Pattern and Process in Macroecology

Pattern and Process in Macroecology

KEVIN J. GASTON
Royal Society University Research Fellow
Department of Animal and Plant Sciences,
University of Sheffield,
Sheffield

TIM M. BLACKBURN
Leverhulme Special Research Fellow
Imperial College (London University),
Silwood Park,
Ascot

Blackwell
Science

© 2000
Blackwell Science Ltd
Editorial Offices:
Osney Mead, Oxford OX2 0EL
25 John Street, London WC1N 2BS
23 Ainslie Place, Edinburgh EH3 6AJ
350 Main Street, Malden
    MA 02148-5018, USA
54 University Street, Carlton
    Victoria 3053, Australia
10, rue Casimir Delavigne
    75006 Paris, France

Other Editorial Offices:
Blackwell Wissenschafts-Verlag GmbH
Kurfürstendamm 57
10707 Berlin, Germany

Blackwell Science KK
MG Kodenmacho Building
7–10 Kodenmacho Nihombashi
Chuo-ku, Tokyo 104, Japan

The right of the Authors to be identified
as the Authors of this Work has been
asserted in accordance with the
Copyright, Designs and Patents Act 1988.

First published 2000

Set by Graphicraft Limited, Hong Kong
Printed and bound in Great Britain
at the University Press, Cambridge

The Blackwell Science logo is a
trade mark of Blackwell Science Ltd,
registered at the United Kingdom
Trade Marks Registry

DISTRIBUTORS

Marston Book Services Ltd
PO Box 269
Abingdon, Oxon OX14 4YN
(*Orders*: Tel: 01235 465500
            Fax: 01235 465555)

USA
Blackwell Science, Inc.
Commerce Place
350 Main Street
Malden, MA 02148-5018
(*Orders*: Tel: 800 759 6102
            781 388 8250
        Fax: 781 388 8255)

Canada
Login Brothers Book Company
324 Saulteaux Crescent
Winnipeg, Manitoba R3J 3T2
(*Orders*: Tel: 204 837 2987)

Australia
Blackwell Science Pty Ltd
54 University Street
Carlton, Victoria 3053
(*Orders*: Tel: 3 9347 0300
            Fax: 3 9347 5001)

A catalogue record for this title
is available from the British Library

ISBN 0-632-05653-3

Library of Congress
Cataloging-in-publication Data

Gaston, Kevin J.
    Pattern and process in macroecology /
    Kevin J. Gaston, Tim M. Blackburn.
            p.   cm.
    Includes bibliographical references (p.   ).
    ISBN 0-632-05653-3
        1. Ecology. I. Blackburn, Tim M.
    II. Title.

QH541.G38   2000
577—dc21   00-029739

For further information on
Blackwell Science, visit our website:
www.blackwell-science.com

Contents

Preface

The first photographs of the Earth to be taken from outer space are said to have had a profound influence on how humanity perceives its place in the Universe. They provided a true visual representation of the broad sweep of the planet we inhabit and how little of it each of us has explored, and, perhaps more importantly, set against the vastness of the cosmos they gave a sense of the vulnerability of this tiny spherical piece of rock. The significance of these images results above all else from the difficulty which humans have in comprehending phenomena at large spatial scales. It is now possible to traverse thousands of kilometres in a few hours, but most people live out their day-to-day existence in thousands of metres. In this sense our lives are perhaps little changed from those of our prehistoric ancestors. Almost inevitably, in consequence we are best adapted to detecting and responding to patterns and processes at such limited scales. After all, it is these which determine who we meet, where we gain food and shelter, and what threatens our existence from moment to moment.

It would be surprising indeed if the scales on which humans typically operate were not strongly to influence the way we conduct many of our affairs. This applies just as much to the way in which science is carried out as it does to other aspects of wider society, despite the attempts of scientists to strive for absolute objectivity. Thus, in ecology, the majority of studies are conducted over small areas—experimental plots seldom span the size of the average back yard—and explanations for observed findings are usually couched in terms of processes which act at the scale of the locality of interest. Just as most natural historians have a 'local patch', whose denizens they come to know, so many ecologists have a 'study plot', the activities of whose denizens both stimulate research questions and supply the answers.

Although humans detect and respond best to patterns and processes at local scales, it is obvious that wider forces have always been at work on their lives. Arguably, attempts to explain, influence and ameliorate the unpredictable vicissitudes of a world beyond the comprehension of people occupying only a small part of it have influenced the establishment of a number of religions. The most obvious manifestation of such wider forces (especially to two Englishmen) is in the effect of large-scale changes and differences in the climate on the fortunes of human societies around the world. For example, the so-called Little Ice Age in the middle of the second millennium AD is thought to have been responsible for the extinction of the small Viking colony established on Greenland around four hundred years earlier. The climatic change meant that the crops on which these people subsisted could no longer be maintained, while essential imports from Europe dried up as the voyage to Greenland became impossibly harsh.

More dramatically, Diamond (1998) argues convincingly that large-scale differences in the climate, geography and biogeography of the various continents have had significant effects on the development of human societies over the last 13 000 years. He suggests that Eurasia was the cradle of agriculture mainly because, for reasons of biogeography, most suitable crop plants and domesticable animals were found in this region. Then, the east–west orientation of this land mass would be particularly amenable to the widespread adoption of this suite of agricultural species, because latitudinal gradients in climate make it easier for organisms to spread within latitudes than across them. Agriculture led to more populous societies, but removed the need for all members to produce or collect food. This in turn opened the way for governing and soldier classes to develop. Large populations were also a natural source of innovations and inventions, including the development of writing, that would have given the societies that possessed them a technological advantage over those that did not. The sum of these and other processes over many millennia was to lead to the current global dominance in wealth and power of peoples of Eurasian origin (Diamond 1998).

The influence of regional patterns and processes on the course of human history is enhanced by the interconnectedness of societies. While developments occurring thousands of kilometres from a local community may seem irrelevant to the day-to-day activities of its members, this interconnectedness effectively embeds all individuals in a broad-scale web of interactions, and makes the Earth a smaller place than in some senses it might otherwise seem. Just how small is shown by the mathematics of so-called 'small-world networks' (e.g. Collins & Chow 1998; Watts & Strogatz 1998). This deals with the number of connections required to link together nodes in a network. For example, if every person in a society knew about 100 individuals, and each of those people also knew about 100 individuals, and so on, then in theory every person in a population of one billion would be connected to every other person through a chain of no more than six mutual acquaintances, or six 'degrees of separation' (Collins & Chow 1998). Degrees of separation will only be this low in a population if connections amongst individuals are random. This is clearly unlikely in human societies, where people tend to know friends of friends. However, it only takes a few well-connected individuals for the degree of separation in a real network to approach the low level found in one that is randomly connected (Watts & Strogatz 1998). The potential this creates for the spread of ideas, innovations, religions, political movements and diseases is readily apparent.

Scarce well-connected individuals typify populations of other animal species as much as they do humans. Thus, while in most bird species the majority of individuals will die within a short distance of where they were born, and will interact with relatively few others, a few individuals will move large distances in their lifetimes effectively connecting disparate populations. Such interconnectedness will enhance any regional-scale forces acting on these species. Thus, adverse climatic conditions in one region will influence populations in another, through differences in the numbers, condition and behaviour of individuals which move between the two. Just as for humans, regional forces are likely to have major impacts on patterns and processes at the local scales on which other species also typically operate. Indeed, the community of plants and animals at a local site will depend on such things as the composition of the regional pool of species potentially available to occupy that site, their abundance, distributions and other characteristics, and on the location and history of the region which have moulded the composition of this pool.

To date, there have been few expositions of the role of regional forces in structuring local species communities and assemblages. In major part, this is an inevitable by-product of the emphasis that has been placed on ecological studies at small spatial scales, and covering relatively brief periods. This book is our attempt to show how an understanding of ecological patterns improves with a broader vision, and some of the implications of the broader view. In short, it is an attempt to forge a link between the micro- and mesoecology which humans find easy to perceive, because this is the scale at which we typically operate, and the macroecology which is more difficult to comprehend, but which is just as important to the structure and function of the natural environment we inhabit.

Just as human affairs in general are influenced by regional forces, the writing of a book is seldom isolated from the influences of a wide range of friends and colleagues. The creation of this one has been no exception. First, and foremost, we express our deepest gratitude to John Lawton and Mark Williamson. John has been a constant source of encouragement and support for our research on macroecology from its earliest days, despite the weight of other demands on his time and attention.

Mark found space in a busy schedule to read and comment on the entire manuscript. His comments were by turn perceptive, challenging and cautionary; they were uniformly helpful. Steven Chown, Richard Duncan, Jeremy Greenwood, Richard Gregory, David Griffiths, John Harte, John Lawton and Phil Warren all read one or more individual chapters, and we thank them for the comments and insights they shared. In addition to all these, other individuals have helped shape our thinking about the topics addressed in this book, or have provided practical help with data or analyses. In particular, we would thank Andy Brewer, Brian Enquist, Sian Gaston, Andy Gonzalez, Paul Harvey, Bob Holt, Peter Kabat, Dawn Kaufman, Ron Kettle, Bill Kunin, Natasha Loder, Brian Maurer, Brian McArdle, Stuart Pimm, Andy Purvis, Rachel Quinn, John Spicer and Diane Srivastava. Linda Birch at the Edward Grey Institute library, and the staff of the libraries at the University of Sheffield and at Silwood Park have helped us trace many an obscure reference. Financial support was provided by the Royal Society, the Leverhulme Trust, the National Environment Research Council and the Centre for Population Biology. Part of the book was written while T.M.B. was visiting the Ecology and Entomology Group, Lincoln University, New Zealand, and he is especially grateful to Richard and Sue Duncan, David and Joy Coombes, Claire Newell, Steve Wratten and other staff and students of the university for their kind hospitality during his stay. It was a pleasure to work with Ian Sherman and Karen Moore at Blackwell Science on the book's production. K.J.G. thanks Sian and Megan for their constant encouragement, especially as he obtains first-hand experience of the scale of the world in which we live. T.M.B. thanks his parents and sister for their support over the years, and Clare for her patience and understanding during the preparation of this book.

K.J. Gaston & T.M. Blackburn
September 1999

1 The Macroecological Perspective

One's ideas must be as broad as Nature if they are to interpret Nature. [Arthur Conan Doyle 1887]

1.1 Introduction

We first visited Eastern Wood, Bookham Common, in the spring of 1998, on an unseasonally cold April morning. Snow had fallen across much of south-east England the previous day, and an overnight frost meant that some snowy patches were left in the grass around the car park. We emerged from our cars through the clouds of our own breath.

The birdwatching at this small (16-hectare) oak woodland in the southern English county of Surrey (Fig. 1.1) was distinctly, though predictably, disappointing, and made even thoughts of a return to our computer keyboards (and mugs of hot tea) an attractive proposition. Between 07.30 and the first drops of rain two and a half hours later, we tallied a mere 25 species (Fig. 1.2), none of which we had not seen on hundreds of previous occasions. The initial flurry of sightings, as we located those species determined enough to sing despite the cold, soon gave way to a slow but steady accumulation of fresh records (about one every 10 minutes) that continued until the weather forced us to cut short our observations. Although Fig. 1.2 suggests that we would have seen more species had we persisted through the rain, the additions would inevitably have slowed even further, and unlike the weather, would soon have dried up completely. Given that this is true, the morning in Eastern Wood offered scant inducement to pay the site a return visit: almost any woodland in southern England would have provided as much ornithological excitement, and we know several sites that would have yielded much more. It is not credible (to us at least) that the poor species list we generated could be a consequence of poor ornithological skills, and so it must in some measure reflect the composition of the avian community of Eastern Wood.

Like so many places, however, the merits or otherwise of Eastern Wood should not be judged on the basis of a single visit. The site is well known to many ecologists as the result of a study by the London Natural History Society (LNHS) of the composition of its breeding avifauna. This work began in 1946

1

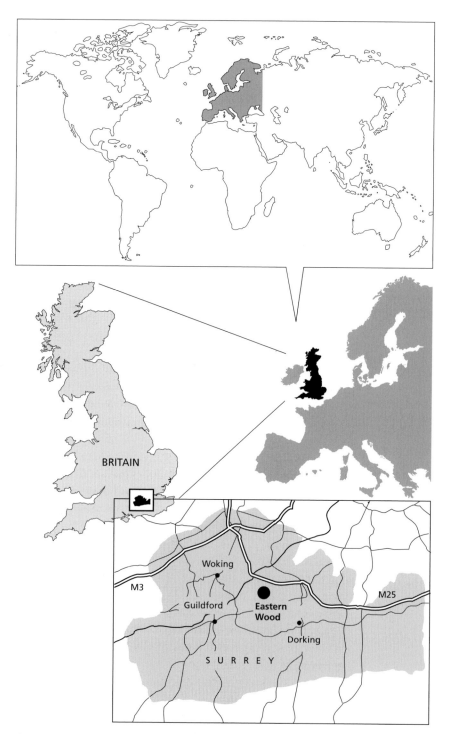

Fig. 1.1 The location of Eastern Wood within Surrey, of Surrey within Britain, and the geographical position of Britain in the world. Lines on the bottom map represent roads; the M3 and M25 are motorways.

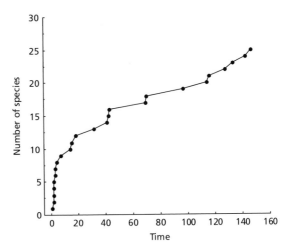

Fig. 1.2 The cumulative number of bird species recorded with time (minutes) in 2.5 hours in Eastern Wood on the morning of 16 April 1998; 0 minutes is 07.30 hours.

with population censuses of the robin (scientific names of bird species cited in the text throughout the book are listed in Appendix I), blackbird and chaffinch (Beven 1976). In 1949, the survey was extended to all species bar the blue tit and dunnock, and from 1950 to 1979 complete censuses of the breeding birds were carried out in every year (except 1957, when there was no census). The data amassed permit a range of features of the assemblage to be quantified, allowing our own observations to be set in perspective. This context reveals a number of interesting patterns.

Perhaps the most notable feature of our morning in Eastern Wood was that while it was not an especially suitable day for censusing birds (the cold discouraging most individuals from raising their voices above the drone of London's air traffic), the tally of 25 species meant that in three hours we recorded more than half (in fact, 56%) of the bird species recorded breeding in the wood over the period from 1949 to 1979 (Fig. 1.3; data for this period used here and elsewhere in the book are referenced and listed in Appendix II). Moreover, our sightings constituted 69% of the maximum number of species recorded breeding at the site in any one of those years (36 in 1972 and 1975). While we could not prove in a single visit that any of the species we observed were actually breeding on the site (although a great spotted woodpecker was watched excavating a nest hole), most of them do breed there annually, and only one that we saw (jackdaw) was not recorded breeding in the wood between 1949 and 1979 (Beven 1976; Williamson 1987). Several of the regular breeders we did not observe are summer migrants that on the date of our visit had probably not yet returned to the site from their wintering grounds. In sum, the short list of species we recorded is a fair reflection of the richness of the breeding avifauna.

A second striking feature of the bird assemblage we encountered in

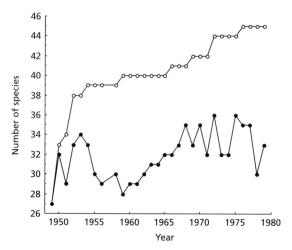

Fig. 1.3 The cumulative number of bird species recorded breeding in Eastern Wood in the period 1949–79 (open circles), and the number of species recorded breeding in each separate year (filled circles).

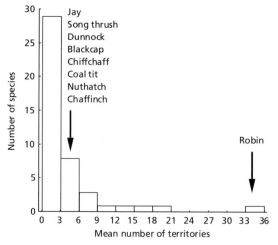

Fig. 1.4 The frequency distribution of the mean number of territories (when present) held in Eastern Wood by breeding bird species in the period 1949–79.

Eastern Wood was that the majority of species were rather scarce at the site. Thus, while the songs of robins and the calls of blue tits provided a near continuous background to our walk, we encountered only singletons of pheasant, green woodpecker, sparrowhawk, willow warbler, nuthatch, bullfinch and chaffinch, and only two coal and two marsh tits. Thus, more than one-third of the species tallied were represented by only one or two individuals. Such a pattern of abundance is demonstrated even more strikingly by the observations of the LNHS survey team (Fig. 1.4). Commonness, as exemplified by the robin, is the exception. Rarity, as exemplified by the nuthatch, is the rule.

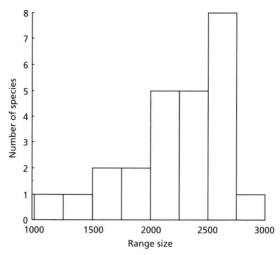

Fig. 1.5 The frequency distribution of range sizes in Britain (number of occupied squares on the 10 × 10-km British National Grid; Gibbons *et al*. 1993) of bird species recorded in Eastern Wood on the morning of 16 April 1998. The maximum range site possible is 2830 squares.

A third feature of the Eastern Wood bird assemblage, albeit one that is much less immediately striking, was that most of the species encountered are relatively widely distributed across Britain. The geographical distributions, or range sizes, of all bird species breeding in Britain have twice been mapped by teams of volunteer recorders in schemes organized by staff of the British Trust for Ornithology (BTO). These projects resulted in two distribution atlases, mapping bird occurrences in the 10 × 10-km squares of the British National Grid over the periods 1968–72 (Sharrock 1976) and 1988–91 (Gibbons *et al*. 1993). If we examine the species we recorded at Eastern Wood using data from the more recent of the two atlases (Gibbons *et al*. 1993), most are found to occupy a high proportion of all possible 10 × 10-km squares (Fig. 1.5). Indeed, only the marsh tit and nuthatch are found in less than 50% of squares, and only six species occupy fewer than 70%. Similar patterns pertain to the entire breeding avifauna of the wood in the period 1949–79 (Fig. 1.6). We did not have to go to Eastern Wood to see the species we did.

One final characteristic of the bird species we observed in Eastern Wood was that most were rather small bodied (Fig. 1.7). Almost two-thirds had average masses of less than 100 g, while the median mass was just over 20 g. This pattern also is not a quirk of the set of species we happened upon on a single visit. The distribution of body masses of all birds recorded by the LNHS as breeding in Eastern Wood is also highly skewed towards small species (Fig. 1.8), with a median mass of 23.4 g.

All these observations attest that, while the weather on the day we first visited Eastern Wood may have been unusual for the site at that time of year, the set of species encountered there certainly was not. We discovered an avifauna composed of a reasonably small number of generally small-bodied species,

5

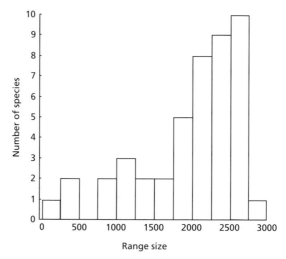

Fig. 1.6 The frequency distribution of range sizes in Britain (number of occupied squares on the 10×10-km British National Grid; Gibbons *et al.* 1993) of bird species observed breeding in Eastern Wood in the period 1949–79.

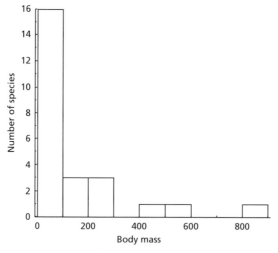

Fig. 1.7 The frequency distribution of body masses (g) of bird species recorded in Eastern Wood on the morning of 16 April 1998.

mostly present in low numbers at the site but widely distributed across Britain. These are all features confirmed by the more detailed and prolonged LNHS survey. With the possible exception of the wide distribution of species, similar observations could also be made about the avifauna of almost any locality in Britain.

In fact, similar observations could be made about the avifaunas of sites the world over (although, subjectively, there are many sites that are far more exciting to visit than Eastern Wood). For example, Holmes *et al.* (1986) censused the

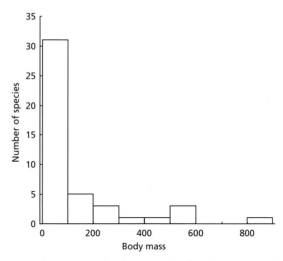

Fig. 1.8 The frequency distribution of body masses (g) of bird species recorded breeding in Eastern Wood in the period 1949–79.

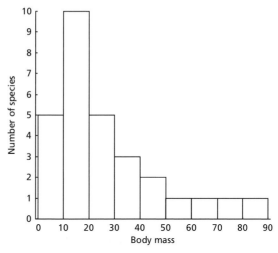

Fig. 1.9 The frequency distribution of body masses (g) of bird species recorded breeding in a 10-hectare plot of deciduous forest at Hubbard Brook, New Hampshire, in the 16-year period 1969–84 (Holmes *et al.* 1986); $n = 29$. Body mass data were kindly supplied by B. Maurer.

avifauna of a 10-hectare patch of deciduous forest at Hubbard Brook, New Hampshire, USA. As in Eastern Wood, most species recorded breeding in the American wood were small bodied (Fig. 1.9), and most were scarce at the site even when present (Fig. 1.10).

Although on opposite sides of the Atlantic, Hubbard Brook and Eastern Wood are not dissimilar in terms of physiognomy. Both are temperate zone deciduous forests. Most of the trees at Hubbard Brook would be clearly akin to

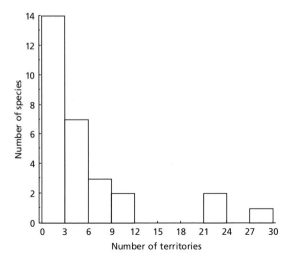

Fig. 1.10 The frequency distribution of average number of territories (when present) for bird species recorded breeding in a 10-hectare plot of deciduous forest at Hubbard Brook, New Hampshire, for the 16-year period 1969–84; $n = 29$. From data in Holmes *et al.* (1986).

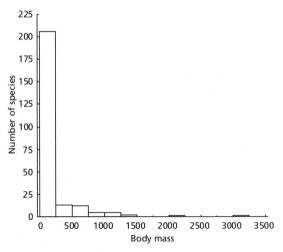

Fig. 1.11 The frequency distribution of body masses (g) of bird species recorded holding territory in a 97-hectare plot of floodplain forest in Manu National Park, Amazonian Peru, in a 3-month census period in 1982; $n = 245$. From data in Terborgh *et al.* (1990).

species familiar in Britiain and the two sites even share a bird species in common (the wren). However, the broad similarity in the structure of the avifaunas of the two sites is not a consequence of this alone. Terborgh *et al.* (1990) censused the avifauna of a 97-hectare patch of floodplain forest in Manu National Park, Amazonian Peru. The species richness of this tropical site was hugely greater than that of Eastern Wood (245 territorial species recorded in

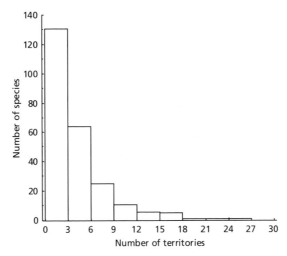

Fig. 1.12 The frequency distribution of number of territories (per 100 hectare) of bird species recorded holding territory in a 97-hectare plot of floodplain forest in Manu National Park, Amazonian Peru, in a 3-month census period in 1982; $n = 245$. From data in Terborgh *et al.* (1990).

3 months versus 45 in 30 years; see Chapter 2), although the most abundant species maintains fewer territories, on average, than the most abundant in the much smaller patch that is Eastern Wood. However, the qualitative patterns are the same. Most species present were small bodied (Fig. 1.11) and in low numbers (Fig. 1.12). Thus, the structure of the avifauna of Eastern Wood is not particularly unusual, excepting that it has been especially well studied.

1.2   Scale and avian ecology

Ecologists have spent several decades attempting to understand what structures animal assemblages at local sites such as Eastern Wood. That is, they have been trying to determine why most species are scarce and a few are abundant, why most are small and a few large, and so on.

With some significant exceptions, answers have predominantly been sought at the spatial scales of these localities, or sometimes at even finer scales. Broader contexts have, by and large, been ignored. We distinguish between cartographic and colloquial definitions of scale, because unfortunately what is small scale according to the former is large scale according to the latter (Curran *et al.* 1997). Here, following convention in this field, scale is used in the latter sense, as a synonym of words such as size and area; small scale refers to a small area, and large scale to a large area.

Reference to *Ibis*, one of the most highly rated journals publishing studies of avian ecology, reveals the extent of the preoccupation with small scales in this field. Local-scale studies never comprised less than 55% of all published

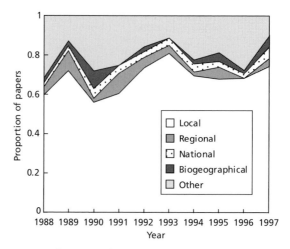

Fig. 1.13 The proportion of papers in the journal *Ibis* in the 10-year period 1988–97 concerning different spatial scales. This classification does not include short communications (comments) or papers in supplements. Studies were classified as 'local' if they were performed over restricted areas (e.g. at well-defined sites), or if they were performed at a few reasonably well-separated sites but this separation was irrelevant to the aim of the study. Thus, the paper by Yamagishi and Eguchi (1996) on the comparative foraging ecology of Madagascan vangids was classified as local, even though it involved work at several well-separated sites. The separation was irrelevant to the study, which could equally well have been carried out at one site had all the vangid species been present. By contrast, Matthysen's (1997) study of geographical variation in nuthatch song types, which was carried out at nine sites in northern Belgium, was classified as regional, because here the site separation was relevant. 'Regional' studies were those concerning scales roughly equivalent to an English or American county, or a restricted part of a country. The 'national' scale refers to studies across regions roughly equivalent to whole countries, whereas 'biogeographical' studies consider multinational, continental or global scales. Studies that could not readily be assigned to any class in this scheme, or for which scale was not relevant, were lumped into the 'other' category. These principally comprised taxonomic, experimental and review papers.

papers in the 10-year period 1988–97 (Fig. 1.13). The second largest category is usually 'other', which includes all those studies that cannot be categorized by spatial scale. Large-scale studies (lumping regional, national and biogeographical categories) generally contributed only between 10 and 15% of the papers published in *Ibis* in any one year, and just over 10% of the papers in total. There is no hint that this proportion has been increasing over the last decade.

This emphasis on small spatial scales is typical of ecology in general (Kareiva & Andersen 1988; May 1994a; Baskin 1997; Lawton 1999). In a similar vein, most ecological studies are of very short duration, usually two or three years at the most (Weatherhead 1986; Tilman 1989; Elliott 1994; Malmer 1994; Baskin 1997). Most focus on just a few species, and there is some evidence that both the proportion of community studies and the number of species per community study have recently been in decline (Pimm 1986; Shorrocks 1993; Kareiva 1994).

The preoccupation with small spatial scales implicitly assumes that it is forces operating within sites which determine their faunal structures. Nonetheless, the success of this approach has been mixed. On the one hand, it has revealed a wealth of important details, including data on the determinants of the reproductive success of individuals and species, patterns of foraging, the temporal dynamics of populations and the influences upon them, and the effects of competition and predation. On the other hand, these studies have largely failed to answer the bigger questions as to why avian assemblages are broadly structured in the ways in which they are. Indeed, arguably this failure was sufficiently acute that, until recently, attempts to resolve these issues had to some marked degree been abandoned. There appears, for example, to have been a significant hiatus in such studies between the late 1960s and the mid to late 1980s.

1.3   A wider perspective

The species richness and the abundance and body size structures of local avian assemblages are among their most apparent features. This is so whether or not the observer is overtly interested in birds and, if interested, whether the context be amateur or professional. Understanding the determinants of these patterns would therefore seem to be a priority. However, it has become readily apparent that the answers lie in a much broader perspective. Local species assemblages are influenced not only by local forces, but also by those at larger scales.

This point has been recognized by a number of ecologists, whether specialists on birds or otherwise (e.g. MacArthur 1972; Maurer 1985, 1999; Brown & Maurer 1987, 1989; Ricklefs 1987; Wiens 1991; Cornell & Lawton 1992; Holt 1993; Ricklefs & Schluter 1993; Brown 1995; Lawton 1999). Thus, in his 1991 Witherby Lecture, Goss-Custard (1993, p. 82) argued that: 'if the populations of such mobile animals [as birds] are to be properly understood, we must address more directly the factors and processes that determine numbers over very large areas and throughout the year, and not just in one locality at a particular time of year. Indeed, because numbers in one study locality are likely to be affected by the size of the greater population of which the local group forms a part, studies on larger geographical scales may be needed even to understand numbers in a particular place.' Similarly, Haila (1988, p. 89) noted that a: 'population decrease on the regional scale is likely to lead to local "extinctions" in small plots, but these cannot be interpreted in terms of a local "equilibrium" '. Much earlier, in their *The Theory of Island Biogeography*, MacArthur and Wilson (1967, p. 182) concluded that: '(g)lobal patterns of distribution also need to be reconsidered. We know that species diversity, relative abundance

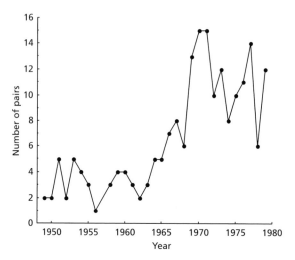

Fig. 1.14 The number of pairs of woodpigeon recorded breeding in Eastern Wood in each year from 1949 to 1979.

and population geometry change with climate. Such variation affects the structure, stability, and energy flow of the plant and animal communities.' Their message is clearly that an understanding of community structure requires consideration of the region within which the community sits. The point was recognized by Holt (1993), who wrote that: 'an important item on the agenda for community ecology will be to grapple with the messy reality that local communities contain species that experience the world at vastly different scales. The structure of a community will surely reflect the interplay of disparate regional processes.'

The avifauna of Eastern Wood well illustrates the significance of events at broader scales. The wood is embedded in a landscape of other woods, agricultural lands and suburban areas. Changes in this landscape and its management change the regional abundances of species, which must in turn impact abundances within the wood. For example, the numbers of breeding woodpigeons in Eastern Wood have shown a marked increase from the early to mid 1960s (Fig. 1.14). Beven (1976) noted this in his summary of changes in the avifauna, but offered no explanation. However, it coincides with the widespread adoption of oilseed rape as an agricultural crop, which provides the woodpigeon with an important source of winter food. Significant preference for this crop in winter seems to decrease winter mortality (Inglis *et al.* 1990), and so may have contributed to the higher breeding numbers seen in Eastern Wood. Other agricultural changes have had similarly marked effects on the structure of British bird communities, albeit mainly for open habitat species such as the grey partridge, skylark and corn bunting, which have not been recorded by the Eastern Wood survey. The recent rise in woodpigeon numbers

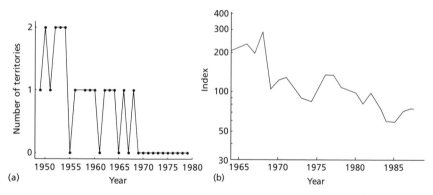

Fig. 1.15 (a) The number of territories of whitethroat recorded in Eastern Wood in each year from 1949 to 1979. (b) Variation in an index of the number of whitethroat breeding on woodland census sites across Britain in the period 1962–88. From Marchant *et al.* (1990) and data in Appendix II.

must be set against sharp declines in the abundances of most other bird species typical of farmland (Gibbons *et al.* 1993; Fuller *et al.* 1995).

A number of the species breeding and wintering in Eastern Wood are migratory. Their presence and the numbers in which they occur are therefore influenced by events many hundreds of kilometres away. For example, while never common, the whitethroat and the garden warbler have both shown steady declines in numbers on the site, to the extent that they became extinct as breeding species in the late 1960s and early 1970s. The whitethroat last bred there in 1968 (Fig. 1.15a). In fact, 1969 was a disastrous year in general for this species, as only one-quarter of the British breeding population returned from the drought-stricken wintering grounds in Sahelian Africa (Winstanley *et al.* 1974). The population has not subsequently recovered to the precrash level (Fig. 1.15b). The population of the garden warbler was also affected, although as this species mainly winters south of the Sahel, the consequences have been less severe and less persistent (Fig. 1.16b). Indeed, the garden warbler managed to re-establish, albeit tenuously, as a member of the Eastern Wood breeding bird community following its initial extinction (Fig. 1.16a). Nevertheless, the decline of the garden warbler and extinction of the whitethroat as breeding species in Eastern Wood seem likely to be direct consequences of events occurring several thousand kilometres from the site (for more examples, see Järvinen & Ulfstrand 1980; Terborgh 1989; Baillie & Peach 1992; Newton 1995, 1998). In a local context, these population changes might at best have been attributable to stochastic effects. In a regional context, they are clearly explicable in terms of specific identifiable causes.

Subsequent chapters will reveal the extent to which features of the Eastern Wood bird assemblage, such as species richness, abundance patterns and body size structure, are influenced by processes operating at large spatial and

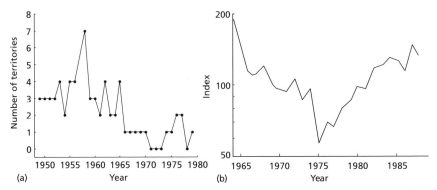

Fig. 1.16 (a) The number of territories of garden warbler recorded in Eastern Wood in each year from 1949 to 1979. (b) Variation in an index of the number of garden warbler breeding on woodland census sites across Britain in the period 1964–88. From Marchant *et al.* (1990) and data in Appendix II.

temporal scales. Some of these effects, such as the role of conditions on the wintering grounds in determining breeding abundances, will probably not be particularly surprising. Others are much more subtle, and there are all shades in between. Of course, many of these regional phenomena are also of interest in their own right, irrespective of their effect on the structure of local assemblages. There is a need to understand why there are general regularities in the distribution and abundance of bird species that transcend specific and regional idiosyncrasies, at least as much as to understand the interactions between individual birds at local sites. Nevertheless, it is the influences of large-scale patterns and processes at the small scale most readily perceived by humans that are most persuasive of their more general importance.

Although large-scale processes clearly influence the community structure of local sites, we do not wish to imply that local processes are not also important. A number of the temporal changes in the avifauna of Eastern Wood may be ascribed to these. For example, a decrease in the willow warbler population (Fig. 1.17) is not mirrored more widely across Britain (Gibbons *et al.* 1993), and seems to be a consequence of changes in management of the wood (Beven 1976). Until the early 1950s, trees had periodically been felled to be sold for timber, and the wood had also been thinned by shrub clearance. The cessation of these practices (in response to requests by the team censusing the bird populations) led to a greater proportion of mature trees, and encroachment of scrub onto open areas and grassy rides. The grassy areas provided nesting sites for the willow warbler, and the simultaneous reduction of both in Eastern Wood is suggestive of a connection. The increase in populations of tits and starlings (Fig. 1.18) seems likely to be a consequence of the greater availability of nest holes in mature and dead trees, which previously would have been felled (Beven 1976). Interestingly, the encroachment of scrub at Eastern Wood should have

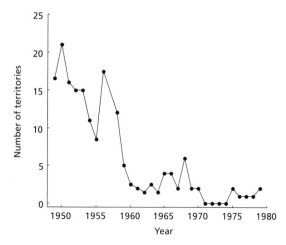

Fig. 1.17 The number of territories of willow warbler recorded in Eastern Wood in each year from 1949 to 1979. From data in Appendix II.

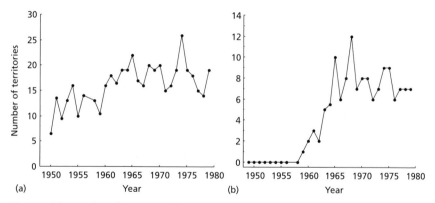

Fig. 1.18 The number of territories of (a) blue tit recorded in Eastern Wood in each year from 1950 to 1979 and (b) starling recorded in Eastern Wood in each year from 1949 to 1979.

benefited the garden warbler, which prefers woods containing such habitat (e.g. Fuller 1982). Indeed, the population of this species did increase there in the late 1950s (Fig. 1.16a). However, its subsequent decline in the late 1960s mirrors its general decline in British woodland (Fig. 1.16b). Variation in the Eastern Wood garden warbler population serves to emphasize that a range of perspectives are required to understand such dynamics.

1.4 The macroecological approach

Recognition of the importance of a regional perspective for understanding the structure and dynamics of local assemblages, and of regional-scale issues in their own right, has in major part stimulated emergence of the field of macroecology (Gaston & Blackburn 1999). Macroecology is concerned with understanding the abundance and distribution of species at large spatial and temporal scales (Brown & Maurer 1989; Brown 1995; Blackburn & Gaston 1998; Gaston & Blackburn 1999; Maurer 1999). It covers the point of intersection of several other fields of biology, including ecology, biogeography and macroevolution.

The philosophy underlying the macroecological approach has been discussed by Brown (1995), Gaston and Blackburn (1999) and Maurer (1999), although the basic ideas have a long and distinguished pedigree. The starting point of this philosophy was typically well described by MacArthur (1972) when he wrote that: '(m)ost scientists believe that the properties of the whole are a consequence of the behaviour and interactions of the components. This is not to say that the way to understand the whole is always to begin with the parts. We may reveal patterns in the whole that are not evident at all in its separate parts.' Macroecology seeks to develop an understanding of ecological systems through the study of the properties of the whole (a 'top-down' approach). This can be contrasted with a more traditional approach, which seeks to develop such an understanding through study of the component parts ('bottom-up'). However, the two are clearly complementary. For example, an examination of the properties of an ecological community might suggest the features of its component species that are of particular importance in its structure, while the behaviours of the species might suggest features of the community that could benefit from more detailed attention. By following both 'bottom-up' and 'top-down' paths, a better understanding is reached than would have been derived from either approach alone. This philosophy is not peculiar to macroecology. A complete understanding of most, if not all, scientific disciplines is likely to arise only by incorporating observations made from a range of viewpoints, or at a variety of scales (Gaston & Blackburn 1999). The focus often changes as a field develops, and important gaps in knowledge are identified. Of course, large- and small-scale approaches are both simply tools for examining and trying to understand the complexity of ecological systems. The systems themselves are continuous over all scales, and the differentiation between large- and small-scale processes is simply arbitrary, manufactured for convenience (for example, see Wilber 1979). Macroecologists recognize that their interests are informed by ecological studies at smaller spatial and temporal scales, but the reverse is also true.

The large-scale approach has advantages and disadvantages. Arguably, one of the principal advantages is that it takes a sufficiently distant view of ecological systems that the idiosyncratic details disappear, and only the important generalities remain (Brown 1995). A fitting analogy is that it attempts to see the wood for the trees. Adopting a distant viewpoint reveals patterns that in some cases would at best otherwise have been difficult to predict, and at worst would have been overlooked. Active research programmes have developed around many such patterns, as will become apparent throughout this book.

Many perceived problems with the macroecological approach are as much imagined (typically by non-practitioners) as they are real (Gaston & Blackburn 1999). Despite a history of studies dating back at least to the middle of the 19th century that today would be called macroecological, it is still very much a fledgling field. The reasons for this probably have much to do with the difficulty of testing macroecological ideas (see next section), but whatever its cause, this immaturity has two particular consequences. One is that many of the analytical tools required for macroecological analyses are only crudely developed. Many macroecological patterns are not simple in form, negating their examination by conventional statistical techniques. For example, some bivariate relationships, such as that between abundance and body size (Section 5.5), and between body size and range size (Section 5.3.3; Brown & Maurer 1987; Gaston & Blackburn 1996a), are what has been described as 'space-filling' (Blackburn & Gaston 1998), or 'polygonal' (Brown & Maurer 1987). How to compare patterns such as these is unclear, although simple methods have been developed to measure the magnitude of the slopes of their boundaries (Blackburn *et al*. 1992; Thomson *et al*. 1996; Scharf *et al*. 1998).

The second consequence of the relative youth of macroecology is that its practice has been unfavourably (and we think unfairly) contrasted with the more traditional small-scale approach to ecological problems. It has been seen by some as an exercise in *post hoc* pattern explanation, with few or no testable hypotheses or predictive theory (Gaston & Blackburn 1999). In part this view is justified, but only to the extent that any scientific discipline must inevitably pass through a phase where the phenomena of interest are identified, and when only tentative theories for these phenomena exist. Put simply, the macroecological approach can be seen as having three phases in its development (Gaston & Blackburn 1999, following Wiegert 1988). The 'What?' stage documents the patterns which the field has to explain. This step is reasonably well advanced, although many of the patterns that have been discovered are not as well known as they should perhaps be. The 'How?' stage documents the anatomy of the patterns, that is how they are structured. This step remains poorly developed, and as such constitutes a significant hindrance to the third stage. This final stage is the 'Why?' stage, which is concerned with the mechanisms underlying the observed patterns. For most patterns a number of hypothesized mechanisms now exist and the challenge is to subject these to rigorous theoretical and empirical testing. However, that the Why? stage is only now being explored in earnest simply reflects the natural development of the field.

1.5   Testing macroecological hypotheses

The need to test theory highlights an additional problem with the macroecological approach, which is that the scale of study largely precludes (for practical and ethical reasons) the use of manipulative experiments in so doing.

The experimental approach has come to dominate the study of ecology (Rousch 1995; Lawton 1996a; McArdle 1996). This is in large part because a well-designed experiment can potentially give more rigorous results than any other form of investigation (McArdle 1996), but also because the results are usually more conclusive. Most macroecological patterns have multiple competing explanatory hypotheses. The inability to use manipulative experiments means that it is hard conclusively to falsify any of these, and so make progress towards an understanding of the mechanisms that underlie the patterns. Thus, macroecology clearly suffers from the inability to exploit this valuable tool as fully as would be desirable.

Nevertheless, manipulative experimentation is not the only way in which hypotheses can be tested. As Diamond (1998) notes, 'the word "science" means "knowledge" (from the Latin *scire*, "to know", and *scientia*, "knowledge"), to be obtained by whatever methods are most appropriate to the particular field', and several others are appropriate to macroecology (Gaston & Blackburn 1999). Of particular relevance to avian ecology is the use that can be made of natural experiments (Diamond 1986). There are many examples of large-scale changes in ecological systems that have been wrought either by natural events, such as earthquakes, fires, volcanic eruptions and, in the longer term, by changes in climate, or by human agency, such as introductions of exotic alien species to faunas, and anthropogenic changes in atmospheric composition. The large-scale effects of these can be used to examine the bases of macro-ecological patterns and hypotheses. Natural experiments have the advantage that the experimental system is likely to be more realistic than that of any laboratory or field manipulation, and thus they are not simply a poor relation of more controlled approaches.

Studies of British birds provide several good examples of the use of natural experiments. For example, the likely underlying causes behind large-scale changes in the distribution and abundance of the corncrake in the British Isles were identified by comparing population changes in different regions, allow-ing connections between the decline and changes in agricultural practice to be drawn (Norris 1947; Green 1996; see also Winstanley *et al.* 1974; Newton 1995). The introduction of several species of British bird into alien environments by homesick expatriates has contributed (albeit unintentionally) to studies exam-ining the factors determining which species can successfully invade avian assemblages, and why (e.g. Moulton & Pimm 1983, 1986; McLain *et al.* 1995; Veltman *et al.* 1996; Duncan 1997). Birds on small British islands have been used to examine patterns of population turnover (the result of population extinction and colonization) through time (Diamond 1984; Pimm *et al.* 1988; Tracy & George 1992; Russell *et al.* 1995; Manne *et al.* 1998).

The methods by which hypotheses are tested shape the practice of macro-ecology in a number of ways. Foremost, as the examples of the previous para-graph illustrate, they force macroecologists principally to employ comparative methodology. This approach involves comparing the distributions of traits among species, or comparing patterns in variables measured for different communities or in different regions, with the aim of identifying causes of variation in those traits or variables. Usually, one or more hypotheses are proposed to explain a pattern of interest, relevant data are then analysed, and the consistency of the results of these analyses with the various hypotheses is assessed.

The comparative approach can be demonstrated using the example of the relationship between the abundances of animal species and the extents of their distributions. In general, widespread species have higher local abundances, resulting in a positive relationship between abundance and range size across species (see Section 4.2 for more details). However, some widespread species are more abundant than others. One possible reason is that species that are more abundant are smaller bodied than rarer species of equivalent range size. This would be expected if smaller-bodied species had lower per capita energy requirements, and so could attain higher local densities for a given range size, and amount of available energy, than could their larger-bodied relatives. Blackburn *et al.* (1997a) tested this hypothesis using data on the distribution, abundance and body size of British birds. They found no evidence that species attaining higher densities were smaller bodied than species of equivalent geographical range size attaining lower densities.

The comparative approach to hypothesis testing in macroecology has two important consequences. The first is that a precise consideration of the null hypothesis becomes paramount. The null hypothesis has been defined as one that: 'entertain(s) the possibility that nothing has happened, that a process has not occurred, or that change has not been produced by a cause of interest' (Strong 1980). Put simply, it is the hypothesis that there is nothing to explain. In macroecology, the consequence of the failure to frame an adequate null hypothesis is the potential acceptance of a pattern as the result of a biological process, when in fact it is no more than an artefact of a particular methodology. For example, the observation that more abundant species tend to use a wider spectrum of habitats than do less abundant species has little biological interest unless it can also be shown that it is not a simple statistical consequence of sample size (by chance, we expect the fewer individuals of the rarer species to be encountered in fewer habitats; Gaston 1994a). This kind of error is always a danger in studies where inferences have to be drawn from patterns in the absence of any experimental control set. In effect, the null hypothesis is the control (Strong 1980). The importance to macroecological studies of a good null hypothesis cannot be overemphasized (see also Colwell & Winkler 1984; Gotelli & Graves 1996). That an observed pattern is a simple consequence of the null expectation should always be the first hypothesis tested by the macroecologist.

The second important consequence of the comparative approach to hypothesis testing is that careful thought needs to be given to the design of the test.

This is best illustrated by pursuing the analogy of an experiment in the previous paragraph. A well-designed experiment must not only include a control treatment, but also ensure as far as possible that no factors other than the one of interest confound interpretation of the results. One way to do this is to set up many replicates of the experimental treatment and control, and randomly to distribute these replicates around the study site. The chance that treatment and control plots differ in respect of factors other than those being experimentally manipulated is then greatly reduced. A problem with the comparative approach is that even when an appropriate null hypothesis has been defined, a test of it may be confounded by other traits that also vary among the species compared. Problems of this sort are particularly likely in comparative tests, because species vary in many traits other than those directly relevant to the hypothesis. Moreover, related species may be similar because they share common ancestry, rather than because they show a common response to a variable of interest. This is certainly true for birds, where up to 25% of the interspecific variation in some traits can be explained by the relatedness of the species in the analysis (Böhning-Gaese & Oberrath 1999). In comparative tests, unlike experiments, one cannot randomize the distribution of the variable of interest among the units of comparison. Thus, the inevitable effect of related species in comparative studies is to increase the likelihood that the results obtained are consequences of confounding variables, rather than those being tested (see Felsenstein 1985; Harvey & Pagel 1991; Harvey 1996).

For example, consider again the relationship between local abundance and range size studied by Blackburn *et al.* (1997a; see above). It is possible that the positive relationship observed between these two variables is a result of the confounding effect of relatedness among species. Most widespread, abundant species could be passerines, say, and most narrowly distributed, rare species non-passerines. This difference could have arisen for a number of reasons (e.g. because passerines tend to be smaller bodied) at the time of the divergence of the passerines from the non-passerines, and have simply been retained in the subsequent radiations of these taxa. However, there may be a general evolutionary trend for abundance to be negatively associated with range size within passerines and non-passerines separately, and within all the subtaxa within these groups (Fig. 1.19). This general evolutionary relationship would then be obscured by a single evolutionary event—the separation of passerines from non-passerines. A positive relationship appears because many passerines and non-passerines are included in the test. Yet, the individual species are not independent examples of the association between abundance and range size. This subtle effect is analogous to the problem of pseudoreplication in experimental design (the use of inferential statistics to test for treatment effects with data from experiments where either treatments are not replicated or replicates are not statistically independent; Hurlbert 1984).

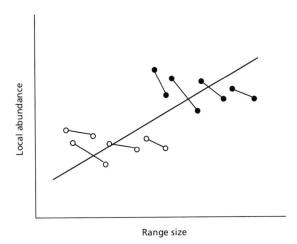

Fig. 1.19 A hypothetical interspecific relationship between local abundance and range size in birds. Open circles represent non-passerine species, and filled circles passerines. Closely related species are joined. A general evolutionary trend for more common species to be less widespread than their rarer relatives would be obscured in an interspecific analysis, because passerines tend to be more abundant and widespread than non-passerines.

To circumvent these kinds of problems in comparative tests, it is necessary to perform analyses that take account of the relatedness of species. Currently, the best way to do this is to base them not on raw species data, but instead on information derived from phylogenetically independent comparisons. This will determine not whether sets of traits are associated across species, but instead whether there are associations between changes in the sets of traits that have occurred since taxa last shared a common ancestor. Comparisons of this sort are not confounded by the phylogenetic relatedness of species, because differences between the taxa must have developed since their divergence. Correlations between traits analysed in this way are therefore evidence for repeated independently evolved trait associations. This method ameliorates the problem of the non-independence of species as points in comparative analyses. Also, because related species will differ in far fewer ways than two randomly chosen species, phylogenetically independent comparisons greatly reduce the likelihood that significant associations are caused by confounding variables. The rationale behind this method is reviewed in detail by Harvey and Pagel (1991), and a programme for implementing it has been developed by Purvis and Rambaut (1995).

Although the phylogenetically independent comparative method helps circumvent the problem of phylogenetic relatedness, the extent to which it does so depends on the extent to which its assumptions are met by the data analysed. In particular, the method assumes a certain model of evolutionary change, and that the hypothesis about the evolutionary relatedness of the species concerned, as represented by their phylogeny, is correct. Several studies have shown that error rates in the estimates of known (simulated) evolutionary relationships provided by comparative methods of this sort depend on the match between the evolutionary process being modelled and that assumed by the method, and on the accuracy of the assumed phylogeny (Martins & Garland 1991; Gittleman & Luh 1992, 1993; Purvis *et al.* 1994; Diaz-Uriarte & Garland 1996; Harvey & Rambaut 1998). Nevertheless, it is often not appreciated that simple interspecific comparisons implicitly assume the hypothesis that all species are equally related (Harvey & Rambaut 1998). As this is patently incorrect, most studies have found that phylogenetically independent comparative methods almost always provide better estimates of true relationships among traits than do interspecific analyses (Martins & Garland 1991; Purvis *et al.* 1994; Diaz-Uriarte & Garland 1996; Harvey & Rambaut 1998), despite criticisms of the approach (Westoby *et al.* 1995a–c; Ricklefs & Starck 1996; Björkland 1997). For this reason, throughout this book, we use phylogenetically independent comparative methods wherever possible and appropriate, coupled with the assumption that the phylogeny of bird species follows that proposed by Sibley and Ahlquist (1990) and Sibley and Monroe (1990, 1993). While this phylogeny has itself been the subject of criticism (e.g. Houde 1987; Sarich *et al.* 1989; Harshman 1994; but see Mooers & Cotgreave 1994), it is probably the best, and certainly the most comprehensive, avian phylogeny available.

In sum, the macroecological approach is a logical development of the far broader research programme of ecology as a whole, and one that is consistent with general scientific philosophy. It certainly has weaknesses, but the same is true of any discipline, and those it does have are likely to be further ameliorated as the field matures. We think it is an approach that is both interesting and informative, and hope to demonstrate as much in subsequent chapters.

1.6   The avifauna of Britain and this book

This book is concerned with the structure of regional (i.e. large-scale) assemblages or communities, and with the influence this structure must have on local (i.e. small-scale) assemblages (or communities). It centres on the birds of Britain, as an exemplary assemblage with which to explore macroecological patterns and processes. In so doing no prior knowledge of this avifauna is assumed. Indeed, no particular interest in it is required.

Britain is defined here as comprising the main island of Great Britain (mainland England, Scotland and Wales) and the islands offshore from this. Thus, Shetland, Orkney, the Outer Hebrides and the Isles of Scilly and Man are all included in our definition, but the Channel Islands are not. This definition does not specifically encompass the island of Ireland, comprising the states of

Northern Ireland and the Republic of Ireland. The question of how to treat Ireland is problematical. Its inclusion is justified on geographical and biogeographical grounds, as it is clearly a member of the group of islands of which Great Britain forms the largest part (Sharrock 1999). Zoogeographically, the position is more ambiguous, as the faunas of Britain and Ireland show enough differences to argue for the exclusion of Ireland, but enough similarities to argue for its inclusion. For pragmatic reasons, we decided to exclude it. Principal among these is that the majority of the data pertaining to bird distributions and abundances in these islands (see below) come from Britain. More trivially, including Ireland raises the simple question of what to call the region as a whole. 'Great Britain and Ireland' is cumbersome and clumsy, while the simple contraction to 'Britain' or 'the British Isles' is clearly insensitive (Sharrock 1999). So, for reasons that are unashamedly practical, we limit our consideration here to the fauna of Britain alone. That is not to say that we do not sometimes include data from Ireland, but where we do, this is made explicit. When we are not explicit, we are referring to Britain.

Having defined Britain, we need to define its bird community. The community is a fundamental ecological concept (Cherrett 1989), but unfortunately not one that is easy to specify. The epistemological status of the community concept has been extensively reviewed by Shrader-Frechette and McCoy (1993), who list more than 25 separate definitions (their table 2.1). All these are based on the idea that a community is a collection of individuals of two or more species, but the emphasis placed on various attributes of this collection has changed as the field of ecology has developed. Early definitions stressed the interactive nature of the component species, with communities perceived as self-regulating entities composed of defined or recognizable groups of species. However, more recent definitions drop this emphasis, and instead adopt the broader concept of the community as the organisms or populations found in a given area. This view dates back to Gleason (1926), who saw the community as an opportunistic collection of species (for discussion in the context of birds, see Bond 1957; Maurer 1985; Wiens 1989).

The analysis of large-scale patterns and processes in ecology encourages a view of the community that coincides with the modern concept. Changes in the avifauna of Eastern Wood have been wrought by factors acting beyond the site, and hence beyond the boundaries of any sensible definition of this community. Yet, these changes (e.g. the increase in the woodpigeon and the decline of the whitethroat) have had no apparent impact on other bird species in the wood. Spatial variation in the distribution and abundance of species means that the avifaunas of no two lowland deciduous woodlands will be exactly identical. Over larger scales, the idiosyncratic responses of species to climatic changes in the Pleistocene (e.g. Graham 1992; Valentine & Jablonski 1993; Coope 1995; Graham et al. 1996) strongly imply that the current composition of communities is largely a stochastic consequence of many individualistic species behaviours. Therefore, we adopt the definition, given by Begon *et al.* (1996), of a community as an assemblage of species populations that occur together in space and time, and use the terms 'community' and 'assemblage' interchangeably.

In taking an opportunistic view of the community, we do not imply that interactions between local populations of different species are unimportant for community dynamics. Rather, we suggest that they are not the primary driving force generating observed species associations. Similarly, while the opportunistic view implies no necessary community structure, in the sense of a 'holistic' community level of biotic organization (e.g. Gilbert & Owen 1990; Shrader-Frechette & McCoy 1993), we do not necessarily deny the existence of some such structure. However, even were there to be no structure, this does not mean that no patterns should be detectable in the attributes of communities. As will be seen in subsequent chapters, there are a number of ways in which pattern may be generated in avian assemblages that do not depend on the existence of any form of community structure.

The definition of a community or assemblage provided by Begon *et al.* (1996) requires its extent to be spatially and temporally specified. Spatial delineation of a community is generally straightforward. For example, we discuss the assemblage of birds in Britain, or the bird community of Eastern Wood, knowing the boundaries of these areas, and that they are reasonably well fixed. Temporal delineation is much harder. The size of the bird communities of Britain and Eastern Wood will be lower if sampled over a month than if sampled over a year, or over 10 years. Over longer periods of time (millennia and beyond), these communities will alter completely as climatic and geological changes exert their effects. As far as possible, we indicate the temporal, as well as the spatial, scale when discussing assemblages throughout the remainder of the book. However, the temporal dynamics of community composition is of itself an important topic, and it is one to which we return in subsequent chapters.

While requiring spatial and temporal qualification, the definition of a community by Begon *et al.* does not include an explicit prescription about its taxonomic composition. The community of Eastern Wood, for example, can include all fungus, bacteria, plant, nematode, insect and mammal populations, and so on, as well as all the populations of birds. However, we limit our attention solely to birds. Subsequently, when we discuss the community of a specified region over a specified time period, we are simply referring to all bird populations that occur within those spatial and temporal boundaries.

There are many reasons why the British avifauna makes an ideal model assemblage. The primary justification for basing a case study around the assemblage is the simple pragmatic one that more is known about it than perhaps any other, at least moderately speciose, regional animal assemblage. As Lawton (1996b) has observed, 'it is inconceivable that any group of organisms other than birds could generate a set of notes entitled: "Turnstones apparently preying on sea anemones" (Donoghue *et al.* 1986), "Turnstones feeding on gull excrement" (King 1982), or the all-time classic "Turnstones feeding on human corpse" (Mercer 1966), culminating in "What won't Turnstones eat?" (Gill 1986).' All these articles appeared in the journal *British Birds*, just one of several British journals publishing exclusively ornithological papers (see also *Ibis*, *Bird Study*, *Bulletin of the British Ornithologists' Club*, *Scottish Birds*, and others). There is thus a huge literature on which to draw. Moreover, since 1933 Britain has been blessed with the BTO, a national organization dedicated to promoting the study of British birds and their habitats. The BTO co-ordinates a large network of amateur fieldworkers, and has collated the huge amounts of information on the abundance and distribution of British bird populations supplied by them. Their work is augmented by that of other government and voluntary conservation organizations, plus the work of academics specializing in avian ecology and biology, and an army of hobby birdwatchers. The efforts of all these groups have resulted in a vast body of data on most aspects of all com-ponents of the British avifauna. The distributions in Britain of scarce birds are particularly well known, as evidenced by the details of where to find 106 of these species provided by Evans (1996). Typical are the directions he gives for seeing lesser spotted woodpeckers in Holkham Park, Norfolk: 'where up to six pairs are known to breed. Park by the hall entrance in the designated car park and walk south through the gates. Turn immediately right inside the gates and walk west for about half a mile to the monument. The area of trees around the monument and the lake to the west will produce calling male Lesser Spotted Woodpeckers between 15th March and 4th May, especially prior to 10.00am.' No other national avifauna is known in such detail.

This last point is important in the context of macroecological studies. As macroecology is essentially comparative in nature, this means that for any given analysis comparable data are usually required for the variables of interest for all species in the assemblage of concern. However, in most assemblages, there is likely to be a subset of species for which the relevant data are not available. For example, an analysis of the body masses of the world's bird species (Blackburn & Gaston 1994a) was hampered by the fact that published body mass data could only be found for two-thirds of described species. Moreover, species may be missing because they are not known to be part of the relevant assemblage, rather than because they are not actually so. Clearly, Blackburn and Gaston's (1994a) body mass analysis could not have included any of the bird species discovered and described during and since 1994, nor any of the species currently awaiting discovery. Missing species are seldom a random sample of an assemblage. They are typically among the least abundant, the most narrowly distributed and the smallest bodied (Gaston 1991a; Blackburn & Gaston 1994b, 1995; Gaston & Blackburn 1994; Patterson 1994;

Gaston *et al.* 1995; Allsopp 1997). Thus, their absence from analyses can significantly bias the results obtained (see, for example, Blackburn & Gaston 1994b). The depth of knowledge of the composition and of the characteristics of the British avifauna is clearly useful in avoiding the biases that missing species potentially can introduce.

The British avifauna is also advantageous for this case study because there is huge interest in the assemblage. The number of birdwatchers in Britain is difficult to estimate. However, the fact that membership of the Royal Society for the Protection of Birds (RSPB), Britain's (and Europe's) largest voluntary conservation organization, has recently topped the one million mark indicates that the avifauna is a source of fascination to a relatively high proportion of the British public. By comparison the combined total membership of the three main British political parties is currently of the order of 900 000. We refrain from drawing the conclusion that the British care more about their birds than they do about their politicians.

The British avifauna also has the advantage that it is a fauna with which both authors are familiar. Although neither of us would regard ourselves as ornithological experts, we have been watching and learning about British birds for longer than we care to consider. When analysing patterns in ecological data that relate to scales well removed from 'normal' everyday experience, it is important to remember, though easy to forget, that the numbers in spreadsheets and the points on graphs represent the biological characteristics of real species. Their interpretation needs background knowledge of the taxon involved, and if that taxon resides within the context of a familiar region then so much the better (see, for example, Futuyma 1998). Thus, while we have not been shy of analysing patterns in faunas with which we have less practical knowledge, our experience of the British avifauna makes it the natural model assemblage for our style of research.

Alongside these pragmatic considerations, the avifauna of Britain offers a number of other advantages. Foremost, the basic macroecological patterns are embodied by the fauna (Table 1.1). Thus, while the British bird assemblage may be small, it generally exhibits the same relationships as more widely drawn avian and other faunas. Latitudinal gradients exist in a number of attributes of the community, despite the relatively limited north to south span of Britain (11° of latitude). In those cases where a broader context is required to understand patterns in the British avifauna, that context is readily available, as avian distributions are well known across most of the western Palaearctic and beyond.

Second, it is an island assemblage (or, more accurately, the assemblage of a group of closely associated islands). Thus, while not a closed system (individual birds and species move in and out of the assemblage), it is a reasonably readily circumscribed one. Moreover, its boundaries can be clearly defined, and are natural in the sense that they represent a real discontinuity in the avian

Table 1.1 Summary of some of the patterns of principal concern to macroecologists, and the main sections of the book in which they are addressed.

Pattern	Section	Summary tables
Species–area relationship	2.2, 2.5.2	
Species richness–isolation relationship	2.3	
Peninsular effect	2.3	
Local–regional richness relationship	2.4	
Latitudinal gradient in species richness	2.5	
Species richness–energy relationship	2.5.3	
Longitudinal gradient in species richness	2.6	
Altitudinal gradient in species richness	2.7	
Species–range size distribution	3.2	3.1, 3.2
Geographical range structure	3.3.2	
Range size–niche breadth relationship	3.3.5	
Extinction–range size relationship	3.3.6	
Speciation–range size relationship	3.3.6	
Nestedness of species occurrence	3.4.1	
Spatial turnover in species identities	3.4.2	
Latitudinal gradient in geographical range size (Rapoport's rule)	3.4.3	3.3
Abundance–range size relationship	4.2	4.1, 4.3
Abundance–niche breadth relationship	4.2.2, 4.3.3	
Latitudinal gradient in abundance	4.2.3	
Species–abundance distribution	4.3	
Species–body size distribution	5.2	
Extinction–body size relationship	5.3.1	
Speciation–body size relationship	5.3.1	
Range size–body size relationship	5.3.3	
Latitudinal gradient in body size (Bergmann's rule)	5.4	5.5
Abundance–body size relationship	5.5	5.6, 5.8

environment, rather than a geopolitical imposition. Indeed, the existence of the boundary is demonstrated by the presence of distinct races of several species that are common British breeding birds, races that are rare breeders outside these isles (e.g. red grouse *Lagopus lagopus scoticus*; pied wagtail *Motacilla alba yarelli*; yellow wagtail *M. flava flavissima*), and perhaps also of one of Europe's few endemic species (the Scottish crossbill; Knox 1990).

Third, Britain has some clear environmental clines (e.g. in temperature, rainfall), which mean that the influence of such variation can be examined over relatively short geographical distances. None of these could be described as dramatic, but they are sufficient to allow an interesting diversity of habitat types, with their associated avifaunas, to be maintained within a reasonably limited area (see Fuller 1982).

There is, inevitably, also a downside to using the avifauna of Britain as the basis for a case study. Throughout, it must be remembered that the region has

Table 1.2 A summary of the areas of different biogeographical regions and biomes, and the percentages of these areas that comprise undisturbed habitat. Areas were classified as undisturbed where there was a record of primary vegetation, no evidence of disturbance, and very low human population density (< 10 person/km^2, or < 1 person/km^2 in arid/semi-arid and tundra communities). From data in Hannah *et al.* (1995).

	Division	Area (km^2)	% undisturbed
Regions	Afrotropical	24 473 218	35.8
	Antarctic	13 506 742	98.4
	Australian	8 255 821	62.1
	Indo-Malayan	9 584 014	10.6
	Nearctic	24 749 723	58.2
	Neotropical	21 550 527	59.9
	Oceanian	933 683	75.1
	Palaearctic—all habitats	59 732 302	51.8
	Palaearctic excl. taiga and Sahara	39 360 429	35.3
Biomes	Temperate broadleaf forests	9 519 442	6.1
	Evergreen sclerophyllous forests	6 559 728	6.4
	Temperate grasslands	12 074 494	27.6
	Subtropical and temperate rainforests	4 232 299	33.0
	Tropical dry forests	19 456 659	30.5
	Mixed mountain systems	12 133 746	29.3
	Mixed island systems	3 256 096	46.6
	Cold deserts/semi-deserts	10 930 762	45.4
	Warm deserts/semi-deserts	29 242 021	55.8
	Tropical humid forests	11 812 012	63.2
	Tropical grasslands	4 797 090	74.0
	Temperate needleleaf forests	18 830 709	81.7
	Tundra and Arctic deserts	20 637 953	99.3

been extensively modified by human activities over the last 10 000 years or so that it has been free of ice. The concept of the 'pristine' habitat must be discarded. However, British habitats are not unique in this respect. Hannah *et al.* (1995) assessed the degree to which different biogeographical regions and biomes have been affected by human disturbance. Their salutary study highlights a number of important points (Table 1.2). For example, the degree of human disturbance of the Palaearctic region is not appreciably higher than that of other regions, and is considerably lower than that of the Indo-Malayan region. However, disturbance varies greatly between biomes within biogeographical regions. Thus, the principal reasons why the Palaearctic is less disturbed than might be expected are because the extensive boreal region has been relatively untouched, and because the figures include the vast barren expanse of the Sahara desert; excluding these biomes considerably reduces the percentage of undisturbed habitat. The principal natural habitat of the British Isles (broadleaf forest) has suffered a high level of disturbance worldwide. Nevertheless, the only biomes that can be considered to be effectively unaltered by human interference are the tundra and Arctic desert. All others have been disturbed to some extent, and the data that Hannah *et al.* (1995) present do not mark British habitats as exceptional (albeit that the extent of habitat degradation here is clearly not good). This is an issue to which we shall return in the final chapter.

Of perhaps more concern with regard to use of the British avifauna as the basis of a case study is that ultimately Britain remains but a small, cool, damp, set of islands on the periphery of continental Europe and the Palaearctic region. Indeed, some might argue that it remains too small an area over which to address macroecological questions. This is evidently not true, in as much as it is plainly a vastly broader scale than that at which ecological processes are often considered, and can provide useful insights into the determinants of local assemblage structure. Moreover, as pointed out above, the British avifauna displays the range of macroecological patterns that would be expected from a larger area. Nonetheless, there is also some sagacity in the statement. Many of the patterns we will examine are essentially global in span, and there are certainly dangers in drawing conclusions from data that encompass only part of the total extent of variation (Blackburn & Gaston 1996a, 1998; Gaston & Blackburn 1996a). To address this criticism, we have endeavoured throughout to make reference not only to how local assemblage structures are influenced by regional ones, but also to how those at the scale of Britain fit with structures at yet broader scales. Nevertheless, while we make every effort to use examples from Britain where those exist in the literature, and to document such examples from available data where they do not, there are still times when we will have to look to other faunas around the world for examples of some patterns, and for tests of the mechanisms suggested to explain them. As informative as the avifauna of Britain is, we cannot expect it to answer all our questions.

Finally, birds are but one part of the sum of all life in the world, and a tiny part at that. The current consensus is that there are about 10 000 extant species of bird (Sibley & Monroe 1990, 1993), although the precise number depends on the complicated issue of exactly how a bird species is defined and may arguably be double this figure (Knox 1994; Martin 1996; Zink 1996). Nevertheless, even 20 000 would be but 'a vanishingly small proportion of the species on earth' (Lawton 1996b). Estimates of the total number of species with which we share the planet vary enormously, spanning the range from 3.5 million to 111.5 million (see, for example, Erwin 1982, 1991; May 1988, 1990, 1994b; Gaston 1991b,c; Lambshead 1993; Gaston & Hudson 1994; Hammond 1995; Hawksworth & Kalin-Arroyo 1995; Stork 1997). These estimates depend on a range of different methodologies, but their large variance reflects the fact that all involve some form of extrapolation from current knowledge of species diversity, and extrapolation beyond the boundaries of knowledge is notoriously inaccurate. The best working estimate is considered to be about 13.5 million species (Hawksworth & Kalin-Arroyo 1995; Gaston & Spicer

1998). Birds comprise less than 0.1% of this figure. Of course, they are an extremely well-known 0.1%, and it is that fact more than any other that tips the balance of pros and cons in favour of their use as a model taxon: notwithstanding the well-known definition of an expert (someone who knows more and more about less and less, until (s)he knows everything about nothing), it is more useful to have a good understanding of a small part of life than a poor understanding of most of it.

So, on balance, and perhaps being inherently more biased than we care to admit (birds are, after all, wonderful creatures), we consider the advantages of the avifauna of this small damp group of islands as the basis for a case study vastly to outweigh the disadvantages. Whatever the conclusion, however, this remains the first detailed analysis of the macroecology of a single animal assemblage. By firmly rooting this analysis in an assemblage that will be familiar to many readers or that is clearly analogous to other assemblages with which they are more familiar, we hope that we can clarify the aims and rationale behind what might otherwise seem to be an abstract programme of research (macroecology).

The need for an exemplary large-scale assemblage, filled by the avifauna of Britain, is matched by the need for an exemplary small-scale assemblage, a local context in which to root the macroecological perspective. We use as our local assemblage that with which we started this chapter, the avifauna of Eastern Wood. There are several advantages to this choice. First and foremost, this avifauna has been extensively and intensively studied for a long period (spawning several previous analyses of its ecological features; Williamson 1981, 1987; Simberloff 1983; Boecklen & Nocedal 1991; Newton *et al.* 1997, 1998). Indeed, censuses of breeding birds were carried out in most years from 1949 to 1997, a period of almost 50 years. This means that much of the caution and scepticism required when viewing the results of surveys based on one or perhaps a few visits (Wiens 1981) can be withheld. Second, the site is a lowland deciduous woodland, and so typical of much of the habitat remaining in lowland Britain not given over to agricultural or urban use. Third, the avifauna harboured by the site is relatively rich for its size and position, providing a reasonable sample size for interspecific analyses. Finally, the site is a nature reserve, and so has been sympathetically managed and kept relatively undisturbed for the period of interest.

Although the avifauna of Eastern Wood has been censused up to the present day, by and large we use data only from the period 1949–79. As this involves ignoring much of the information which has been collected, a few lines of justification for this restriction are appropriate.

A high proportion of the credit for the rigour of the Eastern Wood bird census can be attributed to the work of Dr Geoffrey Beven. Beven had joined the LNHS in 1940, just before being posted to South Africa as a Medical Officer in the RAF, but had taken an active interest in ornithology in the London area for several years before that (Ashby 1990). In 1937, he had noted in a speech to the Society that: 'It has always seemed to me to be quite natural to count birds', and the results of his first area censuses were published in the war years (Beven 1945). He became actively involved in the study of Eastern Wood in 1949.

As noted earlier, the first Eastern Wood census was conducted in 1948, when only populations of the robin, blackbird and chaffinch were assessed. The first complete census was in 1950, but 1949 saw counts of all species bar two. Birds were censused using the relatively labour-intensive technique of territory mapping. Observers are required to visit the site several times during the breeding season and plot the locations of all singing males, and other birds seen. To do this properly requires a high degree of commitment. The BTO recommends at least 10 visits to a site to produce an effective census using this technique. However, such was Beven's level of enthusiasm that he organized a group of volunteers to visit Eastern Wood virtually weekly right through the breeding season, and in some years evidently for much of the non-breeding season as well. The fruits of Beven's labour can be seen in the files of neatly compiled territory maps stored in the LNHS archive in the Imperial College Library in London. An example of such a map is reproduced here as Fig. 1.20. Beven's commitment to his project was clearly extraordinary.

Driven by this enthusiasm, the Eastern Wood bird census ran from 1949 to 1979, with 1957 as the only gap in the sequence. Then, before he could start the 1980 census, Beven suffered a stroke. Although it was not as serious as it could have been, it left him unable to continue his fieldwork in Eastern Wood. There was no census in 1980. Volunteers to continue Beven's work were found for most years after that, except for the period 1985–87. Nevertheless, the continuity was gone, and the effort expended in the censuses declined. The subsequent workers certainly cannot be blamed for failing to match the dedication shown by Beven, but an inevitable consequence of the disparity in census effort was that the apparent composition of the bird community changed when Beven stopped censusing. The number of species breeding in the wood dropped by about 25% in censuses after 1979 (Fig. 1.21), and in all years bar 1984 was lower than the lowest richness recorded by Beven. While this change could be genuine, it seems much more likely to be a consequence of Beven's forced retirement.

Annotations to post-1980 territory number estimates, in an unsteady script that was another consequence of his stroke (he was forced to learn to write with his left hand), evidences Beven's interest in the Eastern Wood bird census right up to his death in 1990. His legacy includes 30 years of some of the highest quality data on avian community composition ever collected in Britain (Appendix II). It is those data that we analyse as our typical local British bird assemblage, data that are exemplary in all senses of the word.

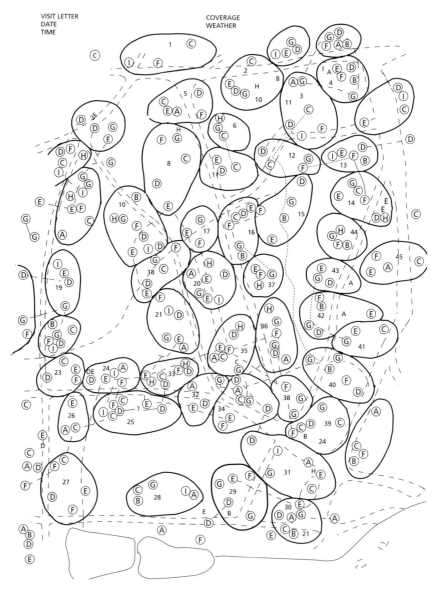

VISIT LETTER
DATE
TIME

COVERAGE
WEATHER

Fig. 1.20 An example of a territory map for Eastern Wood for the robin in 1969, compiled by G. Beven. Letters on the map refer to sightings on different visits. Dashed lines mark paths. From unpublished LNHS records.

1.7 Organization of the book

Each of the subsequent four chapters of the book concern one of the principal macroecological variables, species richness (Chapter 2), range size (Chapter 3), abundance (Chapter 4) and body size (Chapter 5). The reasons why these vari-

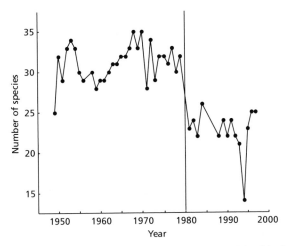

Fig. 1.21 The number of bird species recorded breeding in Eastern Wood in the period 1949–97. Points to the left of the line cover the period when G. Beven was actively involved in the censuses. Data after 1980 from LNHS unpublished records.

ables are of such concern to macroecologists will become apparent as they are introduced. The final chapter (Chapter 6) synthesizes what has gone before and draws out some of the implications. In each of Chapters 2–5, we will take as our starting point the avifauna of Eastern Wood, as an exemplar local assemblage, and explore how patterns at regional scales influence its structure or alter perceptions of this structure. This approach serves to relate the macroecological perspective, which is perhaps often difficult to comprehend, to the everyday experience of local sites. We take the consideration of species richness as our starting point for a similar reason: it is in their contribution to small-scale patterns of species richness that the influence of large-scale processes can most readily be appreciated. An acceptance of their influence in one aspect of avian assemblage structure makes easier their acceptance in others.

Throughout the book our aims are threefold.

1 To draw attention to macroecological patterns in assemblage structure. Some of these patterns are an established part of mainstream ecology, but many more are not, while deserving to be such. In particular, the relationship between abundance and range size (Section 4.2) is a candidate for one of the most general patterns in ecology (Gaston 1996a; Gaston *et al.* 1997a), yet detailed studies of its form are largely non-existent (Leitner & Rosenzweig 1997; Gaston & Blackburn 1999). Attention to these patterns will surely grow as the importance of the regional context in which local sites are embedded comes to be more fully realized.

2 To identify how these macroecological patterns relate to local assemblage structure. We will endeavour to show not only how local assemblages are affected by their regional context, but also how the large-scale patterns are influenced by local behaviours. The connection between large and small scales must be a two-way process, although the relative strengths of the interactions have not been given significant attention.

3 To demonstrate links between different macroecological patterns (as listed in Table 1.1). It has been appreciated for some time that some such patterns must be interrelated (e.g. Harvey & Lawton 1986; Gaston & Lawton 1988a, 1998b), although the degree of variation around them is such that that interrelationship is not always possible to predict a priori. Nevertheless, only recently have links between patterns been drawn more widely (see, for example, Stevens 1989, 1992; Gaston 1994a; Brown 1995; Hanski & Gyllenberg 1997; Harte & Kinzig 1997; Leitner & Rosenzweig 1997). We believe that there is scope for many more links to be formalized, albeit that the current level of knowledge limits those links to a low level of sophistication. Our hope is that we can encourage other researchers to investigate these connections in more detail.

Above all, however, we want to communicate an enthusiasm for macroecology, and to promulgate an understanding of why it is such an important part of the broader programme of research into ecology. As we have stressed, and will continue to do so, a complete understanding of local ecological systems, such as that represented by Eastern Wood and its avifauna, is only likely to be generated if processes are considered acting over as full a range of spatial and temporal scales as possible. That means that attention must be paid to large-scale patterns and processes. If we can sow an understanding of the relevance of macroecology, we will consider this book to have been a success, even if the harvest we reap from any resulting interest is that some of our most dearly cherished beliefs and prejudices ultimately are falsified.

2 Species Richness

There are . . . more species of bird breeding, and also more wintering, in forests than in fields . . . more species of trees in eastern North America than in Europe, and more flies of the family Drosophilidae on Hawaii than anywhere else. There is an even more dramatic difference in the number of species in the tropics than in the temperate . . . Will the explanation of these facts degenerate into a tedious set of case histories, or is there some common pattern running through them all? [MacArthur 1972]

2.1 Introduction

An ecologist setting out to understand the workings of an animal assemblage or community will almost certainly begin by treading a well-worn path. No progress can be made until aspects of the community have been quantified, because without these data there is nothing to explain. A number of features can be measured, but among the questions first addressed will probably be how many species are present, what are their identities and characteristics, and in what numbers does each occur? These issues are fundamental because they define what occurs, and how much of it there is. In this chapter, we focus on the first of the questions, and examine the factors that are likely to determine the answer obtained.

Throughout, we will refer to the number of species in a defined area as the *species richness* of that area (following McIntosh 1967). We think this term is better than the commonly used alternative, *species diversity*, as that has typically been applied to the quantity measured by indices that take account not only of the number of species, but also of the distribution of individuals among them (see, for example, Hurlbert 1971; Magurran 1988). Thus, reference to species diversity hereafter can be taken as meaning a quantity assessed by such an index.

We address the question of how many species can co-exist in a given area before other questions, because the answers are particularly illustrative of the importance of the links between large- and small-scale patterns and processes. No pattern shows more clearly how an understanding of the structure of local assemblages requires a regional perspective. The bird fauna of Eastern Wood provides a prime example.

The avifauna we encountered in Eastern Wood can be viewed as the small-

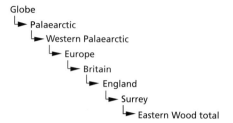

Fig. 2.1 The species richness of the avifauna of Eastern Wood constitutes a subsample of the avifauna from sequentially nested levels above.

est piece in a 'Russian doll-like' set of nested faunas, with the largest representing the global total number of bird species (Fig. 2.1). Each level down can be thought of as sampling a proportion of the fauna from the level above. What we wish to know is what causes each of these samples to assume the magnitude that it does.

2.1.1   Species richness at the smallest scales

The 25 species that we saw on our first visit to Eastern Wood represent almost 80% of the species shown by Beven (1976) typically to breed in the wood in any one year, and just over 50% of the total recorded as breeding in the period 1949–79. In other words, in just three hours in the wood we observed more than a half of the species recorded breeding over 30 years, a high proportion for so little effort. Nevertheless, and despite our best efforts, we failed to observe several species which one might have expected to find in the wood at that season. For example, mistle thrush, treecreeper and starling all eluded us. This failure could have arisen for several good reasons.

Some of those species missed will have been present in the wood at the same time as we were, and had our search continued we would ultimately have found them. This is a straightforward sampling effect, and lies at the heart of any attempt to determine the species richness of an area, be that area small or large. The more time spent, the more likely individual birds are to be encountered, and the greater the number of species that will be recorded. Typically, the rate of increase in the numbers of species observed is initially very high, and steadily declines (Fig. 2.2). In fact, over reasonably short periods of time or when reasonably few individuals are encountered, the pattern may be quite heterogeneous. Thus, on plotting the data for our own short visit to Eastern Wood (Fig. 1.2), we find that the rate of accumulation increased in the period immediately before our departure, suggesting that perhaps a longer stay would have been profitable!

The cumulative growth in numbers of species observed in an area with time constitutes a particular problem for the comparison of estimates of levels of species richness. While the cumulative number may approach an asymptote, it will never actually attain one. Species will forever continue to be added to the

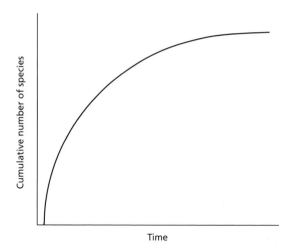

Fig. 2.2 Idealized species accumulation curve, showing the change in the rate at which new species are detected in an area by a given sampling protocol.

avifauna of any given area, albeit that the rate of addition ultimately will be extremely low. Based on this phenomenon, Grinnell (1922) estimated how long it would take before all the bird species recorded from North America had been seen in California. He wrote: '. . . it is only a matter of time theoretically until the list of California birds will be identical with that for North America as a whole. On the basis of the rate for the last 35 years, 1⅗ additions to the California list per year, this will happen in 410 years, namely in the year 2331, if the same intensity of observation now exercised can be maintained. If observers become still more numerous and alert, the time will be shortened.' (p. 375). It remains to be seen whether his prediction will be correct. However, at the present rate of discovery it is an overestimate of the time that will be needed (Bock 1987). In the period 1958–85, the number of bird species recorded in Britain grew at a rate of approximately 2–3 per annum (Fig. 2.3). If this rate continued, it would take about another 3500–4500 years to record all of the world's birds in the country!

The failure to record several of the species that one would have expected to see in Eastern Wood that April morning may not only be a consequence of simple sampling effects, but also of the probability of encountering species which are themselves moving in and out of the area. This is particularly problematical when sampling small areas, over relatively brief periods, for species with very large home ranges, such as many raptors. For example, sparrowhawks have bred erratically in Eastern Wood, a pair being present in some years and not in others (Appendix II). Yet, when they are not breeding there it is likely that the wood falls within the home range of sparrowhawks breeding at other sites (in any one year, Bookham Common as a whole could have held two to three pairs, and at least two further pairs occurred within 1 km of the common in the 1980s; Newton *et al.* 1997). This issue has been found to be par-

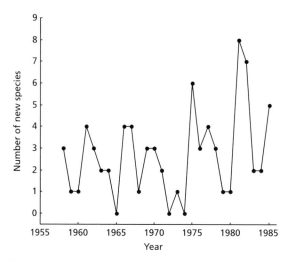

Fig. 2.3 The number of bird species previously unrecorded in the region added to the British list in each year in the period 1958–85. From data in Dymond *et al.* (1989).

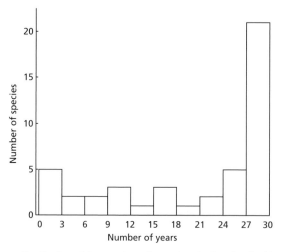

Fig. 2.4 Frequency distribution of the number of years in which individual bird species were recorded breeding in Eastern Wood in the period 1949–79. Recall that the maximum number of years is 30 because there was no census in 1957.

ticularly significant in evaluating the avian species richness of areas of moist tropical forest. Here, species may occur at very low densities and individuals may range over very large areas. Indeed, Terborgh *et al.* (1990) found that relative to the spatial requirements of their temperate zone counterparts, the territory sizes of Amazonian birds were roughly an order of magnitude larger.

Some of the species we would have expected to see in Eastern Wood may simply not have been present there in 1998. Not every species recorded in the wood breeds in every year. Indeed, the number recorded in any one year

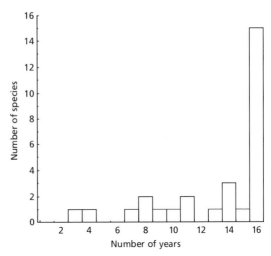

Fig. 2.5 Frequency distribution of the number of years in which individual bird species were recorded breeding in a 10-hectare plot of deciduous forest at Hubbard Brook, New Hampshire, in the period 1969–84; $n = 29$. From data in Holmes *et al.* (1986).

varied from 27 to 36 (Fig. 1.3). Figure 2.4 shows the frequency distribution of the number of years in which each bird species was recorded breeding in the period 1949–79. While most were recorded in most years, almost 50% of species were absent for at least three census years, and only one-third of species bred in every year of the 30. Similarly, the number of bird species breeding on a 10-hectare study area of temperate deciduous forest at Hubbard Brook in New Hampshire, USA, over a 16-year period, varied from 17 to 28, with a mean of 24 (Holmes *et al.* 1986). Of the cumulative total of 29 species, only 15 bred every year (Fig. 2.5). There are a variety of reasons for year-to-year variation in the species richness of an area, some of which we will encounter later. Nevertheless, species that one would expect to find at a site will inevitably be missed because of it.

The species recorded on our first visit comprise a significant fraction of the total breeding avifauna of Eastern Wood in the period 1949–79, which was 45 species. This is clearly an underestimate of the total fauna of the site, because the species richness of an area is affected by seasonality in the composition of the avifauna. Although we have no information on the number of bird species wintering in the 16 hectare of Eastern Wood, data are available for a 25-hectare area of oak wood lying within the boundaries of Northward Hill (High Halstow National Nature Reserve), Kent (Flegg & Bennett 1974). This is a comparable area of comparable habitat in a similar part of the country to Eastern Wood. Knowledge of the avian assemblage is based on censuses over a 12-year period (Flegg & Bennett 1974). The number of species recorded breeding at the site over this period was 43, of which 13 were present in the wood only in summer. This figure is probably quite close to the number of Eastern Wood breeders (Appendix II) that are only summer visitors, which on knowledge of general

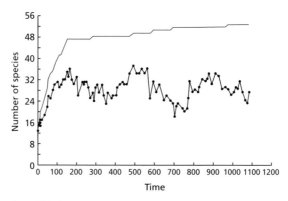

Fig. 2.6 The numbers (filled circles), and cumulative numbers (line), of species of bird recorded with time (days) over a 3-year period on censuses on transects through an upland forest on the Dandenong Ranges in Australia. Reprinted from Mac Nally (1997, with permission from Elsevier Science).

biology alone ought to be around 11 species. However, in addition to 30 resident species at Northward Hill, another 10 species were recorded in the wood only in winter. Thus, the wintering assemblage was only slightly smaller than the breeding one, but there was a significant degree of turnover between seasons. The same degree of turnover at Eastern Wood would add about a dozen extra species to the wood's bird list.

Mac Nally (1997) reports the results of a particularly impressive study of the effects of the temporal frequency of sampling on the sensitivity of monitoring of the avifauna of an upland forest in the Dandenong Ranges in Australia. Ninety-two separate censuses were conducted beginning in the winter of 1993 and continuing through to the end of autumn 1996. The number of species recorded varied markedly from one census to another, from as few as 12 to as many as 37, with this variation superimposed on an apparent pattern of seasonal variation in species numbers (Fig. 2.6). However, the cumulative number of species rose very rapidly, with 47 of the total of 52 species recorded being observed during just the first 17 censuses.

The degree of seasonal variation in the species composition of an area like Eastern Wood depends critically on the latitude at which that area lies. In western Europe, the proportion of species in local avifaunas in summer which are summer visitors increases with latitude, from 29% of breeding species at 35°N to 83% at 80°N (Newton & Dale 1996; see also Herrera 1978). Conversely, the proportion of winter visitors decreases with latitude from 36% of wintering species at 35°N to 8% of wintering species at 70°N and none at 80°N (Fig. 2.7). The overall number of species breeding or wintering tends to decline with increasing latitude (Section 2.5; with the primary exception of breeding coastal birds), as does the number of breeding or wintering species which are migrants (Fuller 1982; Cousins 1989; Newton & Dale 1996). Temporal decreases in the numbers of individual migrant birds in temperate areas,

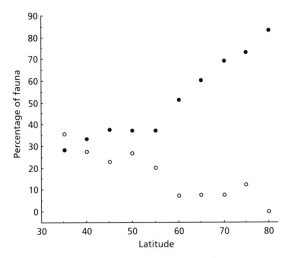

Fig. 2.7 Variation with latitude (degrees) in the percentage of species in an avifauna in Europe that migrate to other latitudes to winter (filled circles) and the percentage in the wintering avifauna that migrate to other latitudes to breed (open circles). From data in Newton and Dale (1996).

particularly neotropical migrants wintering in North America, are presently a cause of much concern (e.g. Leck *et al.* 1988; Robbins *et al.* 1989; Terborgh 1989; Hagan & Johnston 1992; James *et al.* 1996).

2.1.2 *Species richness at larger scales*

As will be continually stressed throughout this book, local sites are embedded in a regional context. Thus, the 45 species breeding in Eastern Wood in the period 1949–79 was about 37% of the 121 bird species that bred in Surrey, the county in which it sits, during the 20th century (data from Parr 1972). This number in turn constitutes just over half of the bird species recorded breeding in the whole of Britain in the 20th century, which at the date of our visit to Eastern Wood stood at 236 (data from Gibbons *et al.* 1996). Of these species, approximately 220 breed on a reasonably regular basis (Appendix III).

Good comparisons with the breeding avifaunas of other countries are difficult to provide, because few peoples have the same detailed knowledge of their avifaunas as do the British. Nevertheless, those comparisons that are possible reveal the British breeding avifauna to be relatively depauperate. One thousand and eleven landbird species alone have been recorded breeding in Colombia, and 835 in Ecuador, out of totals of 1093 and 906 landbird species, respectively (Rahbek 1997). Terborgh *et al.* (1990) carried out a census of a 97-hectare plot in a rainforest in Amazonian Peru over a 3-month period in 1982, and found 245 species holding territory or occupying all or part of the plot.

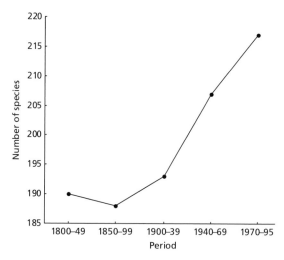

Fig. 2.8 The number of species of bird recorded breeding in Britain in different periods from 1800 to 1995. Note that later periods cover fewer years. From data in Gibbons *et al.* (1996).

Moreover, an additional 74 species visited the plot without breeding there. Thiollay (1994) recorded 248 bird species as regular visitors to a 100-hectare plot of rainforest in French Guiana over a 2-year period. The breeding species richness of local sites within Britain, and of Britain overall, pales in comparison with areas in the tropics.

Although the breeding avifauna of Britain is relatively poor, it is nonetheless higher than it was. Between about 1800 and the present, the total number of indigenous species breeding in Britain has increased by more than 25 (Fig. 2.8). The temporal turnover in the species present has, however, been more marked than this might imply. Over the whole period since 1800, 228 indigenous species have bred, compared with the 218 which did so in the period 1970–95 (Gibbons *et al.* 1996). Four which bred in the period 1800–49 did not breed in 1970–95 (long-tailed duck, Baillon's crake, great bustard, great auk), and 34 which bred in 1970–95 did not breed in 1800–49 (red-necked grebe, slavonian grebe, black-necked grebe, little bittern, whooper swan, gadwall, pintail, common scoter, goldeneye, goosander, common crane, black-winged stilt, little ringed plover, Temminck's stint, purple sandpiper, wood sandpiper, spotted sandpiper, Mediterranean gull, little gull, collared dove, snowy owl, shore lark, bluethroat, black redstart, fieldfare, redwing, Cetti's warbler, icterine warbler, firecrest, brambling, serin, parrot crossbill, common rosefinch, Lapland bunting). Over a similar period, 1850–70, Järvinen and Ulfstrand (1980) showed that Denmark, Norway, Sweden and Finland were colonized by an average of 2.8 bird species and lost 0.6 species per decade and country.

Burton (1995) attributes the pattern of species colonizations in Europe observed over the past 150 years to climate change. He notes that the period from 1850 to 1950 marked a warm phase, but that deterioration in the climate has occurred in the years since. He suggests that the warm period coincided with the northward and westward spread of several species of previously largely southern European distribution, such as the black redstart and serin, while the subsequent cooling prompted colonization by northern species such as the wood sandpiper, snowy owl and redwing. These northern species probably were not recorded breeding in Britain in the 50 years prior to the warm period of 1850–1950 because their small populations were overlooked by the equally small population of birders. Certainly, these species would have bred in Britain in prehistoric times, because the changes in the British avifauna observed over the last 150 years are but a small snapshot of the continual ebb and flow of species across the global landscape in response to climatic changes acting at larger spatial and temporal scales.

Over the last 400 000 years or so, Britain has probably been subjected to four periods of glaciation, separated by short interglacial periods such as that believed to be represented by the current climate (Petit *et al.* 1999). The advance and retreat of ice sheets across Europe inevitably caused major shifts in species distributions. There is good evidence from faunal remains that tundra was the predominant habitat in Britain at the height of some of the glaciations. Arctic bird species would inevitably also have been present. The climate in interglacial periods showed considerable variation. For some periods, temperatures were slightly higher than at present, suggesting that birds currently typical of southern Europe may have been widespread across Britain. At other times, the fauna apparently resembled that of present-day southern Scandinavia. Remains of elements of the avifauna deposited since the last glaciation certainly clearly indicate faunas associated with climates somewhat different to that experienced in Britain today. In particular, bones of both adult and fledgling Dalmatian pelicans identified in Iron Age Somerset reveal a flourishing British colony of a species now confined largely to the eastern Mediterranean. Climatic deterioration may have caused the extinction of this species in Britain, and would inevitably have contributed to many other changes in the avifauna (Burton 1995). Nevertheless, even in the warmest interglacial period, the British avifauna would not even remotely have approached the richness of that seen today in the moist tropics. Neither is it likely to do so in the foreseeable future, despite probable major changes in the composition of Britain's birds as a result of global climate change (Moss 1998).

The wintering avifauna of Britain is no more impressive than that breeding. In January 1991, one of us recorded 138 species of bird during a leisurely day with binoculars and a bicycle at Keoladeo National Park, Bharatpur, India. To produce a list of equivalent length in England on the same (or any) date over the same length of time would have involved equipment and planning worthy of a military operation. The typical entire wintering avifauna of Britain contains only slightly more species than this (Appendix III).

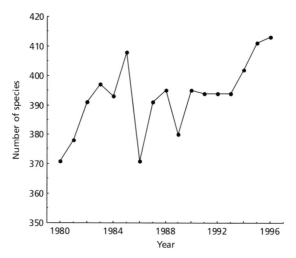

Fig. 2.9 The number of species of bird recorded occurring in the UK in each year in the period 1980–96. From data in Whiteman and Millington (1991) and http://www.uk400.demon.co.uk/yearlist.htm.

Britain may fare poorly in comparison with many other countries in that only 240 species of bird regularly breed or winter in the country (Appendix III), yet the total number of species which was recorded in Britain in the average year between 1980 and 1996 was 393, and fluctuated between 371 and 413 (Fig. 2.9). At the time of our visit to Eastern Wood, the total number of species ever recorded in Britain stood at 550. This figure represents species considered by the British Ornithologist's Union (BOU) to have occurred in Britain in a wild state, and thus excludes those for which a captive origin seems most likely. With an active trade in caged birds, of which a percentage are bound to escape, deciding which occurrences of unusual species in Britain constitute genuine instances of vagrancy is complicated and, to some degree, subjective (it is near impossible to *prove* either as the source of individual occurrences; Simpson 1991; Vinicombe *et al.* 1993; Parkin & Knox 1994; Holmes *et al.* 1998). For example, the same set of records as is judged by the BOU to provide evidence of the wild occurrence in Britain of 550 species is judged by Evans (1997a) to provide evidence for 580 wild species. Nevertheless, the important points are that this number is in the range 500–600, rather than, say, 100–200 or 1000–1100, and that the majority of the total species list for Britain comprises irregular visitors or vagrants. A large number of these have only been recorded at most a handful of times (Fig. 2.10).

In the period 1988–92, there were approximately 5500 records of individual vagrant birds in Britain. This is probably something of an underestimate of the numbers which actually occur. Fraser (1997) uses the phenomenon of 'weekend bias' to calculate these numbers. This refers to the tendency for vagrant birds to be discovered at weekends, when more birders are looking for them.

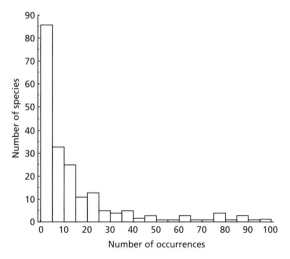

Fig. 2.10 Frequency distribution of the number of observed occurrences between 1958 and 1985 of the 204 species of bird for which less than 100 individuals of apparently wild origin were recorded in Britain in that period. From data in Dymond *et al.* (1989).

Fraser combined the assumption that 90% of all vagrants present on Sunday are detected by observers, with information on how many individuals are found on other days and how long vagrants tend to stay after they have been located, to estimate that almost half of all vagrant individuals to Britain go undetected. Given the frequency distribution of the number of occurrences of the rarest species (Fig. 2.10), some of the undetected vagrants will no doubt belong to species as yet unrecorded in Britain. The British bird list would undoubtedly be higher even than it is if all vagrants that made it to the country were discovered.

Whether the exact figure is 550 or 580, and whether or not this could have been higher had all vagrant individuals been discovered, the number of species recorded in Britain is relatively high for a country of its size and geographical position (intermediate latitudes, edge of a continent). In comparison, Evans (1997a) quotes totals for France and Italy of 515 and 491, respectively. The British list has the benefits of the efforts of the army of resident birdwatchers that make this particular avifauna so good for macroecological studies. Nevertheless, despite this huge contingent of observers, the list still compares poorly with those of many other countries. The species list for the Gambia, for example, stands at around 540 species (Barlow *et al.* 1997), but in a country less than 5% the size of Britain (Anonymous 1997). These figures compare with the 1080 species recorded from Kenya (Zimmerman *et al.* 1996), and the 1695 from Colombia (Hilty & Brown 1986).

Britain sits within the biogeographical region known as the Western Palaearctic. Evans (1997b) gives 938 as the number of bird species recorded from this region (again, the precise figure is debatable, depending on inclusion criteria, but it is the magnitude that is more important). Thus, a high proportion (≈60%) of the species recorded in this region have also been recorded in the small fraction that is Britain. The avifauna of the Western Palaearctic constitutes under 10% of the global total (Sibley & Monroe 1990, 1993). This compares with around 20% of the global total in Africa (Brown *et al.* 1982), and over 30% in South America (Rahbek 1997).

2.1.3 *Making sense of the numbers*

Over the last few pages, we have cited a large array of numbers, constituting species richness estimates for a wide range of areas of different size, scattered across the globe. These estimates are dynamic, increasing with time for reasons that change with temporal perspective. *The* species richness of an area does not, as such, exist. Nevertheless, since the rate of accumulation of species records usually quickly slows to a level at which additions are relatively rare occurrences (e.g. Figs 1.3 & 2.2), it is possible to compare the richness of different faunas, and ask why some are more speciose than others.

In that regard, as we have stepped up the scale ladder illustrated in Fig. 2.1, we have seen that the avifaunas of Eastern Wood, Britain and the Western Palaearctic are all poor relative to other avifaunas for which comparable data exist. However, while tiny in comparison to many local faunas around the

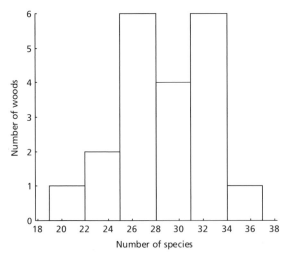

Fig. 2.11 Frequency distribution of the number of species of bird recorded in different woodlands, for a sample of 20 woodlands in southern England, including Eastern Wood. The number of species breeding in Eastern Wood is taken as the arithmetic mean over the 30 years of surveys. Data on the number of species breeding in the other woods are from Woolhouse (1983) and Ford (1987), but only include sites with area in the range 10–34 hectare, and concern a single year.

world, the avian species richness of Eastern Wood is not unusual for a deciduous woodland habitat in southern Britain. Figure 2.11 illustrates the number of bird species recorded breeding in one year in a sample of woods from southern England. Although the richness of Eastern Wood is towards the high end of this sample, it is well within its limits. But, why does it have the richness that it does? And why do other areas the world over have the richnesses they do? These are fundamental ecological questions. The answers encompass a range of different processes.

2.2   Size of area

One of the principal factors determining the number of species likely to be found at any site is its size. That species richness should be related to area is obvious. For example, there are close to 10 000 extant species of bird, but only 550 species have been recorded in Britain and Ireland (314 000 km^2) and 544 in Britain alone (230 000 km^2). Only 312 of the British species have been seen within the current boundaries of the county of Berkshire (1259 km^2; Standley *et al.* 1996), and only six within the current boundaries of T.M.B.'s urban Berkshire back garden (0.000075 km^2). The area within which the Eastern Wood assemblage has been censused is a mere 16 hectare (or 0.16 km^2), set within a larger tract of woodland that covers the approximately 112 hectare which comprise Bookham Common. This area is by no means small in comparison to other woodland patches in the highly fragmented region of southern England (cf. Moore & Hooper 1975; Woolhouse 1983; Ford 1987). Nevertheless, the number of birds recorded breeding in the wood is likely to be dictated foremost by its size. Indeed, the positive relationship between number of species found at a site and its area—the 'species–area' relationship—is one of the most robust and general patterns in ecology (Connor & McCoy 1979; Williamson 1988; Rosenzweig 1995).

More interesting than the existence of a species–area relationship is what form that relationship should take. This has been debated almost since the relationship was first documented (Arrhenius 1921, 1923; Gleason 1922, 1925; Connor & McCoy 1979; Wright 1981; Williamson 1988; Palmer & White 1994; He & Legendre 1996).

Debate about the shape of species–area relationships has centred on the transformation required to linearize them. Such transformation is desirable because linear relationships are easier to understand and compare than are curvilinear ones, and untransformed species–area relationships are rarely linear. In general, species numbers increase with area at a declining rate. Thus, as noted above, the number of breeding bird species in Surrey ($\approx$1850 km^2) is 2.5 times that in Eastern Wood (0.16 km^2), but the total number breeding in Britain ($\approx$230 000 km^2) is only about double that breeding in Surrey. Those studies that have compared the fits of different models to a wide variety of data have shown that species richness and area tend to exhibit a power relationship of the form

$$S = cA^z \qquad \text{(Eqn 2.1)}$$

where S is species number, A is area (km^2), and z and c are constants (e.g. Dony 1970 cited in Connor & McCoy 1979; Stenseth 1979; Williamson 1988). In other words, species number and area are linearly related to each other when both variables are logarithmically transformed, and Eqn 2.1 can be rewritten as

$$\log S = z \log A + \log c \qquad \text{(Eqn 2.2)}$$

This form was first suggested by Arrhenius (1921). Nevertheless, the fit of this model is variable, and was questioned as early as 1922 by Gleason, who thought an exponential model ($S = z \log A + \log c$) was more likely to be an appropriate descriptor. However, while some relationships do indeed seem to be better modelled as exponential (Dony 1970; Connor & McCoy 1979; Stenseth 1979) or other functions (e.g. He & Legendre 1996), it is generally accepted that the power form best describes the majority (e.g. Williamson 1988; Rosenzweig 1995).

Given that the species–area relationship is best modelled as a power function, one can ask what values are typically taken by the parameters z and c. Most attention has focused on the value of the exponent z. A review of this issue has recently been provided by Rosenzweig (1995), who suggests that there is no single answer, the value depending on the areas involved and their interrelationship (see also Williams 1943; Preston 1962). He distinguishes four distinct types of species–area relationship, of which three are of relevance to the macroecology of British birds. These are relationships within biotas, among islands in archipelagoes and between biotas. The fourth type describes the relationship between species number and area for tiny pieces of a biota, a spatial scale too small for present considerations.

The three relevant species–area relationships distinguished by Rosenzweig (1995), and the relationship he postulates between them, are illustrated in Fig. 2.12 (see also Holt 1993). The shallowest slopes are shown by relationships plotted across different areas within a region. Slopes of this sort differ from the other two in that the areas plotted are nested within each other (i.e. smaller areas are parts of the larger areas). An example is given in Fig. 2.13. Rosenzweig (1995) suggests that regression slopes (z-values) from relationships of this type are typically in the range 0.12–0.18, and that in Fig. 2.13 falls just outside this range. Note that although the coefficient of determination (r^2) of this relationship is high, the estimate of its statistical significance is inflated by the inherent non-independence of the individual data points.

Rosenzweig postulates steeper species–area relationships between islands in archipelagoes, with z-values in the range 0.25–0.35. The areas in these plots are not nested: each point refers to a separate island. An example of such a

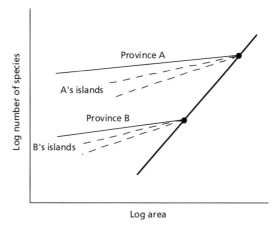

Fig. 2.12 The relationships between number of species and area for different types of areas. These are illustrated for two separate provinces (A and B; solid thin lines) and their associated islands (dashed lines; more isolated island groups have steeper slopes), together with the interprovincial relationship (bold line). Under this scheme, 'islands' are associated with larger land masses, and the main source of their species is from immigration. Real islands that are sufficiently isolated that most of their species derive from *in situ* speciation (e.g. the Hawaiian archipelago, at least prior to human colonization) would be classified as provinces. From Rosenzweig (1995).

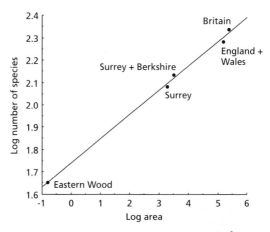

Fig. 2.13 The relationship between number of species and area (km^2) for breeding birds in nested subsets of Britain ($\log_{10} S = 0.11 \log_{10} A + 1.74$; $r^2 = 0.997$, $n = 5$). From data in Parr (1972), Beven (1976), Williamson (1987), Gibbons *et al.* (1996), Standley *et al.* (1996) and Anonymous (1997). Variation between sources means that species numbers are summed over different time periods for different regions: Eastern Wood 1949–79, Surrey 1900–70, Berkshire 1900–94, and Britain (England, Wales and Scotland) 1900–95.

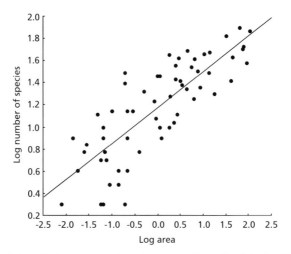

Fig. 2.14 The relationship between number of species and area (km²) for landbird species breeding on islands around the coast of mainland Britain ($\log_{10} S = 0.32 \log_{10} A + 1.17$; $r^2 = 0.66$, $n = 61$, $P < 0.0001$). From data in Reed (1981).

relationship for the landbirds breeding on 61 islands around the coast of mainland Britain (Reed 1981, 1983) is shown in Fig. 2.14. Although the extent to which this set of islands forms a strict archipelago is debatable, its species–area relationship fits well with what would be expected if it were (Rosenzweig 1995). Many authors have drawn an analogy between species–area relationships for 'true' islands (land surrounded by water) and those for habitat islands: that is, for islands of one habitat in a 'sea' of a different habitat. One example of a relationship of this latter type is given in Fig. 2.15, for birds in British woodlands (from data in Woolhouse 1983; Ford 1987). Some of these woodland sites are isolated fragments, while others are embedded in larger woodland tracts. For comparison, we have included data from Eastern Wood. Other species–area relationships for British birds on habitat islands have been published for woodland (Moore & Hooper 1975; Fuller 1982; McCollin 1993) and saltmarsh (Fuller 1982) birds, while Fuller (1982) also plotted the relationship between length and species number for upland and lowland British rivers.

Finally, Rosenzweig (1995) suggests that the steepest species–area relationships should pertain when whole biotas are compared. Figure 2.16 shows an example for the avian diversities of eight separate biotic regions. Rosenzweig does not provide firm figures as to what exponents such relationships should exhibit, but cites a number of examples with z in the range 0.5–1.0. That for the relationship in Fig. 2.16 is 0.66.

Although the avian examples just presented fit well with the pattern of variation in z-values suggested by Rosenzweig to pertain among different types of species–area relationship, it is unclear whether or not this scheme has broader generality. In particular, Williamson (1981, 1988) noted that published

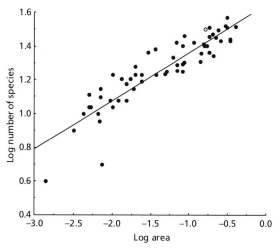

Fig. 2.15 The relationship between number of species and area (km^2) for breeding birds in habitat patches for woodland sites across Britain ($\log_{10} S = 0.28 \log_{10} A + 1.65$; $r^2 = 0.82$, $n = 59$, $P < 0.0001$). Species richness values are averaged over all years censused where necessary. Some of these woodland sites are isolated fragments, while others are embedded in larger woodland tracts. Eastern Wood is indicated by an open circle, with the point immediately to the right having been slightly displaced so that both are clearly visible. From data in Woolhouse (1983), Ford (1987) and Appendix II.

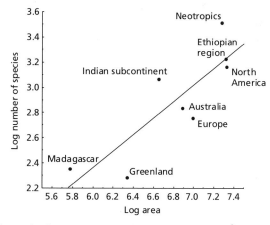

Fig. 2.16 The relationship between number of species and area (km^2) for the avifaunas of different biotic regions ($\log_{10} S = 0.66 \log_{10} A + 1.59$; $r^2 = 0.71$, $n = 8$, $P = 0.008$). From data in Slud (1976), which may be considered comparable, if not up to date.

z-values for species–area relationships among real islands, habitat islands and mainland samples all tended to span much greater ranges of values than implied by Rosenzweig (e.g. 0.05–1.132 for island relationships). However, there are many reasons why parameters of species–area relationships might vary without violating the general patterns illustrated in Fig. 2.12. They are likely to differ for different taxa and in different regions, and will be dependent too on the quality of the data from which they are compiled. A firm test of Rosenzweig's classification would need to control for such factors. Thus, whether or not it is correct remains to be determined.

Despite uncertainties over regularities in the form of species–area relationships, it is nevertheless true that the size of Britain, and the area of local sites like Eastern Wood, provides an initial answer to the question of why their associated avifaunas attain the richness they do. However, this answer simply begs additional questions. In particular, one can ask why larger areas tend to contain more species.

2.2.1   *Why do larger areas contain more species?*

As discussed in the opening chapter, the first hypothesis that the macro-ecologist should consider is the null. The appropriate null hypothesis for the species–area curve is that there is in fact no relationship between species number and area, and that observed is a statistical artefact of variations in sample size associated with areas of different sizes (Preston 1962; Connor & McCoy 1979; Williamson 1988). This idea can be illustrated by comparison of the avifaunas of Eastern Wood and of Britain. Between 1949 and 1979, when the team of ornithologists censused Eastern Wood for breeding birds (Section 1.6), they recorded over 5000 bird territories in the wood (an average of about 168 per year), for 45 different species. By contrast, between 1988 and 1991, a much larger group of ornithologists surveyed the whole of Britain for breeding birds, resulting in the second BTO atlas of British breeding birds (Gibbons *et al.* 1993). These observers submitted a total of 275 732 non-duplicate records of species in the 10×10-km squares of the British National Grid (i.e. no two records refer to the same species in the same 10×10-km square; the total number of records submitted was much higher than this figure), for 219 different species (Gibbons *et al.* 1993). The sampling artefact hypothesis suggests that the number of species recorded breeding in Britain between 1988 and 1991 is so much higher than the number in Eastern Wood from 1949 to 1979 only because the sample size was so much larger for the whole of Britain.

The sampling hypothesis is insufficient to explain most species–area relationships (but for woodland birds in Britain, see Bibby *et al.* 1985). Perhaps the simplest evidence that this is so involves simulation of the relationship. Two distinct simulation approaches can be taken. The first is to model the species–area relationship that would be obtained as a consequence of random sampling from a data set in which the number of species did not increase with area, and compare this relationship with that actually observed. Note that although species number and area are not related in the data set used for this simulation, they will appear to be so, because samples of larger areas include more individuals, which by chance come from more species. The number of species in an ecological sample is a function both of the sample size and of the underlying species–abundance distribution (Heck *et al.* 1975; Simberloff 1979; Haila 1983). The species–abundance distribution describes how individuals are apportioned among species (Chapter 4). Rosenzweig (1995) presents a series of such simulations for sets of data with realistic values of species richness and species–abundance distributions, and shows that the slopes of species–area relationships derived from the random samples fall well below (i.e. are flatter than) those observed in nature (see also Preston 1960; May 1975).

One problem with this approach is that features of the model structure (e.g. the overall species–abundance distribution) must be defined by the investigator. Comparisons with real systems will then only be valid to the extent that the model is an accurate reflection of nature. A better method is to use the data on which a species–area relationship was determined, to test whether the samples on which the species richness estimates from the smaller areas are based have lower richness than equivalent sized samples from the larger area. It has the advantage over the first method that the species richness values and the overall species–abundance distribution are specified by the data, and so their properties do not have to be defined separately.

This second approach can be illustrated with reference to the earlier comparison between the faunas of Eastern Wood and Britain. We can ask whether samples of 5000 pairs from the entire British breeding avifauna tend to include more species than have been recorded from Eastern Wood. Appendix III lists population size estimates for all species of bird considered to be part of the British breeding avifauna, taken from Stone *et al.* (1997). These estimates derive from studies embracing a range of time periods, but should generally be indicative of the normal British population sizes of these species. Summing these estimates, the total number of breeding pairs of all species on this list is just over 62 500 000. This is around the number of pairs of bird we would expect to breed in Britain each year. The mean (± standard deviation) number of species in 100 separate random samples of 5000 from this total is 116.3 (± 3.7). In other words, if the sample of 5000 territories (which we assume is equal to the number of breeding pairs) recorded from Eastern Wood had been drawn at random from the British avifauna, we would have expected to find more than twice as many species as were actually recorded in the wood. Alternatively, we can compare the mean annual species richness of Eastern Wood with the richness of a sample of equivalent size from the British avifauna. An average of 168 territories per year were recorded in the wood in the period 1949–79, and 31.8 (± 2.6) species. One hundred samples of this size from the British avifauna yielded an average of 45.5 (± 3.4) species—significantly more. (The performance of the test of this hypothesis, and tests described in subsequent chapters, might have been different had data been used for population sizes of species in Britain in the period 1949–79. Such data do not exist, but we think it unlikely that the results are greatly affected by this problem, because, although the population sizes of many species have changed, the changes have been relatively minor compared with the more critical between-species differences in population size.)

A shortcoming of this comparison is that some British breeding birds are unlikely ever to breed in Eastern Wood. For example, the guillemot (an auk nesting on sea cliffs) is a common British bird, but even with global environmental change it will be a long time before a pair settle in Eastern Wood (vagrant individuals are only found even wandering markedly inland very occasionally!). We can make the random draw model more realistic by testing whether the species richness of Eastern Wood differs from that of random samples from the community of British birds that breed in deciduous woodland (classified using Ehrlich *et al.* 1994). The total number of British breeding pairs of these 80 species is just over 51 000 000. The mean (± standard deviation) number of species in 100 separate random samples of 5000 from this total is 56.5 (± 2.0), again significantly more than observed. By contrast, comparison of the mean annual species richness of Eastern Wood with the richness of samples of equivalent size (168 territories) from the British woodland avifauna revealed no significant difference. The average richness of 100 random samples was 31.8 (± 2.4) species, exactly the average annual number of species breeding in Eastern Wood.

The simulations just described calculate the expected number of species in small samples from the total British and the British deciduous woodland avifaunas under the assumptions that individuals are randomly and homogeneously distributed across the environment. They are therefore directly equivalent to rarefaction (Sanders 1968; Hurlbert 1971; Simberloff 1979; James & Rathbun 1981; Haila 1983; Brewer & Williamson 1994). This allows us to check the simulation results against the expected number of species in samples from the total British and British woodland avifaunas using mathematical functions for rarefaction. A function estimating the number of species in small samples from a larger sample has been given by Hurlbert (1971), while a different function for the number of species expected in a fraction of a larger area was derived by Coleman *et al.* (1982). Brewer and Williamson (1994) showed that a slight modification of the Coleman function makes the two identical. Using Brewer and Williamson's modified function, the number of species expected in samples of 5000 and 168 territories from the entire British avifauna are 117.2 and 45.8, respectively. The equivalent results obtained by sampling only birds breeding in British deciduous woodland are 56.5 and 31.6 species. These expectations concur with the simulation results.

From the simulation and rarefaction results, we can conclude that Eastern Wood has fewer species than would be expected if it were simply a random sample of the total British avifauna. While the average number of species breeding in any given year is no different to that expected if species are randomly sampled from the British woodland avifauna, the total number recorded in the wood is significantly lower than predicted by this model. Thus, we can conclude that the sampling hypothesis cannot fully explain patterns in avian species richness in Eastern Wood.

Given that a pure sampling hypothesis generally cannot explain species–area relationships, we can turn attention from the null hypothesis, which is also falsified in this particular case, to biological explanations for the pattern. Principal among these is the habitat hypothesis (Williams 1964). Put simply, larger areas might contain more species because they contain more habitats. While, as just argued, the avifauna of Eastern Wood is likely to be composed only of species typical of deciduous woodland and its margins, the avifauna of Britain can include species reliant on open country (e.g. stone curlew, skylark), coniferous woodland (e.g. crested tit, common crossbill), fens (e.g. bittern, bearded tit), rivers (e.g. kingfisher, dipper, grey wagtail), lakes (e.g. grebes, ducks), saltmarshes (e.g. redshank) and moorland (e.g. red grouse, ring ouzel) (Fuller 1982). None of the species listed is likely to breed in Eastern Wood unless major changes in the habitat ensue. As long as species show some degree of habitat specialization, as is clearly true for British birds (e.g. Fuller 1982), Britain as a whole will have more species than Eastern Wood because it encompasses more habitats. In general, the species richness of larger areas may be higher for this reason.

The habitat hypothesis generates two obvious predictions: habitat diversity should be a better predictor of species richness than area, and there should be no species–area relationship (or, at least, not one that differs from the null hypothesis) in cases where increases in area are not accompanied by increases in number of habitats. Several studies support the first of these predictions (see also Johnson 1975; Boecklen 1986). For example, Reed (1981, 1983, 1984) showed that the number of habitat types on British coastal islands was a better predictor of the number of breeding landbirds than was area. Similarly, Haila (1983) noted that the number of bird species in communities of equal size was higher on those islands of the Finnish Åland archipelago with more diversified habitats (but see Martin & Lepart 1989). Rafe et al. (1985) found that a measure of habitat diversity was a better predictor than area of the number of bird species on a selection of Royal Society for the Protection of Birds (RSPB) reserves. Peck (1989) found a positive relationship between the number of bird species using compartments in a forestry plantation in northern England and the number of tree species in those compartments, while the number of tree species was not correlated with compartment size. Further evidence in support of the importance of habitat diversity has been found for a variety of taxa (e.g. Kitchener et al. 1980a,b; Rigby & Lawton 1981; Buckley 1982; Tonn & Magnuson 1982; Fox 1983; Haila & Järvinen 1983; Quinn et al. 1987; Sfenthourakis 1996; Burnett et al. 1998; Nichols et al. 1998).

Nevertheless, a number of studies have failed to support the first prediction of the habitat diversity hypothesis. McCollin (1993) found that the avian richness of woodland fragments in east Yorkshire depended on between-patch (landscape) rather than within-patch (habitat) structure. Ford (1987) showed that area was the primary determinant of richness in Oxfordshire woodland islands, although habitat heterogeneity did explain significant amounts of residual variation in richness, and was the best predictor of richness in his control plots (embedded in larger areas of woodland). Bellamy *et al.* (1996a) found similar results for woodland islands in Cambridgeshire and Lincolnshire, and they have also been obtained for bird species numbers in habitat patches in other parts of the world (e.g. Kitchener *et al.* 1982; Howe 1984; Freemark & Merriam 1986; Møller 1987; Díaz *et al.* 1998).

Teasing apart the relative influences of area and habitat diversity on bird richness is always likely to be difficult. The problem is that area and habitat number are themselves often highly correlated (Harner & Harper 1976; Reed 1981; Rafe *et al.* 1985; Ford 1987; Rosenzweig 1995). Even if the species–area relationship were entirely a consequence of habitat number, the effect of area may be stronger in analyses because area can more easily be accurately assessed: there are many ways in which habitat number can be quantified, and there is no guarantee that the method chosen will reflect heterogeneity in those features of the habitat that influence bird richness (see, for example, Knight & Morris 1996). Given this problem, the analyses reported above are strongly suggestive of an effect of habitat, albeit not conclusive.

The way around the problem of covariation between habitat diversity and area is to examine variation in the number of species when one or other variable is constant: changes in area unaccompanied by changes in habitat diversity should lead to no species–area relationship, or alternatively, differences in habitat diversity in areas of similar size should be associated with differences in species richness. If habitat complexity is used as an indicator of habitat diversity, then there is significant evidence for a positive relationship with avian diversity (e.g. MacArthur & MacArthur 1961; MacArthur *et al.* 1962, 1966; Karr 1968; Recher 1969; Karr & Roth 1971; Moss 1978; Fuller 1982; Tellería *et al.* 1992). Rosenzweig (1995) uses data in Boström and Nilsson (1983) to show that area and avian diversity are unrelated on Swedish peat bogs once the effect of sample size has been removed, arguing that these represent a constant habitat. However, the extent to which that is true is unclear, as Boström and Nilsson themselves argue that variations in species densities are attributable to habitat variation between bogs. Nevertheless, no British birder intent on seeing a large number of species would prefer to spend the day in a square kilometre patch of deciduous woodland rather than a square kilometre containing deciduous woodland, grassland, open fresh water, fen and coastline. The same applies the world over. In sum, habitat diversity seems likely to be a major determinant of the species–area relationship.

Although there is an important role for habitat diversity in determining the number of species found in an area, this seems unlikely to be the sole effect. In particular, the number of studies that show a relationship between area and avian richness once habitat diversity has been accounted for, whether or not habitat diversity is the principal predictor of avian richness, suggest that area is exerting an additional effect on bird species number (Kitchener *et al.* 1982; Howe 1984; Rafe *et al.* 1985; Freemark & Merriam 1986; Ford 1987; Møller 1987; Nilsson *et al.* 1988; Martin & Lepart 1989; McCollin 1993; Bellamy *et al.* 1996a; Díaz *et al.* 1998). A likely candidate for the way this effect may act is through the relationship between area and rates of colonization and extinction. This relationship has been formalized as the equilibrium theory of island biogeography (generally attributed to MacArthur & Wilson (1963, 1967), although the basic theory had been independently developed before their treatment; see Whittaker (1998) for a brief history). This theory has been highly emotive, and has generated one of the largest associated literatures in ecology. A review is beyond the scope of this book (the interested reader is directed to MacArthur & Wilson 1967; Williamson 1981, 1988; Rosenzweig 1995; Gotelli & Graves 1996; Whittaker 1998), and we limit ourselves to a discussion of the salient points.

The theory of island biogeography posits that the number of species on an island results from a dynamic balance between the number of species colonizing from the mainland source pool, and those going extinct after colonization. Colonization rate is hypothesized to decline as the number of species on the island increases, because there are fewer species remaining to colonize, and because the early colonizers will be those best suited to colonization (e.g. good dispersers). Extinction rate on the island is hypothesized to increase with number of species, as each species has its own finite probability of extinction, and because negative interactions between species are more likely when there are more species (although positive interactions may also increase, nullifying this latter effect). Therefore, as the number of species on the island increases, colonization rate declines and extinction rate increases. At some number of species, these two processes will reach equilibrium. This is the number of species predicted to be found on the island (MacArthur & Wilson 1967). Although framed in terms of real islands, this theory has also been applied to habitat islands (e.g. Vuilleumier 1970; Brown 1971, 1978; Johnson 1975; Williamson 1981; Newmark 1987; Brown & Dinsmore 1988; Robinson & Quinn 1988; Lomolino *et al.* 1989; Nores 1995). In fact, since it posits that species derive from the process of colonization, the theory may better apply to habitat islands than oceanic islands, for which speciation is likely to be a more important source of species (Tokeshi 1999).

The equilibrium theory is relevant to the species–area relationship because the processes of colonization and extinction are likely to vary with island size (MacArthur & Wilson 1967). In particular, the total number of individuals, and so the average population sizes of species, will be larger on large islands. Species with larger population sizes are by chance less likely to go extinct

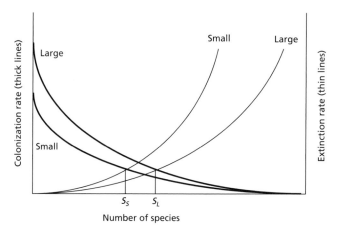

Fig. 2.17 Theoretical relationships between rates of colonization (immigration) and extinction and the number of species inhabiting islands. Functions differ for small and large islands. The points at which the curves intersect define the numbers of species expected on small (S_S) and large (S_L) islands at equilibrium. The colonization and extinction functions are usually presented as being non-linear, but the basic predictions of the model do not change if linear functions are used instead (e.g. Gotelli 1995).

(Chapter 4). Therefore, extinction rates should be lower on larger islands. Larger islands also may have higher colonization rates, because they present larger targets to dispersing individuals (Gilpin & Diamond 1976; Simberloff 1976; Lomolino 1990). Both of these considerations ought to result in higher numbers of species on larger islands (Fig. 2.17). Note that these processes are unlikely to work independently of habitat diversity, in particular because colonization will not succeed in the absence of appropriate habitat; as Haila (1983) notes, the habitat composition of an island determines the size of the source pool of potential immigrants. However, the equilibrium theory may provide an additional effect of area on species richness.

If the numbers of species on islands are influenced by rates of colonization and extinction as proposed, three patterns should be apparent: (i) colonization rate should decline as species number increases; (ii) extinction rate should increase with species number; and (iii) there should be substantial turnover in species composition on the island over time. The first two patterns are assumptions of the model, and the third a prediction.

Figure 2.18 shows the relationship between the number of bird species breeding in Eastern Wood and colonization and extinction rates (number of events the following year). In both cases, the relationship follows that predicted by island biogeography theory: extinction rate increases and colonization rate decreases with species number. These graphs follow analyses by Williamson (1981), but correct minor errors, and include data from four additional years. Williamson found that the regression for extinctions was not stat-

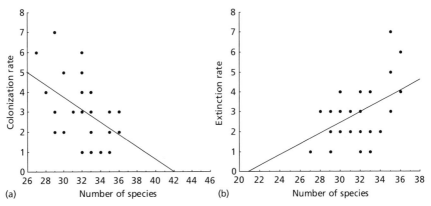

Fig. 2.18 The relationship between the annual rate of (a) colonization (number per year) (slope = 0.31, $r^2 = 0.23$, $n = 29$, $P = 0.0085$) and (b) extinction (number per year) (slope = 0.27, $r^2 = 0.20$, $n = 29$, $P = 0.014$) and the numbers of bird species breeding in Eastern Wood in the period 1949–79. Adding a quadratic term does not significantly increase the amount of variation explained by either regression. There was no census in 1957, and so 1958 is assumed here to follow directly from 1956.

istically significant, but the additional data change this conclusion (Fig. 2.18b). The regression lines in Fig. 2.18 imply that there would be no additional immigration to a breeding assemblage of 42 species in Eastern Wood, and no extinction from an assemblage of 21 species. This range brackets the actual number of species found breeding in Eastern Wood in any one year (Fig. 1.3). The lines intersect at the mean number of breeding species (32), when three species would be expected to colonize and three go extinct. Equilibrium theory implies that even an assemblage of one species should have a non-zero probability of extinction, so theoretically the line in Fig. 2.18b should pass through the origin. Clearly, how the extinction relationship would change in response to larger perturbations in the number of breeding species than occurred naturally in the period 1949–79 is a matter of conjecture. Nevertheless, the Eastern Wood data qualitatively support the predictions of island biogeographical theory.

Additional support for relationships between species richness and the rates of colonization and extinction comes from information on the occurrences of birds on 13 small islands around the coast of Britain. Manne *et al.* (1998) used a maximum likelihood method to fit immigration and extinction functions to these data. They found that avian extinction rates increase consistently with species number on these islands, while immigration rates decrease. Moreover, most of the 13 islands show functions of the concave form predicted by MacArthur and Wilson (1967) (e.g. Fig. 2.17). These results are compromised slightly because avian extinction rates are not significantly higher for birds on smaller islands, as would be required for the equilibrium theory to explain the species–area relationship. Nevertheless, extinction rates are generally higher

Table 2.1 The numbers of bird species which bred in 1968–72 and 1988–91, and did or did not breed in the other period, for (a) Britain and (b) Ireland. Figures in parentheses include species that bred in the wild but which were of introduced or reintroduced origin, or were feral. From data in Gibbons *et al.* (1993).

(a)

		1968–72		
		+	–	Total
1988–91	+	197 (210)	6* (8)	204 (219)
	–	4† (4)	[8 species bred in 1973–87 but in neither other period]	
	Total	201 (214)		

(b)

		1968–72		
		+	–	Total
1988–91	+	130 (133)	7‡ (9)	137 (142)
	–	5§ (6)	[4 species bred in 1973–87 but in neither other period]	
	Total	135 (139)		

* Red-necked grebe, whooper swan, crane, purple sandpiper, parrot crossbill, scarlet rosefinch.
† Great northern diver, black tern, snowy owl, hoopoe.
‡ Black-throated diver, Leach's petrel, garganey, reed warbler, lesser whitethroat, pied flycatcher, common crossbill.
§ Black-necked grebe, Montagu's harrier, greenshank, turtle dove, yellow wagtail.

on smaller islands in this data set. The lack of formal statistical significance may be a function of the relatively limited range of island sizes (Manne *et al.* 1998).

Turnover in species composition is a common feature of British bird assemblages on both real and habitat islands (Table 2.1; e.g. Reed 1980; Williamson 1981; Gibbons *et al.* 1993, 1996; Hinsley *et al.* 1995; Russell *et al.* 1995). In Eastern Wood, for example, 45 species bred at least once in the period 1949–79, but the maximum number in any one year was 36. Most species bred in most years (Fig. 2.4), and 15 species bred in all years, but 20% of species bred in less than one-third of years. The composition of the breeding bird assemblage changed annually throughout the census period (Appendix II). The question is not whether turnover occurs, but whether that which does occur supports the equilibrium theory. Again, the evidence is inconclusive, and can be interpreted either way (cf. Williamson 1981; Rosenzweig 1995). For example, the bird species richness of British woodland habitat islands is often split into a component consisting of woodland specialists, and one consisting of edge or transient species not considered to be a 'true' part of the woodland assemblage (e.g. Ford 1987; McCollin 1993; Bellamy *et al.* 1996a). Turnover can be ascribed largely to the transient species. Conversely, some of the changes to the avifauna of Eastern Wood are the result of broader regional changes in the abundances and distributions of species, while others are the result of habitat changes within the wood. The loss of the whitethroat and the garden warbler as a consequence of drought on their wintering grounds has already been mentioned (Figs 1.15 & 1.16), and they provide obvious examples of the consequences of broader regional changes. Changes within Eastern Wood seem to be directly responsible for colonization by the starling in the late 1950s (Fig. 1.18), and the local extinction of the willow warbler (Fig. 1.17) as a breeding species. Both trends can be ascribed to habitat changes within the wood following cessation of management in the early 1950s, as grassy areas were overgrown by scrub and more trees reached maturity (see Chapter 1). All these examples suggest that changes to bird faunas are predictable from knowledge of ecological processes, albeit acting over a range of spatial and temporal scales. Thus, they hardly provide strong evidence that such assemblages are in the dynamic equilibrium required by the equilibrium theory (Williamson 1981).

On the other hand, some colonization and extinction events do seem to be stochastic, as island biogeographical theory would predict (Rosenzweig 1995; Gotelli & Graves 1996). A likely candidate from the avifauna of Eastern Wood is the treecreeper (Fig. 2.19), which failed to breed in the wood in 1951, despite maintaining between one and five territories in other census years. The treecreeper disappeared from and recolonized Eastern Wood without any marked change in habitat. Such gains and losses are particularly evident on

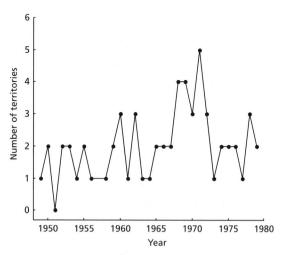

Fig. 2.19 The number of territories of treecreeper recorded in Eastern Wood in each year from 1949 to 1979.

small islands, where population numbers are low and the death of a few individuals can result in local extinction. Indeed, many of the studies of avian colonization and extinction have focused on such islands (e.g. Reed 1980; Williamson 1981; Pimm *et al.* 1988; Russell *et al.* 1995; Manne *et al.* 1998). In fact, for very small sites most of the year-to-year variation in species composition may be a result of stochastic processes. Much of such variation may simply be the result of the continuous change that occurs in the spatial configuration of the populations of most bird species, with new breeders appearing in previously unoccupied sites and old breeders vanishing from previously occupied ones either by dying or by moving their territories (Boecklen & Simberloff 1986; Haila & Hanski 1993; Haila *et al.* 1993, 1996).

At a larger scale, the ebb and flow of some very rare breeding birds in Britain may essentially reflect such stochastic processes. For example, while the range of the marsh warbler has been expanding northwards across Europe this century and the last (Hagemeijer & Blair 1997), it has declined at traditional breeding sites in central England (Kelsey *et al.* 1989; Gibbons *et al.* 1993). Evidence that this small English population is isolated from the greater continental population suggests that it is likely to be vulnerable to stochastic effects (Kelsey *et al.* 1989). By contrast, a newly established marsh warbler population in the extreme south-east of England seems likely to derive from migrant individuals overshooting from the continent. Indeed, even the core European population of the marsh warbler is apparently patchy in nature, suggesting that frequent extinctions and colonizations are a feature of this migrant species (Kelsey *et al.* 1989). Thus, both increases and decreases in British marsh warbler populations over the past 20 years or so seem likely to be in part a result of stochastic events.

In sum, some cases of colonizations and extinctions can be attributed to well-defined changes in habitat or wider population fluctuations. Others seem to be entirely stochastic. Turnover may or may not therefore be indicative of a dynamic equilibrium. Whichever interpretation is preferred, however, all evidence points to the species richness of islands, real or habitat, being affected by the processes of colonization and extinction.

If the relationship between species richness and area was influenced by the processes of colonization and extinction, that could also explain why the parameter z might take characteristically different values in different situations (Fig. 2.12). Recall that z has been suggested to be lower for samples of continuous regions than for islands (lower z-values indicate flatter species–area relationships). In other words, species richness declines more slowly with decreasing area in samples of a continuous region than on islands of equivalent size surrounding the region. If correct, this pattern could be explained by the amount of immigration which areas receive. Many populations in small samples of continuous regions would not be self-supporting if those areas were isolated, but they are maintained because of a constant influx of individuals from the surroundings (the 'rescue effect'; Brown & Kodric-Brown

1977). On islands, this rescue effect is much weaker, and so more populations ought by chance to go extinct. Population extinctions should be more frequent on the smallest islands, because of the smaller average population sizes of their resident species (see above and Chapter 4), and so the species richness of these should be depressed the most. This would result in steeper z-values of islands relative to mainland samples. The slopes for islands would then be shallower than those across biogeographical regions (cf. Figs 2.14 & 2.16), because islands have higher immigration rates than regions. Biogeographical regions are by definition areas that derive the bulk of their faunas from *in situ* evolution, rather than from colonization from other areas (Rosenzweig 1995).

The link between colonization, extinction and species–area relationships has recently been considered from a fresh perspective, namely that of metapopulations. Metapopulation dynamics assume that the landscape is divided into a set of discrete habitat patches, which a species may or may not occupy. At equilibrium, the probability that any one patch is occupied by a species (its incidence) is then a function of its rates of colonization and extinction across the whole metapopulation (e.g. Levins 1969; Hanski 1982a, 1991a; Hanski & Gilpin 1991). Hanski and Gyllenberg (1997) noted that there is a natural link between this metapopulation dynamic perspective and the species–area relationship. The sum of incidences across species gives the expected number of species on islands. Using a simple metapopulation dynamics model, and modelling extinction rate as inversely proportional to the carrying capacity of a species on an island, which is in turn dependent on the product of the species density and island area, Hanski and Gyllenberg were able to derive quantitative predictions for the slope z of the species–area relationship. The value predicted depended on parameter values modelling isolation of patches and variance in species density, but was about 0.1 when 80% of species occurred on an average-sized island, and increased to 0.45 as this percentage dropped to 20%.

As well as generating realistic z-values, Hanski and Gyllenberg's model has other interesting properties. For a given total species pool, it predicts higher values for systems where islands are colonized from a mainland source pool than for systems lacking this source (colonization occurs just among islands). The model predicts that z should increase with the isolation of the islands. The model also provides an explanation for why z-values should be smaller among habitat patches on mainlands than between real islands in archipelagoes. As just described, the normal interpretation of this pattern is that the richness of mainland habitat patches is maintained by the rescue effect. Hanski and Gyllenberg's model suggests instead that the effect is a consequence of the extinction of rare species from the very largest sites, and indeed from the entire metapopulation. However, as this suggests that the lower z-value among mainland habitat patches derives from a reduction in the richness of the largest patches, rather than an elevation of the richness of the smallest patches, we are not convinced that it adds to the model's attractions.

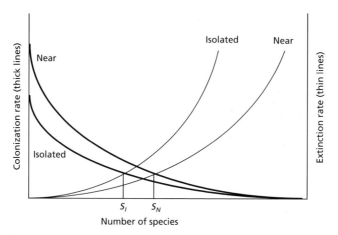

Fig. 2.20 Theoretical relationships between rates of colonization (immigration) and extinction and the number of species inhabiting islands. Functions differ for islands near to or isolated from the source pool of immigrants. The points at which the curves intersect define the numbers of species expected on near (S_N) and isolated (S_I) islands at equilibrium.

2.3 Isolation

If colonization and extinction rates determine, at least in part, the species richness of a site, that suggests an additional feature of sites that should affect their richness. That feature, alluded to in the suggestion of why z-values for species–area curves might be steeper for islands than for mainland habitat patches, is the isolation of the site.

Isolation should influence colonization in a manner similar to area, by making some islands, whether real or habitat, harder targets for immigrants to hit. The further away an island is from a source pool of immigrants, the less likely it is that immigrants will find, and so colonize, it. More isolated islands should therefore have lower immigration rates than islands closer to the source pool. Increased isolation may also increase extinction rates because of a weakening of the rescue effect (Brown & Kodric-Brown 1977). Populations on all islands may at some point in their histories decline to levels where extinction is likely. However, populations on islands close to a source pool of immigrants are more likely to be rescued from extinction by immigrants from this pool. Equivalent populations on more distant islands, without this rescue effect, are more likely to go extinct. This is illustrated in Fig. 2.20, for which the analogy to Fig. 2.17 is obvious. In sum, isolation affects species richness through its effect on rates of immigration.

A number of studies have tested for relationships between the isolation of sites in Britain and their avian species richness. Reed (1981) showed that land-bird richness was lower the more distant offshore islands were from mainland

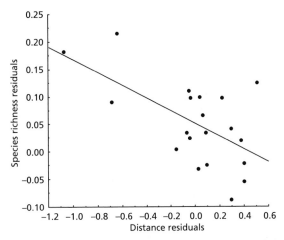

Fig. 2.21 The relationship between species richness and isolation for bird species in Oxfordshire woodland habitat islands, controlling for the effects of woodland area ($y = 0.12x + 0.05; r^2 = 0.36, n = 20, P = 0.005$). The axes are the residuals of regression of $\log_{10}$ species richness against $\log_{10}$ area (hectare), and of $\log_{10}$ distance (km) to the nearest wood > 100 hectare against $\log_{10}$ area. From data in Ford (1987).

Britain, once the effects of habitat diversity and island area had been controlled for. Ford (1987) found that, having controlled for the effect of area, the distance to the nearest wood larger than 100 hectare was the best predictor of avian richness in his Oxfordshire woodland islands (Fig. 2.21). McCollin (1993) noted different effects of woodland patch isolation on species typical of woodland and woodland edges. Woodland species richness declined with isolation, as expected, while edge species richness increased. Edge species in this case are typically those that utilize woodland for breeding but forage in surrounding farmland, and so their richness may better reflect features of the latter habitat than the former. Bellamy *et al.* (1996a) found that area was overwhelmingly the best predictor of the richness of woodland specialist birds in woodland habitat patches in eastern England in censuses for three separate years, but that there was also a significant positive effect of landscape measures related to isolation (although the measure involved differed between years). Other studies demonstrating effects of isolation on avian species richness include those of Hamilton and Armstrong (1965), Greenslade (1968), Vuilleumier (1970), Johnson (1975), Opdam *et al.* (1984), Martin and Lepart (1989), Daniels *et al.* (1992), Nores (1995) and Díaz *et al.* (1998). Johnson (1975) argued that isolation affected habitat diversity as well as bird species richness on the mountaintop islands he studied, so that the relationship he described between habitat diversity and bird richness was in large part determined by isolation. Thus, the species richness of Eastern Wood may be high relative to other woodland patches of similar area (Fig. 2.15) because it is part of a larger area of woodland on Bookham Common, and so is not isolated.

While apparently well established, the idea that species richness and isolation are causally related has been questioned. Most notably, Lack (1969, 1976) argued that it was ecological features of islands, rather than their isolation, that determined the presence and absence of components of mainland avifaunas. The basis of his argument was the contention that the dispersal capabilities of birds are so great that most islands in the world are well within the reach of immigrants. Thus, immigration was unlikely to limit the occurrence on islands of species from the source pool, and their absence was likely to be a consequence of habitat availability and interspecific competition. An example may be provided by the avifauna of New Zealand, where the recent natural colonization of these islands by several bird species seems likely to have been facilitated by habitat change initiated by another recent colonist, humans (e.g. McDowall 1969; Williams 1973). More generally, a number of studies do indeed fail to find evidence for a negative effect of isolation on avian richness (e.g. Helliwell 1976; Kitchener *et al.* 1982; Howe 1984; Reed 1984; Blake & Karr 1987; Møller 1987; Opdam & Schotman 1987; Thiollay 1997).

The regularity with which effects of isolation are identified in studies of British birds is surprising given the scale at which these are carried out relative to the dispersal abilities of most species in the British assemblage. For example, Ford (1987) found a strong effect of isolation on richness in a set of woodland islands (Fig. 2.21) which were, on average, 5.3 km from a wood of more than 100 hectare, and none of which were more than 11.2 km from a wood that large. Yet, a recent study (Paradis *et al.* 1998) of 75 British bird species found that the average geometric mean natal dispersal distance of species in the set was also 5.3 km, while the strong right skew in intraspecific dispersal distances indicated that individuals regularly move much greater distances. At this scale, even isolated woods ought to be readily colonizable by most bird species. One reason why isolation may seem to be important to avian richness is that, in any set of habitats, it is likely to be compounded with other, ecological, differences between sites. Thus, Moore and Hooper (1975) found that while more isolated British woods had lower bird diversities than less isolated woods, they also had fewer plant species, lacked a shrub layer, or were at higher latitudes or altitudes. In fact, the amount of variation in species richness explained by isolation is often small. For example, habitat diversity, island area and isolation together explained between 80% and 95% of the variation in species richness on the islands considered by Reed (1981), yet isolation never contributed more than 3% to these figures. Similarly, Bellamy *et al.* (1996a) could explain between 75% and 80% of the variation in woodland habitat island species richness using measures of area, habitat diversity and isolation, but all measures of isolation combined contributed 6% at most. These percentages are small enough perhaps to be consequences of associations between isolation and other, untested, ecological features of islands.

Despite these objections, however, an effect of isolation on species richness is still the most parsimonious conclusion from the available data. It is repeatedly observed, even in studies that account for habitat differences between islands. For example, Reed (1987) statistically analysed data for birds on islands in the Bahamas and the Gulf of Guinea which Lack (1976) had presented in support of his contention that isolation was not important to the avifaunal richness of islands. In each case, distance from the mainland source of immigrants was the strongest predictor of the richness of the islands, and the only significant predictor for islands in the Gulf of Guinea.

Clearly, Lack was correct in asserting that the vagility of birds is such that many species absent from islands are more than capable of reaching them: the annual arrival in Britain of vagrant individuals from North America and Siberia attests to that (a fascinating examination of the distances involved in such vagrancy is given by Bentley 1995), and one of the authors has observed vagrant individuals of several European species on a remote island in the sub-Antarctic. However, distinction must be drawn between the immigration of individuals and the immigration of 'propagules', defined as the minimum population of a species from which a new colony can be produced. Colonization is only possible in the latter case. Because for birds a propagule probably consists of several individuals, the probability of colonization is concomitantly lower than the probability of vagrancy. Indeed, for all the individual American birds that have arrived on British shores as vagrants, we are aware of only one example of successful breeding here by an American species (the spotted sandpiper in 1975; Sharrock & Sharrock 1976). This cannot only be a consequence of competitive and environmental factors, as American species have successfully established following introduction or escape from collections (e.g. Canada goose, ruddy duck). There must also be an effect of isolation from the source pool of immigrants (North America in this example).

Further evidence for the effect of isolation on richness comes from cases where it is decoupled from the rate of immigration. As this is rarely the case for birds (because of their vagility), we must turn to other taxa. North American mammals provide a good example. Lomolino (1994) compared the relationship between isolation and species richness for mammals on islands of the Drummond Archipelago in Lake Huron, islands off the coast of Maine and islands in the St Lawrence river. Correcting for area, richness is highest for the islands in Lake Huron, although they are no closer to the mainland than are the others. This difference arises because, relative to the other two island groups, the islands of the Drummond archipelago are connected to the mainland by more stable ice bridges in winter, and are separated from the mainland by weaker water currents in summer. The commensurately higher immigration

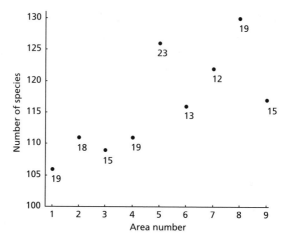

Fig. 2.22 The peninsula effect in the birds breeding in the English west country. Number of species is the number breeding in each of nine areas running from the western tip of Cornwall (area number 1; the seaward end of the peninsula) to Somerset and east Devon (area number 9), with breeding distribution as given by Gibbons *et al.* (1993). The figures by the points are the number of the 10×10-km squares included in each area on which bird distributions are mapped. While these areas are not equal, they do not favour finding a peninsula effect, and there is no relationship between number of squares and number of species ($r^2 = 0.02$, $n = 9$, $P = 0.69$). By contrast, number of species is significantly related to area number ($r^2 = 0.55$, $n = 9$, $P = 0.02$), indicating that richness decreases towards the tip of the south-west peninsula. Multiple regression of number of squares and area number on species richness allows identical conclusions to be drawn (area number: partial $r = 0.85$, $P = 0.008$; number of squares: partial $r = 0.62$, $P = 0.10$).

rate to this island group maintains the higher mammalian richness. Thus, when immigration is independent of isolation, the effect of isolation is lost. Paradoxically, this serves to affirm the importance of isolation on species richness.

A pattern analogous to that for island richness, but which relates to areas that are not completely isolated from the mainland, is the 'peninsular effect'. This describes the tendency for the species richness of terrestrial taxa to decline from the base of a peninsula (i.e. the mainland end) to the tip. The equivalent pattern for aquatic taxa is known as the 'bay effect' (Rapoport 1994). Examples of the peninsula effect exist for some taxa on some peninsulas (e.g. Simpson 1964; Kiester 1971; Brown & Opler 1990; Martin & Gurrea 1990; Currie 1991); it is apparently shown by birds of the Yucatan, Florida and Baja California (Cook 1969; Taylor & Regal 1978). Figure 2.22 shows the effect in the bird species breeding in the English west country peninsula (principally Devon and Cornwall). However, a number of case studies for other regions or other taxa do not exhibit such a trend (e.g. Seib 1980; Due & Polis 1986; Brown 1987), and the generality of the pattern is at best questionable. Where it does occur, it has been suggested most likely to result from a reduction in colonization rates from the base of the peninsula towards its tip, and so to be another consequence of the effect of isolation on richness. Where it does not pertain, such as in the scorpion fauna of Baja California (Due & Polis 1986), this may be because colonization is unimportant. However, plausible alternative explanations invoke autecological and palaeogeographical causes, and the arguments about likely causes, like the patterns and mechanisms themselves, bear good analogy with those relating to islands. In that vein, isolation seems likely to contribute to the peninsular effect, but perhaps only weakly, or as one of several causative factors.

2.4 Local–regional richness relationships

So far, we have considered the effects of area and isolation on the species richness of local sites. Although an understanding of the impact of these factors requires a perspective larger than that of the local site, area and isolation are nevertheless attributes of the site itself. However, implicit in the variation noted in the form of the relationship between species richness and area is another large-scale determinant of local richness: the richness of the region in which the site is located.

To understand why local and regional richness should be related, consider again the multiple forms of the species–area relationship, and in particular, variation in its slope. Values of z are argued to be steep when the areas compared are different biogeographical regions (in the range 0.5–1.0), but shallower when the areas come from the same region (0.13–0.18; Rosenzweig 1995). These different patterns can conveniently be summarized on one plot (Fig. 2.12). From this graphical summary it is immediately apparent that the number of species in equal-sized areas of different provinces ought to be related to the total richness of the province. Simply, local sites should have higher species richness if they are located in richer provinces. (This would be true for areas intermediate in size between local and regional scales even if provincial species–area curves tended to converge.) The association between the species–area relationship and local–regional richness relationships is considered in greater depth by Westoby (1993) and Srivastava (1999).

Although local richness ought theoretically to depend on regional richness in some way, the association may potentially take a range of forms. Typically, two types of local–regional richness relationships are distinguished (Fig. 2.23), albeit these are really only the ends of a continuum of likely possibilities (Ricklefs 1987; Cornell & Lawton 1992). The richness of a Type I local assemblage is proportional to the richness of the region in which it is embedded. For this reason, it is sometimes referred to as following a 'proportional sampling' model; local sites sample a constant proportion of the regional species pool (Cornell & Lawton 1992). The richness of Type II assemblages increases with regional richness up to some maximum level, beyond which local richness becomes independent of that of the region.

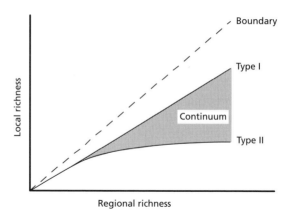

Fig. 2.23 The theoretical relationship between local and regional species richness in ecological communities. Local richness cannot exceed regional richness, setting the boundary. Below that, local richness may increase proportionately with regional richness (Type I), or may reach a limit set by biotic interactions (Type II). Type II communities are often described as 'saturated'. Modified from Cornell and Lawton (1992).

Type I and Type II assemblages are also known as unsaturated and saturated, respectively (Terborgh & Faaborg 1980; Cornell 1985a; Ricklefs 1987; Cornell & Lawton 1992). This reflects the belief that these different types of local–regional richness relationships reflect differences in the structure of the local assemblages contributing to them. The asymptote of Type II relationships is presumed to be set by biotic interactions among the community members, which constrains local richness at a level that is independent of regional richness. The richness of local assemblages in Type I relationships is assumed to be unconstrained by local interactions, and so would go on increasing as long as did the regional pool.

If most assemblages are Type I, we need to understand variation in regional richness when trying to understand variation in local richness. If most assemblages are Type II, then local processes are likely to be the predominant determinants of local richness patterns, at least above some threshold species number. It seems sensible that local communities should become saturated, if only in the most species-rich regions: there ought to come a point beyond which resources are limiting. Nevertheless, Type II relationships are relatively rare. Local and regional richness are positively correlated, with little evidence of an asymptote, for a variety of taxa, including birds, across a variety of regions (Cornell 1985b; Ricklefs 1987; Lawton *et al.* 1993; Cornell & Karlson 1996; Pärtel *et al.* 1996; Willson & Comet 1996; Caley & Schluter 1997; Griffiths 1997; Hugueny *et al.* 1997; Chown & Gaston 1999a; but see also Angermeier & Winston 1998; Thiollay 1998). A review of the literature by Srivastava (1999) found that over two-thirds of studies showed evidence for Type I richness relationships. This will come as little surprise to birders, who appreciate that the

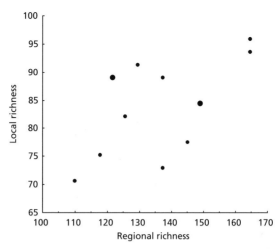

Fig. 2.24 The relationship between local and regional species richness for birds breeding in Britain. Each point refers to the total number of species breeding in a British county (or region in the case of south-east Scotland) and the maximum number of species breeding in a single tetrad (2×2-km square) within that county. Overlap of values for two counties is indicated by the larger points. Ordinary least squares regression indicates a significant linear relationship between the variables ($r^2 = 0.31$, $n = 13$, $P = 0.049$), and there is no evidence of curvilinearity (second-order polynomial regression: $r^2 = 0.31$, $n = 13$, $P = 0.16$). From data in Montier (1977), Harding (1979), Taylor *et al.* (1981), Tyler *et al.* (1987), Sitters (1988), Brucker *et al.* (1992), Guest *et al.* (1992), Thomas (1992), Bland and Tully (1993), Dennis (1996), James (1996), Standley *et al.* (1996) and Murray *et al.* (1998a).

richest birding sites tend to be found in the richest birding regions. Figure 2.24 shows an avian example from Britain.

While most assemblages appear to fit the proportional sampling model, care needs to be taken when drawing conclusions from local–regional richness relationships (see Cresswell *et al.* 1995; Caley 1997; Zobel 1997; Griffiths 1999; Srivastava 1999). There are ways in which Type I relationships can be generated when assemblages are saturated, and Type II relationships when they are not. For example, if local assemblages in richer regions saturate at higher species richness, then a Type I local–regional richness relationship, or something approximating it, might pertain even if local assemblages were saturated. Different geographical regions are likely to differ in a number of factors in addition to the size of their species pool. Many of the variables we discuss in this chapter that are likely to increase the size of the regional species pool may also increase the number of species that can co-exist locally, even if those species are all strongly interacting and local assemblages are saturated. Conversely, there are reasons why unsaturated systems might produce Type II local–regional richness relationships consistent with saturation. For example, overestimating the size of the species pool, combining different local–regional richness relationships in different provinces, or variation in the ratio between the sizes of locality and region, can all lead to Type II curves when local assemblages are unsaturated (Caley & Schluter 1997; Hugueny *et al.* 1997; Srivastava 1999). Type II curves can also be generated in unsaturated communities by the phenomenon of 'pool exhaustion' (Lawton & Strong 1981; Cornell 1985a), where only a fraction of the perceived species pool can actually colonize a site in ecological time.

As Srivastava (1999) notes, it is 'easy to reach the wrong conclusion about species saturation by analysing local–regional richness plots'. For that reason, it is desirable to support conclusions from such plots with other lines of evidence. In particular, Type I plots would be expected to be derived from non-interactive communities, as saturation cannot occur if species do not interact. Studies of fig wasps (Hawkins & Compton 1992), bracken herbivores (Lawton 1982; Lawton *et al.* 1993) and freshwater fish (Oberdorff *et al.* 1998) all support the conjecture that interspecific interactions in Type I assemblages are weak at best. Strongly interacting communities, by contrast, may or may not be saturated, and so observations of strong interspecific interactions do not by themselves provide support for saturation of Type II assemblages (Srivastava 1999). However, Type II assemblages should show evidence of features such as resistance to invasion, resource limitation, density compensation and competitive exclusion. Kennedy and Guégan (1994) cite the latter as evidence for saturation in parasite communities of fish. Thus, while the shape of a local–regional richness relationship is not conclusive evidence for saturation, it seems likely to be a fair indication. If so, then saturation would seem to be unusual.

It may be considered surprising that the richness of birds and other taxa in local assemblages shows little evidence of saturation, especially if one subscribes to a highly structured view of a community. It implies that the number of species co-existing at a site like Eastern Wood is limited only by the richness of the region in which it is embedded (although it will also be subject to the modifying effects of area and isolation). This is to some extent misleading, implying as it does that bird species could (within reason) be added willy nilly to Eastern Wood, and hence to the regional species pool, and be expected to co-exist there. The statement misleads because it imagines Eastern Wood transplanted to a region of greater avian richness without any concomitant changes in the richness of other faunal and floral components of the community. But to the extent that taxa co-vary in their richness (see below), regions richer in one taxon (e.g. birds) are also likely to be richer in others (e.g. insects and plants). This is most obvious if one considers the effect of the area of biogeographical regions on their richness (e.g. Fig. 2.16). The entire community would of course be very different were the site in a different region. To see how different, we only have to look to other continents. That said, it is worth re-iterating that the lack of local saturation implied by Type I local–regional richness relationships itself implies that the number of species co-existing in a community is limited principally by those factors that limit regional richness, and that there is only a

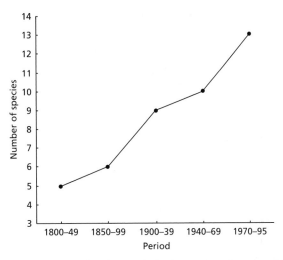

Fig. 2.25 The number of species of birds recorded as having been introduced into the British avifauna at different periods from 1800 to 1995. Note that later periods cover fewer years. From data in Gibbons *et al.* (1996).

minor role for local interactions. This agrees with the opportunistic view of the community espoused in the opening chapter.

Evidence about the saturation of local communities is also provided by introductions. The species richness of many areas has been altered by the addition of introduced species. Ebenhard (1988) surveyed 771 introductions of 212 species of birds which have been reported in the literature. Approximately 20% of these introductions were made on continents, 20% to shelf islands and 60% to oceanic islands. Introduced birds have established wild populations in most countries for which data are available. Ignoring reintroductions of previously native species which had become locally extinct (white-tailed eagle— extinct at the beginning of the 20th century, successfully reintroduced in the 1980s; capercaillie—extinct in the 1780s, successfully reintroduced in the 1830s; Holloway 1996), the total number of introduced breeding species in Britain has increased from five (Canada goose, Egyptian goose, red-legged partridge, pheasant, domesticated rock dove) in 1800–49 to 13 in 1970–95 (further additions being wood duck—not currently considered by some authorities (e.g. the BOU) to be part of the British fauna—mandarin duck, red-crested pochard, ruddy duck, golden pheasant, Lady Amherst's pheasant, rose-ringed parakeet, little owl) (Fig. 2.25; Gibbons *et al.* 1996). In some areas of the world, the numbers of introduced species of bird constitute a substantial proportion of the total avifauna (Table 2.2; e.g. Hawaii, New Zealand). The reasons for the successful establishment of some species post introduction and the failure of others have been much discussed (e.g. Moulton & Pimm 1983, 1986; Lockwood *et al.* 1993; McLain *et al.* 1995; Veltman *et al.* 1996; Williamson 1996; Duncan

Table 2.2 Numbers of native and introduced bird species in a selection of different regions. From Vitousek *et al.* (1997).

Region	Native	Introduced	% introduced
Europe	514	27	5.0
South Africa	900	14	1.5
Brazil	1635	2	0.1
Bahamas	288	4	1.4
Puerto Rico	105	31	22.8
Hawaii	57	38	40.0
New Zealand	155	36	18.8
Japan	248	4	1.6

1997). Some introduced bird species can become very abundant and wide-spread. In Britain, the Canada goose, pheasant and domesticated rock dove come to mind. For birds at least, we know of no evidence to suggest that native species have been driven extinct by introductions. If that is indeed true, introductions would simply enrich local and regional faunas. There are, however, cases where introduced birds have adversely affected populations of natives. For example, hybridization between introduced ruddy ducks and native white-headed ducks in Europe threatens the existence of pure-bred individuals of the latter species (Hughes 1993; Tucker & Heath 1994). Avian malaria carried by resistant introduced birds may have caused the declines of several native Hawaiian species (reviewed by van Riper 1991). Therefore, while enriching regional avifaunas, introduced birds may sometimes depress local species richness.

There is also some evidence that species cannot be added to alien avifaunas *ad infinitum*. Work by Moulton and colleagues (e.g. Moulton & Pimm 1983, 1986; Brooke *et al.* 1995) suggests that later bird introductions may be less likely to succeed than earlier ones. This implies that later introductions may suffer competitive exclusion, the likelihood of which increases with the number of species already successfully added to the fauna, and that eventually a limit to richness would be reached beyond which no more additions would succeed. However, a problem with this conclusion is that a relationship has been shown between introduction success and introduction effort for birds introduced to New Zealand (Veltman *et al.* 1996; Williamson 1996; Duncan 1997; Green 1997). Although these species still show evidence for competitive exclusion after controlling for introduction effort (Duncan 1997), the conclusion that competition influences establishment probability will be impossible to draw from studies comparing the success of earlier and later introductions unless the hypothesis that more effort is expended on early introductions can be falsified. Nevertheless, the wide success of attempts to introduce alien species to avifaunas around the world argues that, at least initially, species are entering unsaturated communities.

The typically Type I relationship between local and regional richness means that to understand the determinants of local richness one must understand the determinants of regional richness. As argued from the opening chapter, the ecology of local assemblages cannot be understood without also incorporating a broader-scale perspective. Therefore, we now turn our attention to what is known about patterns in the species richness of regions, and to consideration of their likely causes.

<table>
<tr><td>2.5</td><td></td></tr>
</table>

Latitude

The effect of isolation identifies the position of a site such as Eastern Wood relative to other habitat patches as an important determinant of its species richness. However, richness may be affected not only by relative position, but also by absolute position in space because species distributions are not spread homogeneously across the planet. Some areas are richer than others. Moreover, the heterogeneity is not random. A number of consistent large-scale trends in richness can be identified. The first we will consider relates to latitude.

Across Britain, there is a marked cline in avian species richness. The highest levels are encountered in the south, and the lowest in the north (see, for

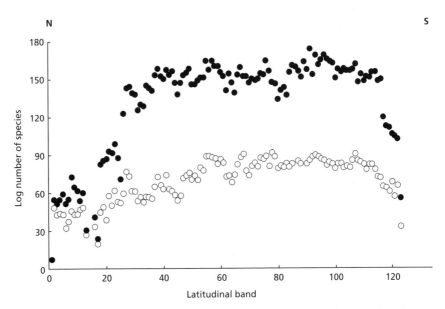

Fig. 2.26 The latitudinal pattern of breeding bird species richness across Britain, based on the mean number of species per 10×10-km grid cell in a latitudinal band (open circles) and the total number of species in each such band (filled circles). The decline in richness in the extreme south reflects the peninsula effect illustrated in Fig. 2.22. From data in Gibbons *et al.* (1993).

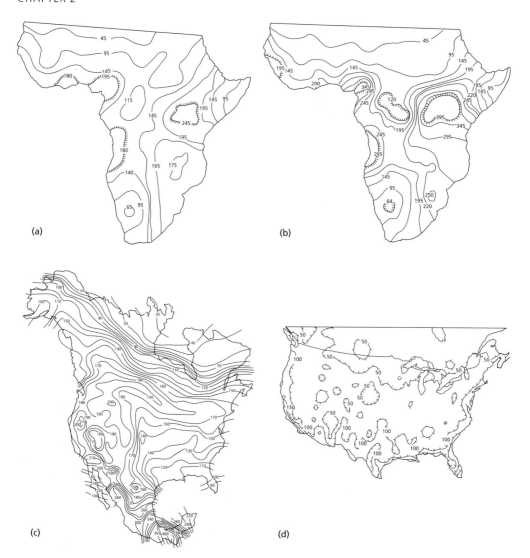

Fig. 2.27 Geographical variation in numbers of species for (a) non-passerines and (b) passerines in the Afrotropics, and (c) breeding and (d) wintering birds in the Nearctic. From Cook (1969, with permission from Taylor & Francis), Crowe and Crowe (1982, with permission from Cambridge University Press) and Root (1988a, with permission from University of Chicago Press).

example, Fig. 2.26, and maps in Sharrock 1976; Fuller 1982; Lack 1986; Turner *et al.* 1988; Gibbons *et al.* 1993; Williams 1996a; Williams *et al.* 1996; Williams & Gaston 1998). This is true not only for the avifauna as a whole, but also for the communities occupying different habitats (Fuller 1982), which means that it is not a simple consequence of a latitudinal cline in habitat diversity. Hence, the high richness of Eastern Wood relative to other British woodlands may be in part because it lies near the extreme south of the region.

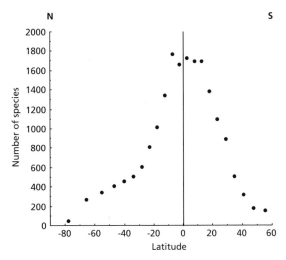

Fig. 2.28 Variation in the total number of bird species recorded (breeding or wintering) in consecutive latitudinal bands across the New World. The unit of latitude is degrees. The Equator is indicated by a vertical line. From data sources in Blackburn and Gaston (1996b).

The cline in richness across Britain is but a subset of a broad latitudinal increase in avian richness, with, for example, the number of resident woodland species increasing across Europe (Hinsley *et al.* 1998). In general, the cline in avian richness stretches from high northern latitudes to the tropics, with a subsequent decline again towards high southern latitudes. Such clines occur for the vast majority of at least moderately speciose groups of animals and plants (for birds, see Dobzhansky 1950; MacArthur & Wilson 1967; Cook 1969; MacArthur 1969; Tramer 1974; Peterson 1975; Haffer 1988; Blackburn & Gaston 1996b; more generally, see table 1 in Stevens 1989). Indeed, the latitudinal gradient in species richness has been described as the bold signature of life on Earth (Lewin 1989). Some avian examples are shown in Figs 2.27 and 2.28.

Figure 2.29 shows the latitudinal richness gradient for the same set of biotic regions for which we earlier showed a relationship between species richness and area (Fig. 2.16). The slope is negative, as expected, albeit not statistically significant. However, the broad scatter around the regression line arises because of the major effect of land area on avian species numbers (Fig. 2.16); controlling for this effect, a significant negative relationship between latitude and richness is recovered (Fig. 2.30). The effect of area on species number is independent of latitude (Fig. 2.31).

Similar latitudinal richness gradients also exist for higher taxa (e.g. Fig. 2.32). Some evidence suggests that the decline in richness with latitude may be faster in the northern than in the southern hemisphere (Platnick 1991; Eggleton 1994). The richness of New World birds is indeed higher in South America south of the Equator than at equivalent latitudes to the north (Fig. 2.32a).

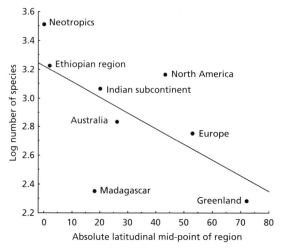

Fig. 2.29 The relationship between number of species and latitude (degrees) for birds in different biotic regions ($y = 0.011x + 3.22$; $r^2 = 0.41$, $n = 8$, $P = 0.087$). From data in Slud (1976), as used in Fig. 2.16.

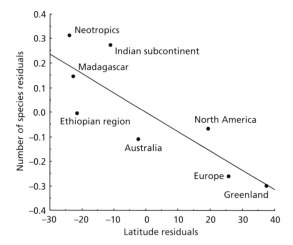

Fig. 2.30 The relationship between number of species and latitude for birds in different biotic regions, controlling for the effects of land area ($y = 0.008x$; $r^2 = 0.70$, $n = 8$, $P = 0.009$). For a given land area, regions centred at low latitudes have higher species richness than regions centred at high latitudes. The axes are the residuals of plots of $\log_{10}$ species richness against $\log_{10}$ area (km^2) and of regional absolute mid-latitude (degrees) against $\log_{10}$ area. Area and richness data were taken from Slud (1976), as used in Fig. 2.16.

However, the difference is slight, and at present the jury is out on the question of its broader generality (Gaston 1996b; Gaston & Williams 1996).

Although the latitudinal richness gradient is one of the most consistent ecological patterns, there are still exceptions. Thus, while most major taxa exhibit a gradient, the same is not true of all their constituent subtaxa. For birds, this is

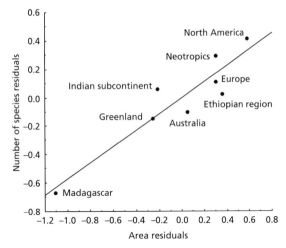

Fig. 2.31 The relationship between number of species and area for birds in different biotic regions, controlling for the effects of latitude ($y = 0.57x$; $r^2 = 0.86$, $n = 8$, $P = 0.001$). For a given regional absolute mid-latitude, larger regions have higher species richness than smaller regions. The axes are the residuals of plots of $\log_{10}$ species richness against regional absolute mid-latitude (degrees) and of $\log_{10}$ area (km²) against regional absolute mid-latitude (degrees). Area and richness data were taken from Slud (1976), as used in Fig. 2.16.

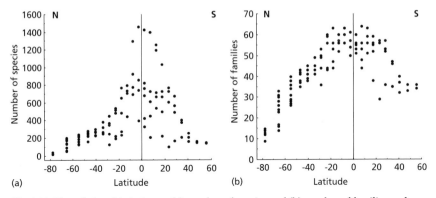

Fig. 2.32 The relationship between (a) number of species and (b) number of families and latitude (degrees) for birds in the New World. Each point represents the number of species or families recorded (breeding or wintering) in each square on the WORLDMAP grid (a cylindrical projection of the world divided into equal-area squares, each approximately 611 000 km², for intervals of 10° longitude, and symmetrical about the Equator; Williams 1992, 1993), and the latitudinal mid-point of that square. The Equator is indicated by a vertical line.

in most cases a trivial consequence of the observation that some taxa are adapted to life at higher latitudes (e.g. penguins and auks). More significant and interesting exceptions exist for other taxa: aphids, sawflies, ichneumonids and bees all show richness peaks at intermediate or high latitudes (e.g. Owen

& Owen 1974; Michener 1979; Janzen 1981; Gauld 1986; Dixon *et al.* 1987; Roubik 1992; Williams 1993; Kouki *et al.* 1994; Williamson 1997). Other taxa may show a gradient in some regions but not in others. Thus, Nearctic and Palaearctic mammals increase in richness towards low latitudes (Pagel *et al.* 1991a; Letcher & Harvey 1994), but Australian mammals do not (Smith *et al.* 1994). The total richness of birds appears to be highest at mid-latitudes within Europe (Mönkkönen 1994; Gregory *et al.* 1998) and Australia (Schall & Pianka 1978; Blakers *et al.* 1984; Pearson & Cassola 1992). No obvious latitudinal richness gradient is shown by birds in Ireland, where centres of highest richness are at mid-latitudes (Gibbons *et al.* 1993). These variations may be useful for distinguishing between potential explanations for richness gradients.

2.5.1   *Why oh why?*

The answer to why there are more species at lower latitudes has long been a puzzle. An obvious starting point is that the pattern has no biological basis at all, but is instead a consequence of random location of species distributions across latitudes. Colwell and Hurtt (1994) have shown that this mechanism can produce species richness gradients as long as there are hard (i.e. impermeable) boundaries that limit species distributions. Because latitudinal gradients tend to have hard boundaries, imposed generally by the Poles, or by land's end for terrestrial species, this is a reasonable condition. Whether this model can explain real latitudinal richness gradients is currently unclear, as it has not yet been widely tested. The only studies that have compared the model against data conclude broad support for its predictions (Lees 1996; Willig & Lyons 1998; Lees *et al.* 1999), but we are not convinced that the results are indicative of a general explanation for latitudinal richness gradients. Thus, while the random model explains reasonable amounts of variation in marsupial and bat species richness across latitudes in the New World (Willig & Lyons 1998), this is hardly surprising given that both real and model gradients peak near the Equator; the pattern of increase with latitude is quite different between the two. Nevertheless, the extent to which random models explain latitudinal gradients awaits further study, and we anticipate that they will play an increasingly major role in studies of species richness gradients.

In addition to the random model, more than a dozen separate explanations for latitudinal richness gradients have been proffered, based on such diverse factors as environmental stability or predictability, environmental patchiness, competition, predation, parasitism, mutualism, productivity, solar energy and latitudinal patterns in geographical range size (numerous reviews and discussions include Fischer 1960; Pianka 1966; MacArthur 1969, 1972; Ricklefs 1979; Shmida & Wilson 1985; Platnick 1992; Rohde 1992; Rosenzweig 1992, 1995; Latham & Ricklefs 1993; Ricklefs & Schluter 1993; Begon *et al.* 1996; Blackburn & Gaston 1996c; Gaston & Williams 1996; Williamson 1997). We do not intend to re-examine all these possibilities here: that process would be a book in itself, especially as it is likely that many of the mechanisms proposed may help to explain the details of richness gradients in some taxa in some situations (Williamson 1997). Rather, we will concentrate on the three attempts to cut the Gordian knot which seem to us to offer the most general promise. While selecting for detailed attention just three of the explanations for latitudinal richness gradients might seem arbitrary, there are good reasons to do so, apart from their popularity in the literature. These will be clearer following Chapter 6.

| 2.5.2 | *Area again* |

One remarkably simple hypothesis is that the tropics has the highest species richness because it has the greatest geographical area; the 'geographical area hypothesis'. This idea originates with Terborgh (1973; see also Schopf *et al.* 1977; Osman & Whitlach 1978), but has principally been championed by Rosenzweig (1992, 1995; Rosenzweig & Sandlin 1997). Large geographical area may translate into high species richness through the effect of area on the geographical range size attainable by species in different regions. Species inhabiting spatially extensive regions can have larger geographical ranges than can those inhabiting more restricted regions. Species with larger ranges are buffered against extinction from accidental causes, because they are more likely to have large population sizes, and against extinction from environmental perturbation, because they are less likely to have their entire population affected. Conversely, species with large ranges may be more susceptible to allopatric speciation through the formation of geographical barriers, which may isolate subsets of their total population (but see Chapter 3). Since speciation rates may be raised and extinction rates reduced in regions of greater spatial extent, these regions should also have higher levels of species richness (Rosenzweig 1992, 1995).

It should be clear that this explanation for the effect of area on species richness is couched in terms quite distinct from that of our earlier discussions of species–area relationships. This is simply because the areas of concern are so large. Four processes ultimately determine variation in species numbers in any area: speciation, extinction, immigration and emigration. For these very large areas, speciation and extinction (at the whole-species level) predominate, and immigration and emigration are less important. For very small areas, immigration, emigration and local extinction predominate, and speciation and species-level extinction are very unlikely.

Considering only the terrestrial environment, the tropics is the most extensive of the biomes (areas of relatively homogeneous environmental conditions), and so should, on the basis of the geographical area hypothesis, have the highest richness, which indeed it does. However, successive biomes (as defined in Rosenzweig 1992) north of the tropics all have about the same land area. If the

geographical area hypothesis is correct, these regions should all have approximately the same species richness. That they do not could result, at least in part,
from the ranges of tropical species extending out into neighbouring biomes.
This effect should be stronger in those biomes closest to the tropics, giving rise
to a 'secondary' latitudinal richness gradient. Rosenzweig (1992, 1995) suggested that if species with partly tropical distributions were excluded, the richness gradient north of the tropics should disappear. By extension, the strength
of the richness gradient across biomes south of the tropics should also depend
on the relative land areas of those biomes once tropical species have been
excluded, although the predictions for both hemispheres may be tempered by
the general decrease in the productivity of the environment at higher latitudes.
Blackburn and Gaston (1997a) tested the effect of removing tropical species on
latitudinal patterns in avian species richness in the New World. They found
that there is indeed a relationship between the land area and the species richness of a biome once predominantly tropical species are excluded.

The geographical area hypothesis potentially explains why we see large-
scale patterns in bird species richness across the globe. It provides one reason why the richness of Eastern Wood is lower than that, for example, for
Kakamega Forest in Kenya. However, it is difficult to use the geographical area
hypothesis to explain the existence of a latitudinal gradient in avian species
richness within Britain, and hence to explain why the latitude at which it lies
contributes to the numbers of species occurring in Eastern Wood. This is not so
surprising. As just noted, the geographical area hypothesis explains large-
scale richness patterns. Additional processes, such as those discussed above,
are likely to fine tune the details within the broad patterns. For example, the
avian species richness gradient within Britain may in part result from the
greater isolation of northern areas from the continental source pool (Section
2.3). The geographical area hypothesis may contribute to the effect in that the
source pool for many southern species (e.g. nuthatch, lesser whitethroat,
hawfinch) may be the European temperate biome, whereas the source of many
northern species (e.g. ptarmigan, dotterel, snow bunting) may be the poorer
boreal biome. Thus, the richness of Eastern Wood may compare unfavourably
with other woodland sites around the world as a consequence of the size of the
area from which it draws its species, but favourably with woodland sites in
Britain for other reasons. However, one of these other reasons also has the
potential to explain large-scale gradients in species richness, and we consider
it next.

2.5.3   Energy

Alongside area, the other strong contender as an explanation of latitudinal
variation in species richness is the 'energy hypothesis'. Put baldly, this states
that higher energy availability in an area provides a wider resource base,

permitting more species to occur there (Tilman 1982; Wright 1983; Turner *et al.* 1987, 1988, 1996; Currie 1991; Wright *et al.* 1993). In fact, there is absolutely no reason at all why *per se* this should be so (Currie 1991; Rohde 1992; Blackburn & Gaston 1996c), and the hypothesis requires some more detailed embellishment to make it clear.

At the upper limit, the absolute amount of life on Earth (hydrothermal vent and other chemolithotrophic communities excluded) cannot exceed that which can be supported by the harnessing of all energy arriving from the sun. Because some energy is conducted and convected away from the region in which it first falls, not all can be converted. Since initial plant, and subsequent animal, conversion efficiencies are less than perfect (i.e. 100%), the absolute upper limit will inevitably be still lower than that expected from levels of solar radiation alone. Effectively, this energy is converted to biomass, or some number of individuals. The way in which this biomass or these individuals are 'divided up' into species then determines the species richness of an area. Energy levels may dictate how much biomass (say) there is to be allocated, but seem unlikely to mediate the allocation process (Blackburn & Gaston 1996c). Nevertheless, it is relatively easy to construct an argument explaining why energy-rich areas should also be species rich, based on simple assumptions about speciation and extinction (Turner *et al.* 1996).

First, speciation is assumed to be a stochastic process, operating in the same way at all latitudes. This would generate equal diversities across the globe. The exact process by which speciation occurs is unimportant for this argument, but it is important that some speciation does happen. If it does not, then one species simply monopolizes all available energy. Reasons why speciation would be expected in any given biogeographical region can easily be posited. Most obviously, no single species can master all ways of life, and different strategies are likely inevitably to lead to speciation. Second, extinction is assumed to be inversely related to population size. Given these two assumptions, more species are expected to persist where extinction is lowest, which will be in areas where species have larger population sizes. These should be areas with higher levels of energy input (Turner *et al.* 1996), where more biomass can be sustained, and hence where larger population sizes may be expected for a given number of species (or, looked at the other way round, where more species may sustain populations above some critical size).

These ideas beg a number of important questions, and we will return to some of these at various places later in the book (especially Chapter 6). At this point, the important issue is whether energy and species richness are related in a way which can explain latitudinal gradients in species richness.

A large number of studies have documented relationships between species richness and estimates of energy availability. As energy availability is reasonably difficult to measure directly, these studies typically use surrogate measures, the principal of which is primary productivity. Productivity is assumed to reflect the input of solar energy because it is only plants that can use this energy to produce biomass. Clearly, energy input and primary productivity are not perfectly related, as there are many parts of the globe that receive plenty of energy with little resultant plant growth, due to the lack of other essential elements (principally water). Nonetheless, their relationship may be sufficiently close for general purposes. The other major surrogates for energy availability are measures of climatic variables.

Studies of the relationship between energy availability and species richness tend to fall into two broad groups. On the one hand, there are those which report hump-shaped relationships, in which species richness peaks at intermediate levels of energy availability or productivity (Grime 1973; Al-Mufti *et al.* 1977; Tilman 1982; Abramsky & Rosenzweig 1984; Owen 1988, 1990; Kerr & Packer 1997; Guo & Berry 1998; Chown & Gaston 1999a). On the other hand, there are studies which report broadly positive relationships, in which species richness peaks at the highest levels of energy availability, or in the warmest climates (Wright 1983; Turner *et al.* 1987, 1988, 1996; Adams & Woodward 1989; Currie 1991; Wylie & Currie 1993a,b; Blackburn & Gaston 1996b; Fraser & Currie 1996). Wright *et al.* (1993) observe that variation in the shape of the relationship depends on the spatial scale of study, with hump-shaped relationships observed at smaller scales and positive relationships at larger ones. They suggest that this scale dependency implies that different factors control richness at different scales, with energy important at large spatial scales.

One notable exception to the generalization by Wright *et al.* about scale dependency is a study by Chown and Gaston (1999a). They mapped the distribution of all Procellariiformes (albatrosses, shearwaters, petrels, storm-petrels, diving-petrels) onto an equal-area grid covering the world's oceans, and found a hump-shaped relationship between the species richness of grid squares and estimates of productivity. Thus, the richest areas of ocean do not hold the highest numbers of seabird species.

Chown and Gaston suggested two reasons why a hump-shaped richness–productivity relationship might occur in this system. First, they noted that highly productive areas of ocean cover smaller geographical areas than do areas of intermediate productivity, while local and regional seabird species richness are positively correlated. Thus, the hump-shaped relationship may arise because larger areas support more species in total, despite having lower productivity per unit area, while regional richness elevates local richness, as we saw earlier (Section 2.4). Second, Chown and Gaston noted that highly productive areas of ocean also tended to exhibit higher temporal variability in productivity. Thus, highly productive areas may have lower species richness because seasonality in production prevents exploitation of those areas by seabirds for substantial periods of the annual cycle. Supporting this second idea is a negative relationship between procellariiform species richness and geographical range size: seabirds with small range sizes cannot persist in areas with highly seasonal productivity, which cannot support their populations all year round.

Additional evidence presented by Chown and Gaston (1999a) seems to suggest that the second of these two explanations for the hump-shaped richness–productivity relationship is the most important in this system. This led them to propose a general explanation for when hump-shaped relationships should occur, and when such relationships should be positive. When productivity and its variance are positively correlated, hump-shaped relationships should pertain, because productive areas will not be able to maintain high species richness all year round. However, positive relationships between richness and productivity should be found when productivity and its variance are negatively correlated.

At the scale of Britain, Turner *et al.* (1988, 1996) have shown that patterns in bird species richness fit well with predictions of the energy availability hypothesis. In particular, they compared predictors of the richness of species between seasons. They argued that if climate (and hence energy) is an important predictor of avian richness, then different patterns of species density should pertain in summer and winter, as the principal temperature gradient runs north–south in Britain in summer, but in winter has a much stronger east–west component. In essence, this is what they found. However, the exact patterns were modified by body size (see Cousins 1989). Winter temperatures affected the distributions of species of all sizes, whereas summer temperatures only affected the distributions of the smallest species (Turner *et al.* 1996). Turner *et al.* suggested that this further supports the energy hypothesis, as all species should suffer from temperature stress in winter, whereas only small-bodied species are likely to be susceptible to such stress in summer.

These energetic arguments might suggest why the breeding bird richness of Eastern Wood is low relative to many places in the world, but high relative to other British woodlands. Latitudinal gradients in energy availability could generate both of these observations, because Eastern Wood is a long way from the Equator (where energy availability ought to be highest), but not as far from the Equator as is most of Britain. Energy may therefore dictate the number of species that Eastern Wood can support.

One obvious deficiency with energetic arguments such as that propounded by Turner *et al.* (1988, 1996) is that while energy does a good job of explaining patterns in the modern-day distribution of British birds, very few of those species are likely actually to have evolved in Britain. Indeed, just a few thousand years ago most of the country was covered by a thick layer of ice. Thus, while the results reported by Turner *et al.* can explain the current distribution of bird richness, they do not necessarily say anything about the factors that drove the evolution of that richness. On the other hand, they do illustrate an association between energy (climate) and richness that needs explanation. That the association might be causal is not improbable.

The fact that Britain was until relatively recently covered by glaciers highlights an important additional consideration with respect to theories of the evolution of species richness. The richness of an area will ultimately depend on the processes of speciation, extinction, immigration and emigration. However, all these processes take place in time. Large, energy-rich areas will be species poor if there has been little time for speciation or immigration to have occurred. In the extreme, of course, the planet was species poor early on in evolutionary history. This has led to a third popular explanation for latitudinal gradients in species richness, that they are dependent on the amount of time available for the processes of speciation and colonization to have occurred at different latitudes.

2.5.4   *Time hypotheses*

Time hypotheses can loosely be split into that concerning ecological time and that concerning evolutionary time (Pianka 1966). The ecological time hypothesis proposes that the low richness of some regions is a consequence of the insufficient period available for species to colonize or recolonize since an earlier ecological upheaval. Adams and Woodward (1989) examined reasons for the well-known difference in tree species richness between North America, Europe and eastern Asia. The tree flora of Europe is depauperate relative to the other two regions. They showed that most of the variation could be explained by differences in productivity between regions: European areas had no fewer tree species than expected on the basis of their productivity. However, Adams and Woodward also noted unexpected regional differences in tree richness in the most productive areas, with these attaining higher richness in eastern Asia. The bird species richness of eastern Asia, eastern North America and the Western Palaearctic mirrors the pattern exhibited by trees. Thus, the avifauna of the Western Palaearctic is depauperate relative to the other two regions (Blondel & Mourer-Chauviré 1998), and this is particularly true of the bird communities associated with forest (Mönkkönen & Viro 1997). Adams and Woodward (1989) and Blondel and Mourer-Chauviré (1998) both attributed these differences to the effects of glaciation in the three regions. They suggested that glaciation resulted in higher levels of extinction in the Western Palaearctic because the predominantly east–west orientation of geographical barriers there (e.g. the Alps, Mediterranean Basin and Sahara Desert) prevented northern species from retreating south as the glaciers advanced, and then prevented southern species from recolonizing as the glaciers retreated. In effect, they are arguing that the numbers of tree and bird species present in high productivity areas in Europe and North America have not regained their preglaciation levels.

In the case of Eastern Wood, avian richness might be low because recolonization of Britain by bird species since the last glaciation is still in progress.

Because this recolonization would inevitably proceed from the south, the process would also explain the richness gradient across Britain. However, while this mechanism can potentially explain small-scale richness patterns, it is unlikely to provide a general explanation for latitudinal richness gradients. The question of what generates the number of species available for recolonization remains.

An answer to this question seems more likely to derive from an evolutionary time hypothesis. This suggests that the richness of regions relates to the length of time available for species to evolve to fill habitats and niches in those regions. As such, it deals with longer spans of time than the ecological time hypothesis, and focuses on speciation, rather than colonization. It proposes that large-scale upheavals, such as glaciation or climatic drying, periodically drive extinct many of the species in a region, and that the floras and faunas of those regions exposed to more frequent upheavals have less time to rediversify. This leads to regional differences in species richness. The mechanism also fits well with the opportunistic view of ecological communities espoused in the opening chapter, as it implies that most communities are likely to be unsaturated collections of species, and thus open to additional species invasions (see also Järvinen & Ulfstrand 1980).

Some evidence for the effect of evolutionary time on richness has been produced by Rohde (1978, 1986, 1992, 1997). He noted differences between the Atlantic and Indo-Pacific oceans in the species richness of both fish and their monogenean gill parasites, with richness higher in the latter region. He reviewed explanations, and concluded that evolutionary time was the most likely. Latham and Ricklefs (1993) argued that the patterns of tree species richness analysed by Adams and Woodward (1989) might be better understood if the effects of evolutionary time on richness were considered.

If the evolutionary time hypothesis is correct, it implies that more time should have been available in the tropics to have allowed so much additional net speciation to have occurred there. Rohde (1992) argues that this is not the case, as short- and long-term fluctuations between warm and cold states have occurred in the global climate for the past 700 million years (Fischer 1981). However, he goes on to suggest that what is important is not the absolute amount of time available at different latitudes, but the effective evolutionary time. This will be a product of absolute time and the rate at which the evolutionary process occurs. Rohde argues that it is this rate that is likely to differ across latitudes as a consequence of the effect of climate, leading to the higher richness of the tropics.

Whether or not an evolutionary time hypothesis based on effective rather than absolute time is an improvement remains to be demonstrated. In particular, it is as yet unclear whether evolutionary rates are indeed faster in the tropics (Rohde 1997), as the hypothesis suggests. Also, quite why fluctuations in the global climate across all latitudes *per se* should count against the

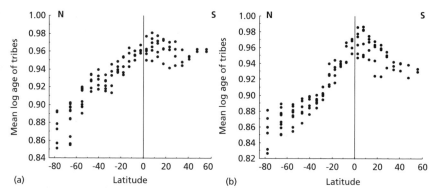

Fig. 2.33 The relationship between the mean age of avian tribes (from data in Sibley & Ahlquist 1990) and latitude (degrees) for birds in the New World. Each point represents the mean of the $\log_{10}$-transformed ages of tribes in each square of the WORLDMAP grid (see Fig. 2.32 for details), and the latitudinal mid-point of that square. In (a) tribes are unweighted, and in (b) are weighted by the number of species of each tribe present in the square. The Equator is indicated by a vertical line. From Gaston and Blackburn (1996b).

evolutionary time hypothesis is unclear to us. What matters is the magnitude of these fluctuations at different latitudes, and they are surely likely to be more severe at latitudinal extremes (e.g. Fischer 1981). Moreover, a fascinating result for birds suggests that there are genuine latitudinal differences in the long-term persistence of taxa. Gaston and Blackburn (1996b) showed that the mean age of tribes and families inhabiting different latitudes in the New World was highest at the Equator (Fig. 2.33). While this pattern can be produced by a variety of different latitudinal patterns of speciation and extinction rates (including, paradoxically, higher extinction rates in the tropics: Gaston & Blackburn 1996b), it is perhaps most likely to show that tropical bird taxa have persisted for longer periods of time, as the evolutionary time hypothesis predicts. That conclusion agrees with other more fine-scale analyses of the distribution of avian species of different ages in South America and tropical Africa, which indicate that lowland tropical forests act as 'sinks' where species accumulate over time (Fjeldså 1994).

2.5.5   A 'primary cause'—holy grail or wild goose?

Implicit in what we have said about latitudinal gradients in richness has been the idea that some explanations can account for some patterns. A good example is the ecological time hypothesis. However, an assumption common to many discussions of the determinants of these gradients is that no mechanism can be supported unless it explains the patterns in all taxa and in all regions. That is, that there is one primary cause of the increase in species richness from high to low latitudes, and any study that unequivocally rules out a mechanism in one region of the globe automatically rules it out in all others. This view is exemplified by MacArthur and Connell's (1966) general statement that: '(w)herever there is a widespread pattern, there is likely to be a general explanation which applies to the whole pattern'.

There is, however, no logical reason why this need be so. To argue for a single primary cause may be to expect a simplicity from ecological interactions for which there is little evidence, and a number of authors have pointed out that observed ecological patterns are likely to be generated by several contributory mechanisms (Wilson 1988; Warren & Gaston 1992; Blackburn & Gaston 1996b,c; Jablonski 1996a; Lawton 1996a; Gaston *et al.* 1997a; Williamson 1997; Gaston & Blackburn 1999). Moreover, Lawton (1996a) has suggested that the strongest and most general patterns are those where all the different mechanisms pull in the same direction: generality is not evidence for primary cause.

In this regard, it is interesting to note the similarities in the geographical area and energy availability hypotheses. The geographical area hypothesis assumes that area influences richness through its effect on geographical range size, which in turn influences rates of speciation and extinction. The energy availability hypothesis explicitly assumes that area influences richness through the effect of energy on population size, which in turn influences extinction rate. As we will see later (Chapter 4), though, there is good evidence that the geographical range size and abundance of species are generally positively correlated. Any factor that increases one of these variables will also be likely to increase the other. Therefore, the geographical area and energy availability hypotheses may be rather closely related. Both depend, in effect, on some factor that is posited to influence the biomass available to be worked on by the processes which ultimately determine how many species there are: speciation and extinction. However, this biomass will be a product of both area and available energy per unit area (Wright 1983; Wright *et al.* 1993): presumably, it is for this reason that small areas tend to be species poor however high their energy input, whereas large areas tend to be species poor if there is low energy input (cf. Madagascar and Greenland; Fig. 2.29). Therefore, both area and energy are likely to be important in determining large-scale patterns in the richness of species. Their effects are likely to be modified to some degree by the effects of effective evolutionary time. Whether effective evolutionary time is greater in the tropics because of latitudinal variation in evolutionary rates or climatic stability, real gradients in evolutionary time across latitudes would serve to enhance the resulting gradients in richness (Wilson 1992).

2.6　Longitude

The latitudinal richness gradient is a widely reported phenomenon, and has received significant attention from ecologists. By contrast, the tendency for richness to vary with longitude has been largely ignored, although it has long been appreciated. Indeed, Fuller (1982; p. 88) noted that the 'tendency for bird

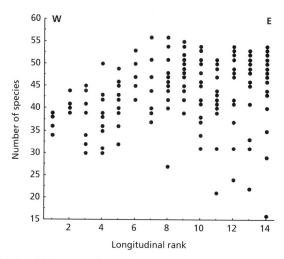

Fig. 2.34 The relationship between the number of forest-associated breeding passerine bird species and longitude across Europe. Each point represents the number of species breeding in a 40 000-km² square, and the longitudinal rank of the square. There is a significant trend towards increased diversity in eastern squares (Spearman rank correlation $r_S = 0.31, n = 173$, $P < 0.0001$), albeit that there is considerable scatter in the data (mainly caused by latitudinal variation in richness). Squares covering Britain are mainly of ranks 3 and 4. From data in Mönkkönen (1994).

communities to change with latitude and longitude is a recurring trend in British ornithology'.

In Britain, highest levels of avian richness are observed in the east (see maps in Fuller 1982; Turner *et al*. 1988; Gibbons *et al*. 1993; Williams *et al*. 1996). Thus, the relatively high richness of Eastern Wood relative to other woodlands of equivalent area fits not only with its latitude within Britain, but also with its longitudinal position. The longitudinal pattern in Britain again seems to be part of a broader cline in avian richness from west to east across Europe. For example, Mönkkönen (1994) presented data on the number of forest-associated passerine bird species in each square of a 200×200-km grid laid over Europe. These show a significant association between species richness and longitude (Fig. 2.34), with the highest richness in eastern grid squares. However, whether this pattern for forest passerines applies more broadly to European birds is difficult to assess, because quantitative results are not given by other studies. Thus, while Cotgreave and Harvey (1994a) found higher levels of avian species richness in the east of the Western Palaearctic, they only report the relationship after controlling for variation in climate. Whether there is a simple trend towards higher richness in the east in their data is unclear. The map of European bird species richness presented by Hagemeijer and Blair (1997) appears to show higher levels of richness to the east, albeit that some missing data make the trend difficult to judge objectively by eye.

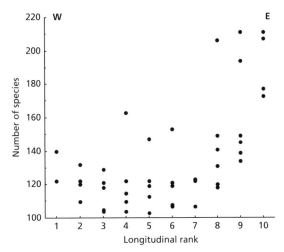

Fig. 2.35 The relationship between the number of breeding landbird species and longitude across Australia. Each point represents the number of species breeding in a 122 500-km^2 square, and the longitudinal rank of the square. There is a significant trend towards increased diversity in eastern squares (Spearman rank correlation $r_S = 0.56$, $n = 46$, $P = 0.0002$), although a significant quadratic term in parametric polynomial regression through these data indicates that lowest richness is at mid-latitudes. From data in Pearson and Cassola (1992).

Assuming an increase in species richness to the east across Europe, longitudinal variation in avian richness is not consistent between continents. Tramer (1974) and Blackburn and Gaston (1996b) showed increases to the west to be most common in North and South American species. By contrast, increases to the east have been demonstrated in Australian birds (Fig. 2.35; Schall & Pianka 1978; Blakers et al. 1984; Pearson & Cassola 1992) and in African waterbirds (Guillet & Crowe 1985, 1986). Cotgreave and Harvey (1994a) showed increases in avian richness in the directions just described on these continents after controlling for climatic variation.

While not necessarily responses to a common cause, longitudinal patterns in species richness have frequently been suggested to provide a golden opportunity to distinguish between hypotheses proposed to explain latitudinal richness gradients. The reasoning is that whatever causes large-scale richness patterns in the north–south direction is likely also to cause the large-scale richness patterns running perpendicular. Moreover, as fewer factors change systematically with longitude (in contrast to the multitude that change with latitude), and because longitudinal gradients run in different directions on different continents, the number of competing explanations for longitudinal gradients ought more readily to be whittled down.

Of the three hypotheses which we suggested were most likely to untie the Gordian knot of latitudinal richness gradients, that based on area is the least likely to explain longitudinal patterns. The geographical area hypothesis is posited to act between biomes, which basically divide up by latitude. It can be extended to encompass differences in area between regions within biomes, but this extension does not do much for its ability to explain longitudinal gradients. In particular, the extensive central desert of Australia is not richer in birds than the smaller eastern coastal forest strip (Fig. 2.35), and the huge expanse of lowland Amazon forest is not richer in birds than the narrow strip of the Andes.

Energy availability and related arguments about productivity are perhaps more likely to contribute to the generation of longitudinal richness patterns. In Australia in particular, productivity is highest along the east coast, where most rainfall occurs (Anonymous 1997). While energy input to central Australia is not lacking, moisture input is. Australian bird richness coincides reasonably well with regions of high productivity (Schall & Pianka 1978; Pearson & Cassola 1992). We have already discussed the relationship between climate and richness in Britain (Turner *et al.* 1988, 1996), which can explain changes in richness gradients from mainly latitudinal to mainly longitudinal in different seasons. Nevertheless, energetic explanations appear to lack generality as predictors of longitudinal richness patterns. In New World birds, for example, gradients in productivity and richness run in opposite directions within latitudinal bands (Blackburn & Gaston 1996b).

If area and energy are insufficient to explain all longitudinal richness trends, what about evolutionary time? In fact, an argument with a strong temporal component has been suggested to explain longitudinal richness gradients across all continents. Cotgreave and Harvey (1994a) found longitudinal richness gradients that persisted after climatic variables had been controlled for statistically (see above). They proposed that these patterns could best be explained by the biogeographical history of the various continents. Regions of high richness tended to be either those that most recently shared a common border with another continent, or, for isolated continents, those regions that had longest exhibited the habitat most characteristic of the continent. Richness will be higher in areas close to a common boundary between regions because of shared species; the region is then the continental equivalent of an ecotone. Cotgreave and Harvey suggest that it is species sharing with Africa and Asia that makes southern and eastern Europe the richest regions of the continent for birds. On isolated continents, the effect of common boundaries ought to have faded with time. Instead, richness should be higher in those areas that have longest exhibited the habitat most characteristic of the continent, because of the evolutionary time hypothesis. Cotgreave and Harvey relate the biogeographical histories of the various continents in qualitative support of both conjectures.

To summarize, it seems that the same mechanisms that have been proposed to explain latitudinal richness gradients can also explain the gradients with longitude. However, their relative explanatory power differs along the different spatial dimensions. We see no contradiction in this conclusion. Just as no single mechanism need explain all latitudinal richness gradients, so no single mechanism needs to be the principal explanation for the gradients in both latitude and longitude. The latitudinal richness gradient is very general and very consistent. The longitudinal richness gradient can be thought of as providing a level of pattern and complexity one scale down. As with species–area relationships at different scales (Section 2.1), there is no reason why different processes cannot contribute in different measure to these related patterns.

2.7   Altitude

Latitude and longitude map two of the three spatial dimensions in which bird species richness can vary. Yet, richness also varies in the third dimension, defined by altitude. In general, the lowest levels of species richness are found at the highest altitudes for any given taxon. For example, Fuller (1982, p. 126) shows that the richness of piscivorous birds breeding on lakes in different regions of Britain declines with the mean altitude of the lake. Terborgh (1977) and Patterson *et al.* (1998) showed that avian richness declines with altitude on elevational gradients in the Peruvian Andes, while Stevens (1992) showed a similar decline for a regional compilation of breeding birds in Arizona. Other studies showing broad declines in avian richness with elevation include Able and Noon (1976), Sabo (1980), Ferry and Frochot (1990), Stotz *et al.* (1996) and Hawkins (1999). A British example is shown in Fig. 2.36.

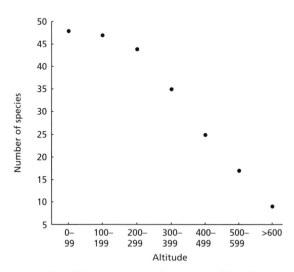

Fig. 2.36 The mean number of bird species breeding per tetrad (2 × 2-km square) in south-east Scotland in tetrads of differing mean altitudinal class. Altitude is given in metres. From data in Murray *et al.* (1998a).

Nevertheless, this generalization of a decline in richness with altitude masks a degree of uncertainty about the normal position of the richness peak. It has been widely assumed that the highest levels of richness are found at the lowest altitudes (Rahbek 1995), the altitudinal gradient thus being an analogue of the latitudinal pattern. However, a reasonable proportion of altitudinal studies find that richness peaks in mid-altitudes (e.g. Rahbek 1995, 1997). Moreover, rather few studies have controlled adequately for differences in area or sampling intensity at different altitudes (Rahbek 1995), both of which may be greater at low altitudes. Area may be greater because land typically slopes more shallowly at low altitudes, so that there is more actual surface area between 0 and 500 m (say) than between 500 and 1000 m. Sampling intensity may be greater because low altitudes are easier to access, and their gentler slopes make them easier to work on. Whatever the reasons, variations in area and effort have the potential to obscure hump-shaped altitudinal richness patterns by inflating the richness of the lowest altitudes. Indeed, when Terborgh (1977) controlled for variation in the effort expended in sampling his Peruvian bird community, the monotonic unstandardized relationship between altitude and richness developed a mid-altitudinal hump. Similarly, Rahbek (1997) changed a monotonically declining relationship between richness and altitude into a hump-shaped one by controlling for the area available at different altitudes. Peak richness for the Arizona bird assemblage examined by Stevens (1992) is not at the lowest altitude. Confusion about the generality of any altitudinal patterns in species richness may also arise through a failure to discriminate between patterns of variation in richness along a particular altitudinal transect, and patterns of variation in the numbers of species in areas within a region in relation to the peak or average elevation of those areas (Gaston & Williams 1996). In sum, the lowest levels of species richness tend to be found at high altitudes, and the highest levels at mid to low altitudes, but whether the 'normal' relationship is monotonically decreasing or hump shaped is currently unresolved.

Assuming a hump-shaped relationship between richness and altitude once area has been controlled for, two likely explanations present themselves. The first is that the pattern is an artefact. On any altitudinal gradient, species altitudinal distributions are constrained to fall between zero (assuming that the lowest possible altitude is at sea level) and the highest point on the gradient—no species can have an altitudinal distribution that extends beyond one or both of these limits. If species altitudinal distributions are chosen at random from a feasible set of values and placed at random on the gradient, then the highest numbers of species are expected by chance alone to be found at mid-elevations (Colwell & Hurtt 1994; Rahbek 1997; Lees et al. 1999). This argument is identical to the random model for latitudinal richness gradients discussed earlier (Section 2.5.1). The second possibility is that a hump-shaped altitudinal richness gradient is analogous to the hump-shaped relationship between richness and productivity in the horizontal dimension discussed earlier. Productivity is perhaps the more likely of the two explanations. The artefact hypothesis predicts that richness should peak at mid-elevations, whereas the peak is generally at low elevations, even when not at the lowest. However, Rahbek (1997) cites evidence that the relationship between productivity and altitude may itself not be linear, and sometimes show a peak at mid-elevations. If so, a direct productivity–richness relationship may pertain, and its variation be explained by variation in the form of productivity–altitude relationships.

2.8   Summary

The number of species recorded at a local site is not bounded, but depends on the time period considered. In part, temporal changes in richness are a consequence of sampling effort, with fewer of the species present being overlooked in longer, more thorough surveys. However, allowing for sampling effort, recorded levels of richness would be expected to increase through time due to faunal changes associated with such factors as seasonality, colonization and extinction, and, ultimately, speciation. Attempts to understand patterns of species richness need to take account of the time period over which the richness data have been assembled.

That said, the determinants of the species richness of a site like Eastern Wood are reasonably well known. They present a complex amalgam of processes acting at a variety of spatial and temporal scales. Consideration of processes at all scales will be vital for a complete understanding of the causes of observed richness patterns at local sites.

The habitats of which the site is composed determine in the first instance which species can make a living there. However, how many species do make such a living depends additionally on a broader set of factors.

At the smallest scales, the size of a site affects how many species it is likely to support. Size seems to exert its influence here through its positive relationship to the number of different habitats that the site encompasses, together with its effect on rates of immigration into and extinction from the site. More habitats mean more species. Larger sites are more likely to be located by immigrants, and support larger populations of those species that do locate the site, which as a consequence are less likely to be driven extinct by the vagaries of chance. The likelihood of colonization is also affected by the isolation of the site, with fewer immigrants locating isolated sites. Species in isolated sites will also have higher extinction rates because stochastic population declines are less likely to be rescued by influxes of immigrants. Thus, small, isolated sites will tend to contain fewer bird species than larger, less isolated sites, even if the sites are composed of equivalent habitat.

The size and isolation of local sites affect how they sample the avifauna of the region in which they reside, but features of that region are just as important.

Therefore, sites that are embedded within species-rich regions are likely to sample more species, and so themselves be richer, than are sites of identical size, isolation and habitat embedded within species-poor regions. Regional species richness is itself most obviously related to the latitude at which the region sits, although there are additional effects of longitude and altitude. The reasons for latitudinal variation in species richness are contentious, but are most likely to relate to variation in the area and productivity of regions, which together determine how much life (e.g. biomass) a region can support, and to variation in the effective amount of time available for the processes that divide that biomass up into different species. Thus, speciose regions ought to be those that are large, productive, and have had greater effective time for their floras and faunas to diversify. Deficiencies in any of these variables ought to reduce a region's richness. The least well-understood link in this argument at present is that relating to diversification: why should there be more species where levels of biomass are higher, and not just more individuals of a few generalist species?

This chapter has been unashamedly biased. It has emphasized large-scale processes as determinants of species richness, but has largely ignored many of the small-scale processes, such as competition, apparent competition and predation, that help determine how many species can co-occur together at any given site. We believe that this bias is wholly justified. It is large-scale processes that determine in the first instance how many species may be expected at a site. Small-scale processes fine tune their interactions. We hope that this chapter has convinced readers of the importance of a large-scale perspective for understanding ecological patterns, because it is for patterns of richness that this perspective is most readily appreciated. For the remainder of the book we move on to consider patterns for which the large-scale perspective is equally useful, but less well appreciated.

3 Range Size

In spite of these discouraging remarks, . . . I shall try to demonstrate that the geographical range of species and other taxa can be studied, and that such a study can give us valuable information for the better understanding of the game of Nature, that is, ecology. [Rapoport 1982]

3.1 Introduction

Implicit in the treatment of patterns in species richness in the previous chapter is an observation about the distributions of species. Not all those species that occur in a region occur at all sites within that region. For a set of sites, such as woodlands in southern Britain, one can envisage a presence–absence matrix of the bird species that they contain. Along one axis would be listed the identities of the sites, along the other the names of all the species, and each site by species combination (each 'element' of the matrix) would be represented by a one or a zero depending on whether the species was present or absent at that site. Summing across rows would give the numbers of species which had been recorded from a given site, its species richness. Summing across columns for a given species would give the numbers of sites at which it had been recorded, a measure of its occupancy. Species richness is a feature of the site, whereas occupancy is a feature of the species. An example of a matrix of this sort for the breeding birds of Berkshire is given in Appendix IV.

This example matrix gives a very simple, but graphic, illustration of the interrelatedness of patterns of species richness and patterns of occupancy (see also Ryti & Gilpin 1987; Hanski & Gyllenberg 1997). However, it also reveals the different ways in which variation in the species richness of sites could be achieved. All species could have the same overall level of occupancy, and differences in richness could result from the particular combinations of sites at which different species occurred. Alternatively, species could differ in their levels of occupancy. Of course, as every birder and natural historian knows, there are widely distributed and narrowly distributed species, and every shade in between. Indeed, in the opening chapter it was observed that the majority of species which have bred in Eastern Wood are relatively widespread across Britain, whilst only a few are particularly restricted in their occurrence (Fig. 1.6). This observation is, however, only one of many that can be made about the occupancy of bird species. This chapter reviews a number of other patterns, and examines how these relate to the composition of the avifaunas of local sites.

That large-scale patterns in the occupancy of sites by species should affect the composition of a local avifauna may not be obvious to some readers. It is easier to believe that features of local communities will determine patterns of species occupancy across sites. Indeed, these features must contribute significantly to occupancy. For example, the presence of some species but not others in Eastern Wood can be attributed to the suitability of the habitat. Woodland specialists require the presence of woodland, and their distribution across the landscape must ultimately be dependent on the distribution of woodland. The absence of many species from Eastern Wood, and from surrounding sites, can be ascribed to this or similar reasons alone. Nevertheless, this 'bottom-up' view of the interplay between local and regional patterns of occupancy is not sufficient.

The richness of local sites like Eastern Wood depends on how the wood samples the avian assemblage of the landscape in which it sits, which depends in the first instance on the area of the wood (Section 2.2) and its isolation from similar sites (Section 2.3). The site samples the fauna of the landscape in a probabilistic manner. Whether or not a species is found at the site is dependent not only on the ecologies of the species and the site, but also on factors that determine the likelihood that the species will encounter the site. Principal among these factors is the distribution of the population of the species across the wider landscape. Many of the species that could potentially breed at a given site may fail to do so because of their pattern of distribution (Andrén 1994a). For example, Hinsley et al. (1996) showed that when the population sizes of most woodland bird species in a region of eastern England are low, whether or not they are found breeding in any given wood depends largely on its size. Woods that could be occupied by these species, and indeed are occupied when their regional population sizes are high, go unused. Similarly, Newton (1993) suggests that with more than 80% of the goldfinch population of Britain wintering in Belgium, France and Spain, the number breeding in Britain is probably more dependent on conditions in continental wintering areas than in Britain itself. Again, we showed in Chapter 1 how the whitethroat probably became extinct as a breeding species in Eastern Wood not because of changes at the site, but because of a wider population decline precipitated by drought on the wintering grounds. Thus, more species could breed at Eastern Wood than do in any one year, as is clearly evidenced by the difference between the total number of species breeding at the site between 1949 and 1979 compared to the number breeding in any one of those years (Fig. 1.3). Patterns of occupancy across the wider landscape of Britain and beyond affect the species composition of the avifauna of Eastern Wood.

Just as the species richness of the avifauna of Eastern Wood depends on how features of the wood allow it to sample the regional avifauna, so the composition will depend on patterns in the occupancy of those bird species that determine which are most likely to be sampled. A number of such patterns may be important; the most obvious is the size of the geographical ranges of the species.

3.2   Species–range size distributions

3.2.1   *Range size measures*

Some initial comments are required as to what is meant by the range size of a species. At first glance, this might seem self evident. However, as anyone will know who has ever compared where a species occurs in the field with where it is supposed to be found according to the map in a field guide, 'the' range size of a species is a nebulous concept. Rapoport (1982) wrote that: 'geographical areas of distribution are the Chinese-lantern shadows produced by the different taxa on the continental screen: it is like measuring, weighing, and studying the behaviour of ghosts.'

Two broad proximate objectives in the measurement of the range sizes of species can be recognized. These are to quantify either the area between the outermost limits to the occurrence of a species, or the area over which it is actually found. These two quantities have been termed, respectively, the extent of occurrence and the area of occupancy (Gaston 1991d, 1994b). The former is what is commonly depicted in field guides, and the latter is what is illustrated in the best distribution atlases, epitomized for birds in Britain first by the work of Sharrock (1976) and then by Lack (1986) and Gibbons *et al.* (1993). The area of occupancy will tend to be smaller than the extent of occurrence, because species do not occupy all areas (or habitats) within the geographical limits to their occurrence (no species is continually distributed in space). Indeed, the finer the resolution at which occurrence is mapped, the smaller will be the measured area of occupancy (e.g. Kunin 1998; Cowley *et al.* 1999), because greater areas over which the species does not occur will become apparent. In a similar way, many species do not occur throughout a patch of habitat, such as Eastern Wood; there are, for example, edge specialists and interior specialists. However, because both extent of occurrence and area of occupancy can be measured in many ways, the magnitude of the difference between the two can vary considerably, and need not always be evident.

Both measures will be influenced by the duration of a study, with extent of occurrence fluctuating as range limits move back and forth, and area of occupancy fluctuating as regions or habitat patches are colonized and relinquished. Area of occupancy will tend to be influenced to a greater degree for a mobile group such as the birds. Summarizing occurrence data over multiple years may lead to a disproportionate overestimation of the ranges particularly of the rarer species. For example, although the great grey shrike was recorded in 238 squares in Britain over the recording period of the winter bird atlas

(Lack 1986), it was recorded in only 71 squares in 1981–82, 124 in 1982–83 and 85 in 1983–84.

In the present context (as in ecology in general; Gaston 1991d), we are primarily interested in the areas of occupancy of species. These can be measured over a range of spatial extents. A species has an area of occupancy for a woodland as much as it has one for an entire continent. This raises an issue of what term to use to describe the spatial distributions of species. In the limit, the total area of occupancy of a species (across the globe) is its geographical range size, which has been the distribution measure of principal interest to macroecology. However, this is a confusing term when applied to smaller spatial scales. 'Distribution' is general, but is likely to generate confusion with statistical meanings of the word (which is the meaning we largely reserve for this term, as in the title of this section), while terms such as 'extent' or 'distributional extent' will inevitably tend to cause the confusion of area of occupancy with extent of occurrence. To use 'occupancy' itself would probably be the most economical solution, but this does not have general currency in the literature. Instead, we use the term 'range size' to refer to the occupancy of a species. This has the advantage that it has been used widely in the literature in this regard (see Gaston 1994a), while it can be prefixed with modifiers to produce the standard terms for the total spatial distribution of a species (geographical range size) and the temporal pattern of distribution of an individual (home range size). Therefore, for the rest of this book, we use range size to indicate the area of occupancy of a species. No single scale is implicit in this term, but modifiers or some indication of scale will often accompany its use.

3.2.2   *Patterns in the distribution of range sizes*

How the elements of a species by site matrix of the type shown in Appendix IV are filled is a prime example of an ecological pattern that is dependent on spatial scale. Converting the rows of the matrix into a frequency histogram of the range sizes of species not infrequently yields a distribution that is bimodal when the extent of the area over which range sizes have been recorded is relatively small, with the peaks in species number in both the smallest and the largest range size classes (Raunkiaer 1934; Goodall 1952; Hanski 1982a–c; Gotelli & Simberloff 1987; Williams 1988; Collins & Glenn 1990, 1991, 1997; Tokeshi 1992; Gaston 1994a). Figure 3.1 shows an example for the occurrence of breeding bird species in the 25 tetrads comprising the 10×10-km square SU87 in Berkshire (from the data in Appendix IV).

The frequency of bimodality appears to increase with reduction in the extent of the area across which are spread the sites at which occupancy is assessed (Brown 1984; Collins & Glenn 1991, 1997; Gaston 1994a). Figure 3.2 shows how range sizes are distributed when the entire ranges of bird species in Berkshire are examined. The sample area is now all of the 391 tetrads that cover the

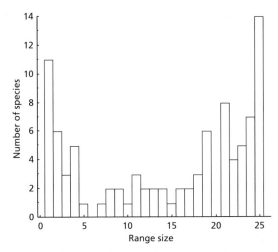

Fig. 3.1 The frequency distribution of the range sizes (number of tetrads—2 × 2-km squares—occupied out of a total of 25) of breeding bird species in National Grid square SU87 in Berkshire. From data in Standley *et al.* (1996).

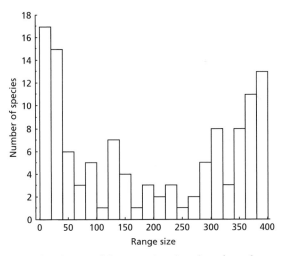

Fig. 3.2 The frequency distribution of the range sizes (number of tetrads occupied) of breeding bird species in the whole of Berkshire (391 tetrads); $n = 118$. From data in Standley *et al.* (1996).

county, rather than a sample of 25. This distribution still has peaks corresponding to the smallest and largest range size classes, as does Fig. 3.1. However, while 11 and 14 species were found in the minimum (1) and maximum (25) possible number of sample tetrads, respectively, only two and four species were recorded from the minimum (1) and maximum (391) possible number of Berkshire tetrads. Bimodality tends to be more pronounced when sites are more similar, and more species can occur at all sites. Habitat heterogeneity increases with scale, and the likelihood of bimodality decreases concomitantly.

In fact, firm statements about the effect of scale on patterns of bimodality in species–range size distributions have proven difficult to substantiate because bimodality is notoriously difficult to test for statistically. A method has been provided by Tokeshi (1992), albeit one that is designed more to recognize bimodality than to provide a definitive statistical proof of it (Collins & Glenn 1997). This involves calculating the probability (P_c) that more species occur in the extreme left- and right-hand classes of the distribution than expected by chance from the null hypothesis of random occurrence, using the equation:

$$P_c = \sum_{i=s_l}^{S-s_r} \sum_{j=s_r}^{S-i} \frac{S! h^{i+j} (1 - 2h)^{S-i-j}}{i! \, j! (S - i - j)!}$$ (Eqn 3.1)

where S is the total number of species, s_l is the number in the left-most class, s_r is the number in the right-most class and h is the frequency interval (where h and $1 - h$ sum to 1). If $P_c < 0.05$, then there are more species in the extreme classes than expected by chance, and one can then go on to calculate whether more species occur in the extreme left- or right-hand classes of the distribution separately using the equation:

$$P_h = \sum_{i}^{S} \binom{S}{i} h^i (1 - h)^{S-i}$$ (Eqn 3.2)

where i here is the number in the left-most (or right-most) class. If this probability is less than 0.05 for both classes, then the distribution can reasonably be called bimodal. Applying these tests to the distributions in Figs 3.1 and 3.2 shows both to be significantly bimodal.

Once we expand consideration to the whole of Britain, no bird species occurs in all possible squares on the British National Grid (Gaston et al. 1998a), despite the fact that ranges are mapped on a coarser grid than within subregions such as Berkshire (10×10 km, rather than 2×2 km). The species–range size distribution for the whole British avifauna is strongly right-skewed. Most species are narrowly distributed and a few are widespread. This is true both of the breeding and wintering assemblages (Fig. 3.3). Such a pattern is common to other taxa across Britain (Gaston & Lawton 1988b; Gaston 1990; Hodgson 1993; Quinn et al. 1996, 1997a; Sanderson 1996; Blackburn et al. 1997b; Gaston et al. 1998a; Cowley et al. 1999), and assemblages of birds and of other taxa in other regions (e.g. Enghoff & Báez 1993; Gaston 1994a; Vaughn 1997; Hecnar & M'Closkey 1998). Bird species have a higher median range size in Britain than do a wide variety of other taxa for which data exist (liverworts, vascular plants, molluscs, dragonflies, moths, butterflies; Gaston et al. 1998a). This does not appear to be an artefact of better recording of the distributions of this group. Indeed, the pattern holds despite the records used to map range size for most

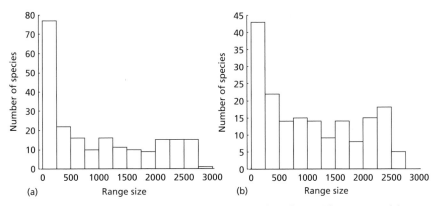

Fig. 3.3 The frequency distributions of range sizes (number of 10 × 10-km squares of the British National Grid occupied) for bird species (a) breeding ($n = 217$) and (b) wintering ($n = 176$) in Britain. From data in Appendix III.

other groups having been accumulated over much longer periods than those for birds. Skewed distributions such as these sometimes become approximately normal when the data are logarithmically transformed. However, logarithmic transformation here tends to be too strong, skewing the distributions to the left rather than making them approximately symmetrical (Gaston *et al.* 1998a). The logit transformation appears to be the most appropriate for normalizing range sizes at regional scales (Williamson & Gaston 1999).

Britain comprises only a proportion, and often a small one, of the entire geographical ranges of the species which occur there. There is only one bird species endemic to Britain. Studies embracing the entire geographical ranges of all, or at least the majority, of the species in a taxonomic assemblage tend to document species–range size distributions which are unimodal with a strong right-skew. That is, as for birds across Britain, most species have relatively small range sizes while a few have relatively large ones (Gaston 1994a, 1996c, 1998; Brown 1995; Brown *et al.* 1996). This observation has been made not only for extant assemblages (birds—Anderson 1984a; Schoener 1987; Pomeroy & Ssekabiira 1990; Blackburn & Gaston 1996a; Gregory *et al.* 1998; Maurer 1999; others—Willis 1922; Freitag 1969; Anderson 1977, 1984a,b, 1985; Pielou 1977a; Rapoport 1982; McAllister *et al.* 1986; Russell & Lindberg 1988a; Pagel *et al.* 1991a; Gaston 1994a, 1998; Roy *et al.* 1995; Hughes *et al.* 1996; Gaston & Chown 1999a; Hecnar 1999), but also for palaeontological ones (e.g. Jablonski 1986a, 1987; Jablonski & Valentine 1990; Roy 1994). Figure 3.4 shows examples for the distributions of geographical range sizes for extant birds of Europe, the New World and Australia. In virtually all published species–range size distributions at such scales in which range size is untransformed, the left-most range size class is also the modal one. These distributions tend toward an approximately normal distribution when geographical range sizes are subject to a

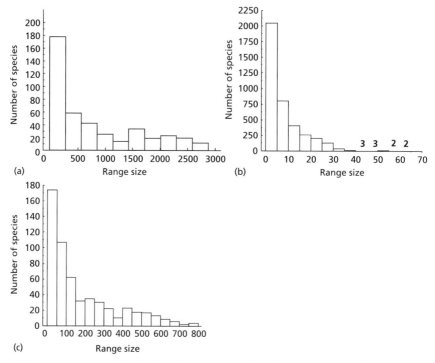

Fig. 3.4 The frequency distributions of range sizes for breeding bird species in (a) Europe (number of approx. 50 × 50-km grid squares occupied), (b) the New World (number of 611 000-km² grid squares occupied) and (c) Australia (number of 1 × 1-degree grid squares occupied). Numbers above columns indicate the number of species in that range size class. For New World birds, the maximum possible range size is 68.55 units. From Gregory *et al.* (1998), data sources in Blackburn and Gaston (1996a) and data in Blakers *et al.* (1984).

logarithmic transformation (Fig. 3.5; e.g. Anderson 1984a,b; McAllister *et al.* 1986; Pagel *et al.* 1991a; Gaston 1994a, 1996c; Blackburn & Gaston 1996a; Brown *et al.* 1996; Hughes *et al.* 1996). However, they appear also consistently to acquire a mild to moderate left-skew under such a transformation, reminiscent, but less marked than, that documented for ranges across less extensive areas (Gaston 1998), and similar to that observed for species–abundance distributions at large scales (Section 4.3). The logit transformation may be the most appropriate here also (Williamson & Gaston 1999).

Although species–range size distributions for birds and other groups of organisms tend to differ in shape at different scales, shifting from bimodal to unimodal as the scale increases, the extents of the range sizes of species at different scales tend nonetheless to be significantly positively correlated. For example, Fig. 3.6 shows that the total number of tetrads occupied by a bird species in Berkshire is well predicted by the number of tetrads (out of 25) in British National Grid square SU87 in which it is found. The number of tetrads occupied in Berkshire is in turn positively correlated with the range size of a

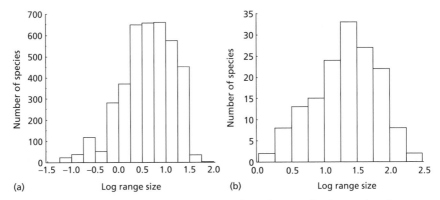

Fig. 3.5 The frequency distributions of $\log_{10}$-transformed geographical range sizes for (a) all bird species breeding in the New World and (b) wildfowl species. From Blackburn and Gaston (1996a) and Gaston and Blackburn (1996c).

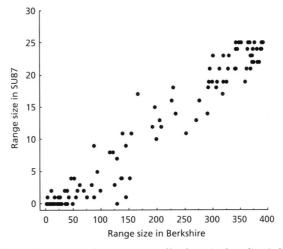

Fig. 3.6 The relationship between the range sizes of bird species breeding in National Grid square SU87 in Berkshire (number of tetrads occupied; $n = 25$) and their range size in the whole of Berkshire (number of tetrads occupied; $n = 391$) ($r = 0.975$, $n = 118$, $P < 0.001$). From data in Standley *et al.* (1996) and Appendix IV.

species in Britain (Fig. 3.7). Gregory and Blackburn (1998) showed that bird species with wide ranges in Britain also tend to be distributed widely across the Western Palaearctic, to have larger global ranges and to be found at a wider range of latitudes. Indeed, combining their data with those in Appendix IV reveals that bird species that are widely distributed across the Western Palaearctic tend to be found in more tetrads in Berkshire than do species with small Western Palaearctic ranges, albeit that the relationship is weak (Fig. 3.8). Correlations between the range sizes of bird species at regional scales and biogeographical or global scales have also been reported for India by Daniels *et al.* (1991), and in studies of a variety of other groups and regions (Gaston 1994a).

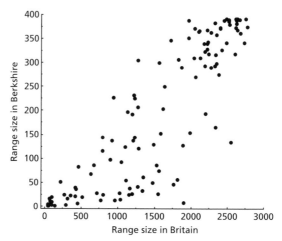

Fig. 3.7 The relationship between the range sizes of bird species in Berkshire (number of tetrads occupied; $n = 391$) and their range sizes in Britain (number of 10×10-km squares of the British National Grid occupied; $n = 2830$) ($r = 0.86$, $n = 118$, $P < 0.001$). From data in Appendices III and IV.

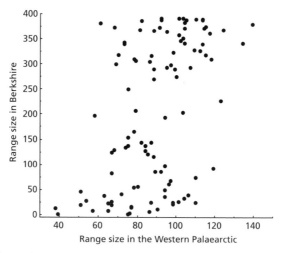

Fig. 3.8 The relationship between the range sizes of bird species in Berkshire (number of tetrads occupied; $n = 391$) and their range sizes in the Western Palaearctic (number of $153\,000$-km^2 grid squares occupied; $n = 139$) ($r = 0.47$, $n = 104$, $P < 0.001$). From data sources in Gregory and Blackburn (1998) and Appendix IV.

Patterns in the frequency distribution of avian range sizes at all scales reveal that many species in assemblages occur at only a very limited proportion of all possible sites. This suggests that many are unlikely to occur in particular local assemblages, such as that at Eastern Wood, not because the wood is an inappropriate habitat, but instead because by chance the individuals that are present in the landscape do not coincide with the site. For this reason, it seems likely that those species that are found in local assemblages will generally be those widespread in the landscape. We have already seen (Section 1.1) that those found to breed at Eastern Wood in the period 1949–79 tended to be widespread across Britain. We can assess this difference formally by comparing the breeding range sizes of bird species breeding in Eastern Wood with those of British species not breeding in Eastern Wood, using the data in Appendix III. The range sizes of the Eastern Wood species are significantly larger (two sample t-test: $t = 7.1, n = 217, P < 0.0001$).

As pointed out in the previous chapter, some British breeding birds are unlikely ever to breed in Eastern Wood. Thus, the above test is probably more realistic if the British range sizes of those bird species that breed in Eastern Wood are compared with the British range sizes of bird species that breed in deciduous woodland but not Eastern Wood. The range sizes of the Eastern Wood species are still significantly larger (two sample t-test: $t = 6.6, n = 80, P < 0.0001$).

Even using the more realistic woodland species pool, however, some pool species are unlikely to occur in Eastern Wood. The pied flycatcher, for example, is very much a bird of western, not eastern, woods. Although species range sizes tend to be correlated across scales, the use of an inappropriate species pool could bias the test (Chapter 6, e.g. Harvey *et al.* 1983; Colwell & Winkler 1984; Gotelli & Graves 1996). A better comparison would be of the woodland birds of Surrey, the county in which Eastern Wood is situated. Unfortunately, range size data for birds across Surrey are not available. However, if we make the not unreasonable assumption that the distribution of species across Surrey is likely to be quite similar to their distribution across the neighbouring county of Berkshire (and Eastern Wood is only a few miles from the Berkshire border), we can use the range sizes in the latter given by Standley *et al.* (1996; Appendix IV) to compare deciduous woodland species that do and do not breed in Eastern Wood. The range sizes in Berkshire of bird species breeding in Eastern Wood do turn out to be significantly larger than those of woodland species that breed in Berkshire but not at Eastern Wood (two sample t-test: $t = 5.3, n = 64, P < 0.0001$).

Although it is clear that the bird species recorded breeding in Eastern Wood tend to have larger than average range sizes, it is not clear whether or not the Eastern Wood avifauna differs from a random sample of the British avifauna in terms of its range sizes. The tests that we have just performed are unweighted by the range sizes of the species: if 45 species were picked at random from the British breeding avifauna, they would have smaller range sizes, on average, than the species breeding at Eastern Wood. However, if, as we have argued, the probability that Eastern Wood samples species from the wider environment depends on their range sizes, range sizes need to be accounted for when testing whether or not the Eastern Wood species are a random subset of the British avifauna. We can perform such a test by comparing the range sizes of the Eastern Wood avifauna with those of many random samples of 45 species selected without replacement from the British avifauna, weighting the probability that a species is selected to be proportional to its British range size (using the data in Appendix III). Using the entire British species pool, there is a marginal tendency for Eastern Wood species to have larger range sizes than expected (1000 random assemblages: $P = 0.012$ that the arithmetic mean range size of Eastern Wood species is no different from the random expectation; $P = 0.054$ for the geometric mean). However, using the more realistic woodland species pool, there is no tendency for Eastern Wood species to have larger range sizes than expected (arithmetic mean range size, $P = 0.92$; geometric mean, $P = 0.89$). Thus, we can conclude that although the bird species breeding in Eastern Wood tend to have larger than average range sizes compared to other British birds, their mean range size is exactly as would be expected if Eastern Wood sampled species at random from the woodland species pool on the basis of their range sizes (Blackburn & Gaston, in press).

3.3 Determinants of species–range size distributions

The species–range size distributions illustrated in Figs 3.1–3.5 show how the range sizes of bird species vary at different scales. The obvious question that next arises is why range sizes show the pattern of variation that they do. In other words, why at large regional scales should most species be narrowly distributed and a few very widespread, and at smaller scales should most species be either narrowly distributed or widely distributed? This is a significant issue: to answer it is to address why there are so many rare species, at least from one perspective on what constitutes rarity (Gaston 1994a).

A number of mechanisms have been suggested as determinants of species–range size distributions (Table 3.1). These are not necessarily mutually exclusive, although their interactions have been poorly explored. In some cases they may simply constitute different levels of explanation. Here, we will address these mechanisms in a sequence which approximately moves from those more pertinent to small scales to those more relevant to large ones.

In some cases, the shapes of species–range size distributions—as indeed with many other frequency distributions—may be generated by peculiarities in the ways in which the data have been assembled or displayed. These determinants will not be addressed here, but it is important to consider them when constructing such frequency distributions (for discussion see Nee *et al.* 1991a; Gaston 1994a, 1998; also Loder *et al.* 1997).

3.3.1 *Random sampling*

If the individuals of each species in an assemblage are dispersed randomly across the same area independently of other species, then a bimodal species–

Table 3.1 Factors postulated as determining or contributing to the form of species–range size distributions.

Factor	Process
Random sampling	Random and independent dispersion of individuals
Range position	Location of study area relative to the range limits of species
Metapopulation dynamics	Metapopulation structures
Vagrancy	Narrow range sizes of vagrant species
Niches	
breadth	Species with broader niches attaining larger range sizes
position	Species exploiting more widespread environments and resources attaining larger range sizes
Speciation, extinction and temporal dynamics	Relationships between range size and likelihoods of speciation and extinction, and the long-term dynamics of range size

range size distribution will be generated when the species–abundance distribution is approximately log-series or log-normal (Section 4.3; Preston 1948; Williams 1950, 1964; McIntosh 1962; Papp & Izsák 1997; see also Gleason 1929) or presumably takes the form of any of a number of long-tailed distributions. This is because there are many relatively rare species, resulting in a left-hand peak to the species–range size distribution, and high classes of occupancy include a wider range of abundances than do the lower classes, resulting in a right-hand peak also. Such a mechanism is most likely to be met at very small spatial scales, smaller than those under consideration here, because it is then that the assumptions are most likely to be met. Nonetheless, such a model provides a reasonable fit to some regional data (e.g. Hanski 1982b).

3.3.2   *Range position*

The shapes of species–range size distributions can, in some instances, be regarded as a consequence of the location of a study area relative to the positions of the range limits of species in an assemblage. Plainly, if a study area is at the periphery of the range of a species, then that species is likely to extend only some way across the area, while if the study area is at the core of the range of a species, then that species is likely to extend across the whole area. Moreover, if, as many have argued (Hengeveld & Haeck 1981, 1982; Brown 1984; Lawton 1993; Maurer 1994, 1999; Brown *et al.* 1995), the occupancy of a species becomes patchier towards its range limits, then even if no species actually reaches its geographical limit within a study area, nonetheless the proximity to the range edges of species may determine how widespread they are across the area. Britain is peripheral to the global geographical ranges of a number of bird species, and these contribute disproportionately to the numbers of species in the lower classes of the species–range size distribution for the region as a whole (Gaston *et al.* 1997b).

This mechanism for generating the shapes of species–range size distributions is essentially a second-order explanation, in that it does not explain why some species penetrate further into a region than do others, or why occupancy becomes patchier towards range limits. By definition, it cannot explain the shapes of frequency distributions of the entire geographical range sizes of species (see also Section 4.2.2: Range position).

3.3.3   *Metapopulation dynamics*

Metapopulation dynamics models have been used in attempts to understand the determinants of the biogeographical limits to species occurrences (e.g. Carter & Prince 1981; Lennon *et al.* 1997). For example, Lennon *et al.* (1997) show that relatively abrupt edges to ranges can be generated not only by discontinuities in environmental variables, but also by the interactions of extinction and recolonization which underpin metapopulation models. Such models have also provided a popular framework for examining the determinants of species–range size distributions (for a general review of metapopulations, see Hanski & Gilpin 1997). This work has centred on a class of models first introduced by Levins (1969, 1970), which relate the proportion of sites occupied to the probabilities of immigration and extinction (analogous to population models in which the rate of change in abundance is expressed as the difference between birth and death rates). Four such models have been explored in considerable detail, expressing the dependency or independency of immigration and extinction rates on regional occurrence (for details see Levins 1969; Hanski 1982a, 1997a; Gotelli & Simberloff 1987; Gotelli 1991).

If the probability of local immigration i is dependent on regional occurrence (the proportion of sites occupied, p) and the probability of local extinction e is independent (Levins 1969):

$$dp/dt = ip(1 - p) - ep \qquad\qquad\qquad \text{(Eqn 3.3)}$$

then if $i > e$ there is a single internal equilibrium at $p^* = 1 - e/i$. For a range of parameter values, a stochastic version of this model, in which e is a random variable, predicts a unimodal distribution of occupancy for one species at different times. This can equally be interpreted as a unimodal distribution of occupancies for different species at one time (Hanski 1982a).

If the probability of local immigration and the probability of local extinction are both dependent on regional occurrence (Hanski 1982a), such that the model incorporates a 'rescue effect' (emigrants from surrounding populations reduce the probability of local extinction):

$$dp/dt = ip(1 - p) - ep(1 - p) \qquad\qquad\qquad \text{(Eqn 3.4)}$$

the outcome is unstable. If $i > e$, then $p = 1$, and if $i < e$ then $p = 0$. A stochastic version of this model predicts a bimodal distribution of occupancy, but with $p = 1$ the more common of the two outcomes (Hanski 1982a; Tokeshi 1992).

If the probability of local immigration is independent of regional occurrence (it incorporates a 'propagule rain'), and the probability of local extinction is also independent of regional occurrence (Gotelli 1991):

$$dp/dt = i(1 - p) - ep \qquad \text{(Eqn 3.5)}$$

there is an equilibrium point at $p^* = i/(i + e)$. For most parameter values a stochastic version of this model predicts that there is a unimodal distribution of occupied sites.

If the probability of local immigration is independent of regional occurrence but the probability of local extinction is dependent (Gotelli 1991), that is the model incorporates a propagule rain and a rescue effect:

$$dp/dt = i(1 - p) - ep(1 - p) \qquad \text{(Eqn 3.6)}$$

there is an equilibrium point at $p^* = i/e$. For most parameter values a stochastic version of this model predicts that there is a unimodal distribution of occupied sites.

Finally, Tokeshi (1992) has suggested one further model, which has been less widely remarked, and which allows a rescue effect, but the rate of extinction is negatively related to p:

$$dp/dt = ip(1 - p) - e(1 - p). \qquad \text{(Eqn 3.7)}$$

Here, the equilibrium for a species depends on the relationship of e/i to p, but will either be $p = 1$ or $p = 0$. If e and i show stochastic variation, then $p = 0$ is more common.

The Levins' model and its derivatives make a number of simplifying assumptions. Suitable habitat occurs in discrete patches, and these patches are identical and infinite in number, individuals can move equally well between any pair of them, and, in the multispecies context, they can potentially be occupied by breeding populations of all species. The real world is plainly much more complex, and it remains a contentious issue how adequate this caricature is for predicting what will actually occur (Harrison 1994; Hanski 1997a,b). Nonetheless, these models convey some important messages as to the possible determinants of the shapes of species–range size distributions. Foremost, they illustrate how the kinds of distributions observed can readily be generated from simple patterns of colonization and extinction, and that such patterns do not typically result in species–range size distributions which are of a fundamentally different character.

Some of the limitations imposed by assumptions of the Levins-type model have been addressed by the incidence function approach (Hanski 1994, 1997a,b). Here, the number of patches is finite, patches vary in size, and interactions between patches are localized in space. Three of the key, and testable, simplifying assumptions that are typically made in exploring such models are that the size of a local population increases with patch size, that the probability of extinction of local populations increases with decreasing patch size, and that the probability of colonization of patches decreases with their increasing isolation from other patches. For British birds, empirical studies have found support for some of these. For example, Bellamy *et al.* (1996b) showed that population size was related to woodland area for eight species of bird breeding in small woods in Cambridgeshire and Lincolnshire (see also Newton 1998, p. 68), and that, in general, smaller populations were more likely to go extinct in any given year. Whether or not the woods in their study were isolated enough to hinder dispersal among them by such mobile species as birds is debatable, but Bellamy *et al.* nonetheless found no evidence that woodland isolation affected the probability that species recolonized them. Recolonization does not exactly equate to colonization, because individuals may relocate to sites that are already occupied, but this result does question an important assumption of the incidence function approach. Also, extension of this approach to multispecies systems has, as far as we are aware, been limited (but see Hanski & Gyllenberg 1997), and it remains unclear what predictions it makes about the shapes of species–range size distributions.

The extent to which birds, and indeed most groups of organisms, exhibit metapopulation dynamics remains an open question. If metapopulation dynamics comprise local extinctions and colonizations, resulting in population turnover, then many and perhaps most bird species probably fail to exhibit such dynamics (those which are rare or occupy particularly fragmented landscapes or discrete colonies are perhaps most likely to conform, e.g. Hinsley *et al.* 1995; Oro & Pradel 1999). Here, so-called mainland–island systems, in which one or more (mainland) populations persist essentially indefinitely (Harrison 1994), are excluded. Such systems are probably common among birds. If a broader view is taken, in which any assemblage of local populations with migration among them is considered a metapopulation, regardless of the rate of population turnover (Hanski & Gilpin 1997, p. 1), then the answer is almost certainly that birds do exhibit metapopulation dynamics; if this were not so, then we would not observe recolonizations of bird species at Eastern Wood, following their loss from the site. Under this view, a metapopulation approach is less useful when dispersal rates are very high and dispersal distances are long relative to between-patch distances, because the greater population is less likely to be divided into discrete independent units, as metapopulation models assume. Such a circumstance may perhaps be likely to pertain to birds at high latitudes. Birds in Britain can disperse widely relative to the size of the country, although most individuals tend to remain close to their natal site (Paradis *et al.* 1998). Nevertheless, Paradis *et al.* (in press) found that most bird species showed only weak synchrony in temporal population fluctuations across sites in Britain, while Koenig and Knops (1998) found no evidence of spatial autocorrelation in the abundances of birds at sites across the US state of California. Both results suggest that bird populations at local sites may show the demographic independence required by metapopulation dynamics.

3.3.4 *Vagrancy*

Raunkiaer (1934) argued that bimodality in species–range size distributions was produced because species found to occupy all sites were those best adapted to live under the prevailing conditions, while species found to occupy only a few sites were adapted to, and common in, other conditions. This highlights the potential role of vagrants in structuring such distributions.

Vagrant species have explicitly been excluded from most of the species–range size distributions which we have mentioned. Whatever the scale, their inclusion would obviously serve primarily to swell the left-hand range size class, and often markedly so. For example, Appendix III lists the 240 species that we considered to be a normal part of the British breeding or wintering avifaunas. As discussed in that appendix, different inclusion criteria would give different species lists, but not lists that would differ greatly from ours in species number. Yet, a total of at least 550 species have been recorded in Britain in an apparently wild state (Section 2.1.2). Thus, around 300 species, or 55% of the total British avifauna, may be considered vagrants. Of these, about 250 species are so rare in Britain that the provenance of every individual record is assessed to ensure that correct identification has been made and that wild origin is likely. Only a handful of individuals, at most, of these species occur in Britain in any one year, and most of these are found at a few migration hotspots. Britain does not form part of the normal range or migration routes of these species (indeed, many are migrating in the contrary direction to that in which they would be expected to move), and their area of occupancy in Britain, tenuous as it is, is small.

The influence of vagrants on species–range size distributions can be regarded as one manifestation of a phenomenon which has elsewhere been termed the 'mass effect' (Shmida & Wilson 1985). This is the increase in species richness which results from the influx of transient individuals of species that cannot maintain viable populations in an area (in some definitions these species nonetheless become 'established'—however, this usage apparently refers to the vegetative growth of plants, which will not otherwise be present, at least as mature individuals, and does not imply a viable population). The importance of mass effects to levels of species richness has been argued by Shmida and Wilson (1985) to peak at meso-scales, which they defined as spanning areas of approximately $10-10^6$ m^2. In comparison, Eastern Wood covers 1.6×10^5 m^2, and it is probably fair to say that very few of the bird species breeding there could maintain viable populations in the area if it were a closed system, subject, of course, to the definition of 'viable' (Soulé 1987; Burgman *et al.* 1993). Williamson (1981) suggests that five species (great tit, blue tit, wren, robin, blackbird) could do so, assuming that populations of 10 pairs or more were viable. Nevertheless, the extent to which the majority of the avifauna of Eastern Wood can usefully be thought of as transient is debatable. Most components of the avifauna breed there in most years (Fig. 2.4), and that many other species that breed there do so sporadically is likely largely to be a consequence of their pattern of space usage (Section 2.1). While the total avifauna of Eastern Wood could be inflated by transient individuals of species known normally not to breed or winter at the site, few bird species breeding in Eastern Wood seem likely to be genuine transients. The list of breeding species at the site, and hence pattern in the structure of the avifauna, is unlikely to be greatly affected by the mass effect.

3.3.5   *Niches*

As we will discuss later at greater length, the shapes of species–abundance distributions at regional scales have been interpreted as being consistent with niche breakage models, in which species abundances are considered to be associated with particular processes of niche division (Section 4.3.3; Nee *et al.* 1991b; Tokeshi 1996). Species apportioned larger fragments of total niche space by the division process are assumed to be more abundant. Models in which successive niche division tends to be more likely for species with larger niches or higher abundances appear to fit particularly well. Tokeshi (1996) has associated this with the greater likelihood that both higher abundances and larger geographical range sizes are more likely to generate new species, thus implicitly identifying a potential link between the forms of species–abundance and species–range size distributions. In practice, it is difficult to conceive of the generation of species–range size distributions in terms of niche apportionment, if for no other reason than that, at all but the finest resolutions (e.g. the volume filled by an individual organism), physical space is shared by species rather than used by some to the exclusion of all others. Nonetheless, it remains plausible that species with higher abundances should use more space (see also Section 4.2), and hence that such a model could reasonably approximate the shapes of species–range size distributions even if the details of the mechanism are flawed.

The fundamental premise underlying the model is that range size is correlated with niche breadth, such that possession of a broader niche enables a species to become more widespread. This is an idea with a long history, which has been vigorously championed recently by Brown and colleagues (Brown 1984, 1995; Brown *et al.* 1995). Nonetheless, although it is intuitively very appealing, the evidence for a positive relationship between niche breadth and range size is remarkably poor. Most of the analyses which have been published are confounded by the failure to account for differences in sample size between restricted and widespread species; this problem particularly frustrates attempts to draw some general conclusion in the present context from studies of the relationship between the size of the geographical range of a species and the number of habitats or breadth of environmental conditions which lie within that range. If sites were occupied by species at random across a landscape, then, because environments tend to exhibit positive autocorrelation (sites closer together are more similar than sites far apart), by chance the larger the number of occurrences the broader the niche would appear to be (assuming that niches were measured on the basis of environmental variables). Thus, what we need to know is whether the niche breadths of restricted species are narrower than those of widespread species when they are determined on the basis of similar numbers of occurrences.

A formal framework for addressing this question is provided by the concept of niche pattern, the three-dimensional relationship between the niche breadths, niche positions and abundances of the species in an assemblage (Shugart & Patten 1972). Here, niche breadth is the range of environmental conditions in which a species is observed to survive (and perhaps also reproduce). This can be extended to include axes of resource exploitation. Niche position, while it has been interpreted in several ways, is usually regarded as a measure of how typical the conditions in which a species occurs are of the full universe of conditions under consideration. Counter-intuitively, the greater the value of niche position, the more atypical are the conditions.

Gregory and Gaston (2000) used extensive census and environmental data to examine the relationship between niche breadth and range size for breeding birds in Britain. The census data came from a volunteer-based survey of 1830 1 × 1-km squares in 1996, the environmental data from ground surveys and satellite imagery. Eighty-five species, which occurred on over 100 1 × 1-km squares in the survey, were chosen for analysis. Thirty-four land-use or environmental variables were used to generate four ordination axes by Canonical Correspondence Analysis (CCA), and estimates of niche breadth were derived from these axes. Gregory and Gaston found no evidence of a systematic relationship between niche breadth and range size. This result held when controlling for the phylogenetic non-independence of species.

Similar results (although without controlling for phylogeny) have been documented for the habitat volumes (environmental tolerances) of a suite of plants from southern Western Australia (Burgman 1989). Thus, on the basis of current evidence at least, it seems unlikely that species with larger ranges do have broader niches. Unfortunately, no amount of analysis ultimately can preclude the possibility that there is a positive relationship between range size and niche breadth. Niches are, following Hutchinson's (1957) definition, *n*-dimensional. There are therefore many components to the requirements of an individual species, and these will be hard to combine to provide a measure of niche breadth that is meaningful and comparable between species. In addition, with so many niche axes along which species may be varying, any test may simply fail to measure the relevant ones (Colwell & Futuyma 1971). This is especially true for migrant species, for which range size may be affected by factors changing outside the focal region. The best that can be done is to attempt to capture some of the major niche axes. Ideally, this probably needs to be performed at more than one spatial scale, as species may be responding to the environment in complex ways.

If we accept that there is not a general relationship between niche breadth and range size, which we strongly suspect is the case, this does not mean that niches are irrelevant to the range sizes of species. The conditions and resources which a species can exploit plainly play a substantial role in determining where it can occur. This suggests perhaps a rather simpler way in which niches can in some systematic fashion determine the range sizes of species. Range sizes may be determined not by niche breadth, but by niche position—the availability of the environments and resources which are exploited.

Gregory and Gaston (2000) found some support for such a mechanism, based on their analyses for breeding birds in Britain. They observed that, contrary to niche breadth, niche position and range size were negatively correlated across all species, and that this relationship was robust to controlling for phylogenetic non-independence of the data points. Duncan et al. (1999) showed that the current range sizes in New Zealand of birds artificially introduced from Britain in the 19th century are correlated with the amount of habitat in the new environment suitable for the species. In a similar, but more anecdotal vein, Fuller (1982) argues that the most abundant and widespread species of breeding birds on saltmarshes in Britain are those which utilize the typical and common features of the marshes, while the least abundant and poorly distributed species are those which are restricted by their preference for special habitats. As a broad generality extending to many other habitats, such findings would make good intuitive sense to most birders. If, on average, individual species expand their range toward the limits of the geographical distribution of the resources which they exploit, then species exploiting more widespread resources will themselves be more widespread.

Consideration of the relationship between niches and range sizes leads almost ineluctably to the issue of what factors limit the range sizes of species. Traditionally, a distinction has been made between abiotic and biotic factors, with the former typically regarded as being the more significant at higher latitudes (Dobzhansky 1950; MacArthur 1972; Kaufman 1995, 1998). This follows in major part from the observation that the northern limits of the ranges of species in the northern hemisphere are often found to be coincident with particular combinations of climatic conditions (e.g. Voous 1960). For example, Root (1988b) investigated the association between environmental factors and distributional boundaries for 148 species of wintering landbirds across North

America. The comparisons revealed frequent correspondence between range limits and environmental factors. Average minimum January temperature, mean length of frost-free period, and potential vegetation frequently associated with northern range limits. Less than 1% of all the associations observed were expected to occur by chance. Likewise, abiotic factors have frequently been found to act as predictors of the distributions of birds (e.g. Avery & Haines-Young 1990; Carrascal *et al.* 1993; Gates *et al.* 1994; Gibbons *et al.* 1995; Lloyd & Palmer 1998; Venier *et al.* 1999). Burton (1995) gives many examples of species whose ranges are apparently, or have been postulated to be, limited by climate. For example, the distribution of Cetti's warbler seems to be affected by the severity of winter weather (Bonham & Robertson 1975). Hard winters in south-east England in the period 1984–87 caused the main British population of this species to shift to the south-west (Burton 1995).

Also suggestive of a role for abiotic factors is the observation that the distributions of species are often seen to change through time in broad synchrony with changes in environmental conditions. Thus, the occurrence, or the more southerly occurrence, in Britain of many species during particularly harsh northern winters will be familiar to local birders. The smew (a species of duck) is an interesting case in point, with juveniles and females being driven south first, followed by males (which can tolerate colder conditions; Elkins 1995). Over longer periods, the distributions of many bird species in Europe and beyond have waxed and waned as climatic conditions have fluctuated (Burton 1995).

Unfortunately, neither the coincidence of range limits with climatic factors nor the synchronous change in climatic factors and the position of range limits need necessarily imply a direct effect of climate on the distribution of a species (Spicer & Gaston 1999). Rather, climate may, for example, be influencing the distribution of resources, competitors, predators or parasites, to which the distribution of a species is in turn responding. Even where climate is the predominant determinant of range limits, the manner in which it operates may be difficult to discern. Thus, for example, the argument as to whether range limits, particularly northern ones, of birds are coincident with their threshold energy requirements has persisted for a long time (e.g. Salt 1952; Cox 1961; Root 1988a,c, 1989; Castro 1989; Repasky 1991). Most recently, it has been proposed that a physiological ceiling on metabolic rate, preventing the demands of thermoregulation from being met as temperature declines, constrains the northern distributional limits of wintering landbirds in North America (Root 1988a,c, 1989). However, this interpretation has subsequently been challenged (Castro 1989; Repasky 1991).

The roles of biotic and abiotic factors will in practice often be blurred. Ranges of birds may be limited because, for example, the prevailing climate prevents sufficient growth of resources, or prevents their growth over a sufficient period for successful reproduction. Thus, the whooper swan takes 130 days to complete its breeding cycle. Within that part of its range north of the Arctic circle only half the summers are ice free for this length of time, and at these latitudes breeding is often unsuccessful (Elkins 1995, p. 83). Its smaller relative, the Bewick's swan, can breed further north, because it has a cycle of only 100–110 days.

Examples have been documented of more direct roles of competition and predation in limiting the ranges of bird species, although these seem to be relatively scarce in the literature. Thus, Terborgh (1985) argues that the elevational limits of two-thirds of the species in his study of Andean birds were attributable to competitive exclusion (direct or diffuse) and one-sixth to ecotones. He further argues that evidence from temperate regions suggests, in comparison, a much greater importance of ecotones and much reduced importance of competition, likely as a result of more drastic structural changes across ecotonal boundaries. Dekker (1989) argues that the ranges of megapodes and large carnivores show little overlap because their mound-building breeding behaviour makes megapodes highly susceptible to predation. The two only co-occur in areas where megapodes use burrows instead of mounds. The situation here seems to be that although the carnivores are generalists, and would typically switch to other food sources when particular prey species become rare, megapode mounds are easy to locate and may thus be predated even when they occur at very low numbers.

One attempt to elucidate the relative importance of biotic and abiotic factors as limits to range sizes is the work of Kaufman (1998), who argues that the form of range edges should differ depending on the type of factor delimiting them. Abiotic factors should generate more regular range edges than biotic factors, because variation in abiotic factors is likely to be more regularly distributed across space. As abiotic factors have been hypothesized to be more important at higher latitudes (Dobzhansky 1950; MacArthur 1972), three basic predictions follow.

1 Outside the tropics, the high-latitude or 'polar' boundary is more consistent with latitude and more geometrically regular than the low-latitude or 'equatorial' boundary of a species range.

2 Within the tropics, there is no difference between northern and southern boundaries of a species range.

3 Across zones, temperate–polar boundaries are more consistent with latitude and more geometrically regular than tropics boundaries across the group of species used.

Kaufman (1998) used two metrics of range edge consistency to test these predictions for a sample of New World mammals. Her results were in agreement with the abiotic–biotic hypothesis.

Ultimately, multiple factors, both abiotic and biotic, doubtless influence the position of the range boundaries of individual species, and hence the shape of the species–range size distribution, with their relative importance varying in space and in time. In practice, however, understanding why particular species of bird exhibit the geographical occurrence that they do remains largely wanting.

3.3.6   *Speciation, extinction and temporal dynamics*

At large geographical scales, the shapes of species–range size distributions must ultimately be a consequence of patterns of speciation, global extinction, and the temporal dynamics of the range sizes of species through their lifetimes (Gaston 1998; Gaston & Chown 1999a; Webb *et al.* 2000). Speciation adds new ranges, extinction removes ranges, and through time ranges move within the span of possible range sizes. Unfortunately, at present too little is known about the fundamental processes concerned (the nature of some of which remains extremely contentious) to be able to draw any strong conclusions as to how this actually occurs, and hence which processes are the most significant.

If we assume that speciation is predominantly allopatric, then Rosenzweig (1975, 1978, 1995) argues that on a purely probabilistic basis, species with larger geographical range sizes are more likely to undergo speciation, because the likelihood of their ranges being bisected by a barrier is greater than for a small range size. Differentiating, as Rosenzweig does, between two kinds of barrier, 'knives' (which have beginnings and ends) and 'moats' (which surround their isolates), then strictly this assertion is only true of moats. Very large geographical ranges will tend to engulf knives, such that they do not engender speciation, and the probability of division will have a peak at intermediate range sizes. Rosenzweig (1995) argues that this is unlikely to occur because there are no, or virtually no, species with geographical ranges so large that reducing them would make them an easier target for barriers. However, this view depends critically on the frequency distribution of barrier sizes (Gaston 1998). If most barriers are small to intermediate in size, relative to the range sizes of widespread species, then intermediate-sized ranges may indeed have a higher instantaneous probability of speciation. Such an effect would be enhanced because barriers seem far more likely to take the form of knives than of moats. However, if species with larger range sizes persist for longer, and ancestors persist beyond speciation (which is incompatible with some models of speciation), then they may nonetheless leave behind more descendants despite the pattern of instantaneous rates (Gaston & Chown 1999a).

While there are a number of models of the long-term temporal dynamics of the range sizes of individual species between speciation and extinction, whether any general patterns exist remains almost entirely unknown (Gaston & Blackburn 1997a; Gaston & Kunin 1997; Gaston 1998; Gaston & Chown 1999a; Webb *et al.* 2000). At one extreme, approximately maximal range sizes may be attained rapidly after speciation and then essentially maintained until shortly before extinction. Such a model has been championed by Jablonski (1987). It concords with some features of the patterns of extinction of molluscs, particularly the fact that the distribution of range sizes of species that originated in the two million years preceding the end-Cretaceous extinction (whose geological durations were thus truncated) is statistically indistinguishable from that of the species originating in the previous 14 million years (for other evidence for long-term stability of ranges, see Riddle 1998). At the other extreme, range sizes may follow no consistent temporal trajectory from one species to another, rather responding idiosyncratically to the environmental pressures peculiar to any given time, place and organism. Such a model would seem in keeping with the continual adaptation and change that may result from responding to the demands of the Red Queen (Van Valen 1973a; Ricklefs & Latham 1992); with such a dynamic, geographical ranges would have to attain an average size through their lifetime greater than that they acquired at speciation, if average range sizes across species were not to have declined markedly through evolutionary time (for which there is little evidence; Gaston 1998). Between these two extremes lie a variety of other models, including systematic increases in range size with time (the 'age and area' hypothesis; Willis 1922) and a number of cyclic patterns (e.g. the taxon cycle; Wilson 1961; Ricklefs & Cox 1972, 1978). There is at least some limited evidence for most of these models, but no consensus on which, if any, is the more general. The situation is also somewhat complicated because where ancestors persist beyond speciation, such events may have some impact on the long-term dynamics of ranges, reducing the range sizes of ancestors.

The instantaneous likelihood of extinction is almost certainly a decreasing function of range size. Larger ranges on average comprise more individuals (Section 4.2), reducing the likelihood that demographic and environmental stochasticity will wipe out local populations. The greater geographical spread reduces the likelihood that all individuals will experience adverse conditions simultaneously. While logically sound, empirical evidence that categorically demonstrates a negative relationship between likelihood of extinction and range size is, however, difficult to obtain.

There appears to be abundant empirical evidence for a positive relationship between time to extinction and range size for various palaeontological species assemblages (Jackson 1974; Hansen 1978, 1980; Stanley 1979; Koch 1980; Flessa & Thomas 1985; Jablonski 1986a,b; Buzas & Culver 1991; Jablonski & Raup 1995; Erwin 1996; Flessa & Jablonski 1996), although the correlation is not always especially strong and may break down during periods of mass extinction (Jablonski 1986a; Norris 1991). However, a component of this pattern is doubtless an artefact resulting from the fact that more widely distributed species tend also to be locally more abundant (Section 4.2) and thus are more likely to be recorded in the fossil record (Russell & Lindberg 1988a,b). Moreover, it is unclear what contribution to the pattern is made simply from the ways in which geographical range sizes are assessed, which commonly serve to integrate occurrence records across time, and so obscure the distinction between range size and changes in range location (see Gaston & Chown 1999a for discussion).

Turning to more contemporary data, negative correlations have been documented between likelihood of local extinction and range size on a number of occasions (Hanski 1982a; Gaston & Lawton 1989; Gaston 1994a; Mawdsley & Stork 1995). However, analyses based on the entire geographical ranges of species have had to rely on estimations of which species are presently regarded as facing the greatest likelihood of extinction rather than actual extinctions. While these also tend to support the tendency for small range sizes to increase these risks (Laurila & Järvinen 1989; Gaston & Blackburn 1996c; Mace & Kershaw 1997; see also Johnson 1998a), there is concern that range size measures themselves may play too great a role in the estimation of extinction risks for these analyses to be entirely secure. Nonetheless, it would be surprising indeed if, in general, risk of extinction did not genuinely decline with increasing range size.

Particularly because of the uncertainties surrounding the long-term dynamics of range sizes, we are not in a position to determine how in practice the processes of speciation, range size dynamics and extinction fit together to yield the species–range size distributions observed for extant assemblages. It does seem likely that most species have small range sizes at origination, and just prior to extinction. Thus, small ranges ought to predominate. At present, however, we can add little to this simple statement. Ultimately, though, speciation and extinction must interact in such a way as to generate the right-skewed distribution.

3.3.7   *Synthesis*

In sum, the determinants of species–range size distributions may be viewed from a number of perspectives. All probably comprise some grain of truth, at least at one of the scales of interest (Table 3.2). Within Surrey, in which Eastern

Table 3.2 A qualitative assessment (on a scale of 0–4) of the available empirical evidence in support of the role of the factors postulated as determining or contributing to the form of species–range size distributions.

Factor	Assessment
Random sampling	+++
Range position	++
Metapopulation dynamics	+++
Vagrancy	++
Niches	
breadth	+
position	+++
Speciation and extinction	+

Wood resides, differences in the range sizes of bird species probably closely reflect differences in resource usage, and the dynamic interactions between the populations in different habitat patches. Moving to the scale of Britain, the significance of the latter probably declines relative to the former. At yet broader scales, the stamp of speciation and extinction dynamics becomes most evident.

3.4   Patterns of range overlap

We began this chapter by noting that variation in the species richness of sites could result if all species had the same overall level of occupancy, but occurred at different sites in different combinations, or if species differed in their levels of occupancy. We showed that the latter pattern predominated, and then dealt with the question of why some species are widely distributed across landscapes, while the majority are narrowly distributed. The question of the size of ranges is, however, only one aspect of how patterns in species ranges help determine the composition of faunas at local sites like Eastern Wood. Just as variation in richness has the potential to arise from a variety of range size distributions, so also does it have the potential to arise from a variety of patterns in the overlap of ranges. It is these patterns that we next consider.

The question of range overlap primarily concerns three issues: nestedness, spatial turnover and geographical variation in range size. The degree of nestedness in a fauna quantifies the relationship between the composition of high and low richness sites occupied by the constituent species (Patterson & Atmar 1986). Perfect spatial nestedness exists when the species present in any particular fauna are also present in all larger faunas, and where the species absent from any particular fauna are also absent from all smaller ones. It must be stressed that nestedness here refers to the composition of the faunas, and not to the spatial configuration of the sites: it is a trivial observation that faunas are spatially nested when sites are. Spatial turnover describes the extent of species replacement along a spatial gradient (also known as β-diversity; Whittaker 1972). 'Turnover' is also used to describe temporal changes in the species composition of sites (e.g. Section 2.2.1), so we add the qualifier here to distinguish these terms. Spatial turnover among species can be viewed as the antithesis of nestedness, and departures from nestedness have been used to quantify the prevalence of spatial turnover (Williams 1996b). Superimposed on patterns of nestedness and spatial turnover is geographical variation in range sizes, and in particular variation with respect to latitude. This variation potentially interacts with the form of species–range size distributions, as well as with the degree to which species replace each other across space, and so may affect variation in the composition of local sites in a variety of ways.

3.4.1   *Nestedness*

Given the choice between visiting a species-rich or a species-poor site of a given habitat in a particular region, such as lowland deciduous woodland, most birders would, all else being equal, choose the former. Having visited the richer site, there would probably be little to be gained in terms of seeing additional species by also visiting the poorer. This is a trivial observation for a birder, but one with clear implications. It suggests that avian communities commonly are nested. Species present in any particular fauna tend also to be present in larger faunas. Nestedness is a composite property of a suite of biotas (Patterson 1990), and is a phenomenon that has recently been the subject of significant attention (e.g. Patterson & Atmar 1986; Ryti & Gilpin 1987; Patterson 1990; Simberloff & Martin 1991; Wright & Reeves 1992; Atmar & Patterson 1993; Cook 1995; Cook & Quinn 1995, 1998; Kadmon 1995; Boecklen 1997; Warburton 1997; Rohde *et al.* 1998; Wright *et al.* 1998; Brualdi & Sanderson 1999).

While it seems intuitively obvious from the pattern of species richness across local sites that many faunas are nested, problems arise when trying to show that this is the case. A number of related metrics have been proposed and used to test for nestedness in animal assemblages (see Wright *et al.* 1998). All measure nestedness on the basis of those absences or presences that would be unexpected were the incidence matrix (e.g. a matrix of the sort depicted in Fig. 3.9, of presences and absences of species in habitat patches) perfectly nested. It is how the metric is derived from these presences and absences that distinguishes them. The simplest measures simply sum the unexpected absences (Patterson & Atmar 1986) or unexpected presences (Cutler 1991). More complicated measures involve quantifying the number of 'steps' required (by filling in unexpected absences and erasing unexpected presences) to transform a given matrix into a perfectly nested one (Cutler 1991), or by counting the number of times that the presence of a species at a site successfully predicts its presence at equally rich or richer sites (Wright & Reeves 1992). However, a problem with these measures is that they are correlated with matrix size or with matrix fill (the proportion of occupied elements) (Wright *et al.* 1998). These correlations persist even after standardization to remove the association with matrix size. This makes it difficult to compare the nestedness of different faunas.

One metric for which these problems are less severe is the statistic T, which measures matrix disorder, or 'temperature' (Atmar & Patterson 1993). This is calculated as the sum of squared deviations from the matrix diagonal (i.e. a line running from bottom left to top right in Fig. 3.9) of unexpected (in a perfectly nested matrix) presences and absences, divided by the maximum value possible for the matrix, multiplied by 100. Wright *et al.* (1998) found this measure to be independent of matrix fill, and only weakly related to matrix size. It

Species

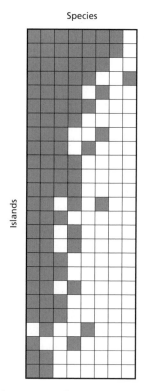

Islands

Fig. 3.9 An example of an incidence matrix showing a significantly nested distribution of species occurrence. Each column represents a species of British breeding landbird, and each row an island off the coast of mainland Britain. An element is filled when a species occupies an island. Forty-eight per cent of possible elements are filled, and the matrix nestedness temperature (Atmar & Patterson 1993, 1995) of 8.3° shows that this matrix is significantly more nested than expected from the null hypothesis of a random distribution of filled cells ($P < 0.0001$). Islands and species are ordered in the matrix to minimize its temperature (Atmar & Patterson 1993). From data in Reed (1980).

also has the advantage that it allows the identification of idiosyncratic species or habitat patches: that is, species or patches that contribute most to matrix disorder by disrupting the nested pattern. For simplicity, we restrict ourselves to this metric as a measure of nestedness.

The choice of metric is the more straightforward to overcome of two problems in measuring nestedness. Less obvious is what should be the appropriate null hypothesis against which to test the significance of the nested pattern. Clearly, significance needs to be tested against some pattern of random allocation of occupied elements among species and habitats. However, there are a number of ways in which such allocation could be performed. Wright *et al.* (1998) list four different null hypotheses that might logically be used. The first three are similar in requiring that the species richness of each habitat patch is equal to that actually observed. They differ in the way species are randomly allocated to these islands up to the value required: species are either selected with uniform probability (R0), with a probability proportional to their incidence in the original matrix (R1), or with a probability proportional to the square of their incidence divided by the sum of the squares of all incidence values (R2). R2 corrects the tendency for R1 to under-represent widespread species (Wright & Reeves 1992; Cook & Quinn 1998). The fourth null hypothesis (R00) places no constraint on the number of species in each patch or on the size of each species distribution, but instead simply allocates occurrences up to the total in the matrix. Thus, the first three null hypotheses place increasing constraint on how occurrences are distributed at random across the incidence matrix, with concomitantly greater nested structure in the random assemblages generated, while the fourth null hypothesis imposes the fewest constraints, and produces the least nested random assemblages (Wright *et al.* 1998).

Problems arise testing nestedness against these different null hypotheses because none is a priori obviously the correct one. Different authors prefer to use different null models for different reasons. Thus, Wright and Reeves (1992) used R0, arguing that while this null makes the unrealistic assumption that all species are equally likely to occur at a given site, differences in incidence nevertheless are part of the phenomenon of nestedness, and so should not be subsumed into the null model (see also Cook & Quinn 1998). By contrast, Patterson (1990) affords no significance to nestedness revealed by R0 but not revealed by the more conservative R1. However, Cook and Quinn (1998) note that the null distributions produced under R1 are biased with respect to that actually being modelled, increasing the probability of type I statistical error.

The issue of which null model to use would not be a problem if results from all models agreed, but unfortunately they do not. The greater constraints imposed by R2, for example, make it much less likely to detect significant nestedness than either R0 or R1 (Cook & Quinn 1998; Brualdi & Sanderson 1999). Wright *et al.* (1998) reviewed 279 incidence matrices for nested subset structure. They found that statistically significant nestedness was common when using R0 or R1. However, when R2 was used, most incidence matrices were identified as significantly *less* nested than expected by chance (although they did not test most nestedness metrics against R00, this should give similar results to R0). Similar conclusions were reached by Brualdi and Sanderson (1999).

Given that there is no correct null hypothesis, we take a pragmatic approach and use R00. This is the null that Wright *et al.* (1998) claim, albeit with little justification, is the most suitable to be used with T. Moreover, a program (Atmar & Patterson 1995) to calculate T and associated probability values under the null of R00 is readily available on the Internet (http://www.fmnh.org/research_collections/zoology/nested.htm). We assume that a fauna is nested if its temperature is significantly higher than expected from null hypothesis R00, as calculated using Atmar and Patterson's (1995) nested temperature calculator program.

Species

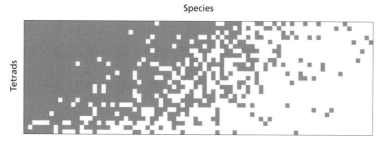

Fig. 3.10 The incidence matrix for the 93 species of bird breeding in the 25 tetrads comprising the 10-km grid square SU87 in Berkshire (matrix fill (the proportion of occupied elements) = 51.6%, $T = 24.12°$, $P < 0.0001$). Tetrads and species are ordered in the matrix to minimize its temperature. Note that all bar one of those species present in all rows of the matrix is automatically removed by the program with which this figure is plotted (Atmar & Patterson 1995). From data in Appendix IV.

Using this approach, quantitative data strongly bear out the impression that British bird assemblages are nested. For example, Reed (1980) presented information on the occurrence of eight species of landbird across 25 islands around the British coast. These are the data shown here as Fig. 3.9. The distribution of these species across these islands is highly significantly nested. Reed's data formed one of the 279 incidence matrices tested by Wright *et al.* (1998), whose analyses of these matrices confirm that significant nestedness is generally a common feature of assemblages of a diversity of taxa living at a variety of latitudes.

Nestedness is also a feature of other subsets of the British avifauna. For example, bird species distributions in the 25 tetrads comprising the 10×10-km grid square SU87 in Berkshire are also nested (Fig. 3.10), although the matrix temperature is much higher than for Reed's island birds. We also examined nestedness more widely across Britain, using a transect of adjacent 50×50-km grid squares stretching from south to north across the country. The first square on the transect was placed with its bottom left-hand corner on square SU41 on the British National Grid. By staggering subsequent squares 10 or 20 km to the west, it was possible to produce a continuous transect of seventeen 50×50-km squares, each entirely covering land, or including only a small percentage of sea. A similar transect was used by Harrison *et al.* (1992) to study spatial turnover (Section 3.4.2). We then plotted the presence or absence of each British breeding bird species in the squares on this transect, using the maps in Gibbons *et al.* (1993). Figure 3.11 shows the resulting incidence matrix. Its temperature is much higher even than that for the birds in Berkshire (Fig. 3.10), but is still significantly more nested than expected by chance.

The minimum requirement for nestedness to develop in an assemblage is that islands or habitat patches differ in their suitability, and species differ in their abilities to colonize and persist on them (Ryti & Gilpin 1987). Nestedness

Species

Fig. 3.11 The incidence matrix for the 179 bird species breeding in seventeen 50×50-km quadrats placed in a continuous transect running south–north across Britain. (Matrix fill = 52.9%, $T = 40.83°$, $P < 0.0001$.) The British breeding avifauna was defined as that listed in Appendix III; however, for this analysis we excluded seabirds, birds that predominantly breed at the coast, such as eider, most terns and rock pipit, and species for which Gibbons *et al.* (1993) do not provide accurate distribution data (e.g. red kite, osprey) or full-sized maps (e.g. serin). Quadrats and species are ordered in the matrix to minimize its temperature.

will tend to be caused by any factor that favours the assembly or disassembly of communities from a common species pool in a consistent order (Mikkelson 1993). Several candidate factors have been suggested, but four seem most likely to be important: random sampling, area, isolation and habitat (Wright *et al.* 1998). Interestingly, this list includes several of the factors suggested principally to generate patterns in local species richness (Chapter 2), and the ways in which these factors might cause both patterns are essentially the same. This similarity is useful, because some of the same data bear on both patterns.

Passive sampling of individuals from a defined species abundance distribution (e.g. a log-normal) can generate nested faunas, if habitat patches are considered to be saturated with species before all species are allocated to them (e.g. Cutler 1994). Indeed, passive sampling may create simulated faunas that are more nested than real ones, because the model makes no allowance for the interaction between species and environment (Wright *et al.* 1998). Wright *et al.* (1998) made no attempt to test whether random sampling could cause the patterns of nestedness observed in their real data, although they did observe that the R1 null model could be thought of as a passive sampling model with incidence replacing abundance (see also Andrén 1994b). In fact, as incidence and abundance are usually well correlated in animal assemblages (Section 4.2), this correspondence is potentially very close. Nevertheless, we think that passive sampling is unlikely alone to explain nestedness in real faunas. Our belief derives from the analogy between the passive sampling models for species–area relationships and nestedness patterns. Low richness habitat patches in both cases are those that sample fewer individuals, and hence by chance fewer species. Models of species–area relationships explicitly assume that the patches sampling fewer individuals are smaller, while the nestedness models make no a priori assumptions as to why different patches sample different numbers of individuals. Aside from this, there is no difference between the mechanisms. Therefore, because simulation models have demonstrated that passive sampling usually produces unrealistic species–area relationships (Section 2.2.1), there is little reason to believe that passive sampling produces observed patterns of nestedness. The same argument can also be invoked to suggest that observed species–range size distributions are unlikely generally to be a consequence of passive sampling (Section 3.3.1).

The distribution of habitat is also an unsatisfactory mechanism for nestedness. This explanation assumes that species use certain habitat types, and that species nestedness is a consequence of habitat nestedness. As such, it simply moves the explanation down a trophic level—we then need to know why habitats are nested. Nevertheless, habitat may play a role in nestedness patterns, if it interacts with area.

Area may generate patterns of nestedness through the minimum area requirements of species. Species with large minimum area requirements will only be found on large islands or habitat patches, because only these will be able to support population sizes large enough to buffer against the risks of stochastic extinction (Chapter 4). Because both large and small areas will presumably be able to support viable populations of species with small area requirements, nestedness should result. This explanation effectively assumes that nestedness results from the disassembly of a common species pool through the effect of extinction. It is the mechanism generally assumed to produce nestedness in the faunas of landbridge and mountaintop islands, as waves of extinction follow the reduction in habitat area brought about through the isolation of these elements following climate change (Brown 1971, 1978; Lomolino 1996). Alternatively, we saw in Chapter 2 that large areas may support more species partly because they support a wider range of habitats. If small areas only contain the commonest habitat types and their associated species, while large areas also encompass rarer habitats and species, then nestedness would again result.

There is evidence for an effect of area on nestedness. Wright *et al.* (1998) found that nestedness was positively correlated with the range in the sizes of islands contributing to the incidence matrices in their collection. Moreover, they also found that nestedness was higher for incidence matrices derived from landbridge islands. Thus, the best evidence to date suggests that the area effect is a consequence of extinction, presumably acting through the minimum area requirements of species.

Nevertheless, the likely effect of area through extinction does not rule out the effect of habitat diversity on nestedness. As discussed in Chapter 2, it is difficult to disentangle the effects of habitat diversity from other area effects. However, nestedness occurs in faunas occupying areas that do not differ in size, such as the breeding birds of the 25 Berkshire tetrads shown in Fig. 3.10. Some factor other than area is causing nestedness in this avifauna, and habitat variation seems the most likely. Unfortunately, the effect of habitat variation will be difficult to assess, as it will depend not only on habitat number, but also on the size of individual habitat patches. Thus, the association between area and extinction probability may exert an effect even in areas of equal total size. However, because habitat does contribute to variation in species richness with area (Section 2.2.1), it seems likely also to contribute to variation in nestedness with area.

The fourth factor proposed to explain patterns in nestedness is isolation. Isolation should cause nestedness because only species with the highest colonization ability would be able to reach the most remote islands or habitat patches. Kadmon (1995) examined floras on islands in an American reservoir, and found that faunas were significantly nested, and that this was explained by variation in island isolation, but not island area. Moreover, the isolation effect was due largely to species with poor dispersal abilities. Species that could easily disperse to all islands contributed little to the nestedness of the system. Similarly, Conroy *et al.* (1999) suggested that isolation was the most likely cause of nestedness in the mammalian fauna of the Alexander Archipelago, Alaska.

Nores (1995) examined nestedness as part of a study of the insular biogeography of the high Andean avifauna of nine mountains in the Sierra Pampeanas, north-western Argentina. These mountains are isolated from the main spine of the Andes, with mountaintop habitats of high-altitude grassland quite unlike the matrix of Chaco woodland and Monte steppe out of which the mountains rise. The avifauna of the high-altitude grasslands is highly nested, with a matrix nestedness temperature $T = 10°$ ($P < 0.0001$). Nores contends that this must have arisen through patterns of colonization, and not extinction, because most of the Sierra Pampeanas islands could never have been connected, either to the Andes or to each other. Thus, they would never have developed a common fauna for extinction to disassemble.

Lomolino (1996) tested three nearshore island and two montane forest archipelagoes for the influence of island isolation and area on nestedness. He ordered the islands in each archipelago first by size, and then by isolation, to see whether this difference created nestedness in associated mammalian faunas. Ordering islands by area always generated nestedness, whereas ordering by isolation generated nestedness in two of the five faunas. Wright *et al.* (1998) found no strong support for an effect of isolation in their extensive analyses. As for patterns in species richness, then, isolation seems to influence nestedness in some cases, but may have an effect that is subsidiary to that of area.

Thus far, we have treated area and isolation as separate explanations for nestedness. The effect of area is generally described as acting on extinction rates, and isolation on colonization rates. However, colonization and extinction rates are likely to be intimately associated, as seen when discussing causes of species richness patterns (Section 2.2.1). It is probably too simplistic to assume that significant area or isolation effects provide evidence for extinction or colonization, respectively, as the principal cause of nestedness patterns. The 'rescue effect' of periodic colonizations is likely to affect extinction-dominated systems such as mountaintop or landbridge islands (Davis & Dunford 1987; Cook & Quinn 1995), while colonization-driven systems will also be affected by extinction (e.g. Patterson 1990; Conroy *et al.* 1999). Both processes will interact to generate nestedness patterns, as long as species vary predictably in their colonization and extinction probabilities. There is evidence that they do (e.g. Cook & Quinn 1995; Hinsley *et al.* 1995; Kadmon 1995; Bellamy *et al.* 1996b).

In fact, it is possible that non-nested systems may be those where neither of these processes act, but where incidence patterns are determined in large part by speciation. Faunas dominated by speciation will have large numbers of endemics, which have been shown to reduce nestedness (Cook & Quinn 1995). Thus, observed differences in nestedness of landbridge and oceanic island faunas (Wright & Reeves 1992; Wright *et al.* 1998) may not, as has been suggested, be differences between extinction- or colonization-dominated systems, but instead differences between systems dominated by extinction/colonization or speciation (Cook & Quinn 1995).

Wright *et al.* (1998) suggested that the different factors proposed to cause nestedness can usefully be thought of as a series of probabilistic filters, screening species with particular characteristics from particular islands or patches. Area and isolation screen species by their extinction and colonization tendencies, while habitat and passive sampling screen them by their habitat preferences and abundance, respectively. It is a view with which we generally concur. It was emphasized at the start of this chapter that patterns of occupancy across the wider landscape of Britain affect the species composition of avifaunas of sites like Eastern Wood, because the site samples the fauna of the landscape in a probabilistic manner (Section 3.1). Features of the site determine the number of species sampled, while their identity is determined by features of the species themselves. Incidence matrices (e.g. Figs 3.9, 3.10; Appendix IV) show clearly that richness and range size are intimately linked. Therefore, that the same set of factors proposed to filter species to produce patterns of nestedness should also be of primary importance in generating richness patterns is hardly surprising. Because the richest sites will include most species in an assemblage, while widespread species will occupy most sites, this view suggests that nestedness may perhaps be an inevitable, 'second-order' consequence of the factors that cause variation in species richness and range size.

3.4.2 *Turnover*

While patterns in the nestedness of species have received a reasonable amount of attention, patterns of spatial turnover in species identities at a comparable scale have received very little. This is rather surprising given how central this topic is to an understanding of geographical patterns in species richness. One possible reason for this state of affairs is the multiplicity of measures of spatial turnover, and the lack of any general agreement as to which are the most appropriate and when (for discussions, see Wilson & Shmida 1984; Magurran 1988; Harrison *et al.* 1992; Blackburn & Gaston 1996d; Gaston & Williams 1996; Williams 1996b).

All measures of spatial turnover attempt to quantify the extent to which faunas occupying different sites are themselves different, a quantity that should increase as the distance separating areas increases, and the size of the areas decreases (Harte & Kinzig 1997). Turnover measures can be broken down into two broad groups: those that quantify how different sites are, and those that quantify how similar they are. The latter, of course, are inversely related to spatial turnover. The most commonly used similarity indices all adopt the same basic approach, in effect calculating the proportion of all species present in two sites that are present in both (for example, see Magurran 1988). The more sophisticated indices include information on the relative abundances of the species. Differentiation between assemblages is also known as β-diversity (Whittaker 1960, 1972). Approaches to measuring this quantity are more varied, although all depend solely on incidence data. The simplest was introduced by Whittaker (1960), and quantifies the extent to which the entire species richness of a set of samples, S, exceeds the average richness of those samples, α:

$$\beta_w = (S/\alpha) - 1 \qquad\qquad\qquad\qquad \text{(Eqn 3.8)}$$

Other indices incorporate the numbers of species that are gained and lost along environmental gradients (e.g. Cody 1975; Wilson & Shmida 1984), the numbers of species with overlapping distributions, or information on the number of samples along a transect in which each species occurs (Routledge 1977). As for nestedness, no one index has logical precedence over any other. Comparative studies of their performance (Wilson & Shmida 1984; Magurran 1988) typically reveal strengths and weaknesses in all measures. For example, while Whittaker's β_w performs well in most circumstances, it was conceived to deal with cases where α is approximately constant across samples, and so cannot distinguish cases of true spatial turnover from cases where species are simply lost along a gradient (Harrison *et al.* 1992). Replacing average α with the maximum α for the set of samples ameliorates this problem (Harrison *et al.* 1992). Nevertheless, interpretation of studies of β-diversity needs to be cautious, as results can depend on the metrics used (for example, see Wilson & Shmida 1984; Harrison *et al.* 1992; Blackburn & Gaston 1996d).

Presently, there are not enough comparable studies of the phenomenon to draw any general conclusions about large-scale patterns in spatial turnover. Those studies that do exist document an inconsistent set of patterns. For example, Harrison *et al.* (1992) found that β-diversity along latitudinal and longitudinal gradients for 15 taxa in the British flora and fauna, including birds, was generally low. Using a version of β_w modified to allow comparisons along transects composed of different numbers of samples, they showed that, for all taxa except bees, β-diversity increased as the latitudinal distance between sites compared increased. By contrast, birds were one of the few groups that showed the same result on the longitudinal transect. The latitudinal results, however, were largely a consequence of species dropping out along the transect, rather than actual species replacement. Using the modification of β_w that replaces average α with the maximum α for the set of samples (= β_2), Harrison et al. showed that only birds exhibited a significant tendency for sites further apart both latitudinally and longitudinally to hold more differentiated faunas. These analyses were based on bird distributions from the first BTO British breeding bird atlas (Sharrock 1976), but the results are very similar when data from the second atlas (those in Fig. 3.11) are used.

British taxa do not show a strong tendency for β-diversity to be higher between more distant sites largely because species ranges are not distributed evenly across the country. Instead, latitudinal richness gradients mean that most ranges are concentrated in the south (for birds, see Sharrock 1976; Fuller 1982; Lack, 1986; Turner et al. 1988; Gibbons et al. 1993; Williams 1996a; Williams et al. 1996). Assemblages then differ because of species loss, not species replacement (Harrison et al. 1992). Such a pattern is indicative of nestedness. British birds are a notable exception to this generalization: northern and southern sites genuinely differ in their avifaunal composition more than adjacent southern or northern sites, while the same is true for eastern and western sites. These results in turn reflect genuine differences in the composition of avifaunas in different parts of the country. For example, the pied flycatcher and redstart are typical of western deciduous woods, the crested tit and Scottish crossbill are found only in Scottish coniferous forests, while the hawfinch and nightingale are typical of south-eastern deciduous woods. These changes are reflected in the relatively high nestedness temperature calculated for the incidence matrix derived from these data (Fig. 3.11).

Gregory et al. (1998) used a different approach to examine spatial turnover of birds more widely across Europe. They calculated β_w for pairs of adjacent 50×50-km grid squares on several transects running north–south across the continent, and then plotted turnover against the latitude of the square. Turnover was highest at the northern and southern edges of Europe, but uniformly low otherwise. Gregory et al. ascribe the northern peak in β_w to species losses, and the southern peak to species gains. Thus, there seems to be no clear pattern of true spatial turnover for birds across Europe. The same analysis can be performed using the incidence data for the transect of 50×50-km quadrats laid south–north across Britain (Fig. 3.11). The results are shown in Fig. 3.12. β_w shows a tendency to increase with latitude along the transect, albeit that the relationship is not quite statistically significant. However, β_2 calculated by Harrison et al. shows no significant variation with position on the transect

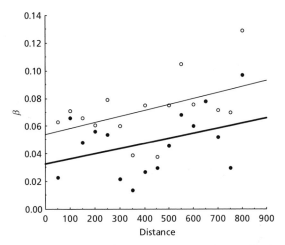

Fig. 3.12 The relationship between the β-diversity calculated for breeding bird species between pairs of consecutive 50 × 50-km quadrats on a transect south–north across Britain, and the distance (km) of the common edge of those quadrats from the southern end of the transect. The open circles and thin line are for Whittaker's β_w ($r^2 = 0.22$, $n = 16$, $P = 0.064$), and the filled circles and thick line are for the modification of β_2 by Harrison *et al.* (1992) ($r^2 = 0.15$, $n = 16$, $P = 0.14$). See text for details of the calculation of β_w and β_2.

(Fig. 3.12). Thus, the higher values of β_w between adjacent squares in northern Britain are due to species loss, rather than true turnover. The results for Britain mirror those for European birds in that there is no point on the latitudinal gradient at which spatial turnover is particularly high.

The lack of latitudinal variation in β for European birds reflects the relative homogeneity of this fauna. We would expect to see more definite patterns in β-diversity across a more heterogeneous region. In fact, we can test for just such patterns for the birds of the New World. Blackburn and Gaston (1996a,d) mapped the distributions of almost 4000 New World bird species on a large-scale equal-area grid. The grid squares were arranged in latitudinal bands, allowing spatial turnover to be compared between pairs of adjacent bands. Figure 3.13 shows how β_w and β_2 calculated for birds in pairs of adjacent latitudinal bands vary with latitude. β_w is highest between the northernmost pair of bands, but shows no clear pattern of variation otherwise. Once again, the peak in β_w represents species loss, rather than species turnover, as only two species inhabiting the most northerly band do not also occur in the band directly to its south.

In contrast to β_w, β_2 shows a clear pattern of latitudinal variation. It is generally highest at low latitudes, and indeed shows a significant curvilinear relationship to latitude (second-order polynomial regression: $r^2 = 0.46$, $n = 21$, $P = 0.004$). The contrast between the values of β_w and β_2 between the

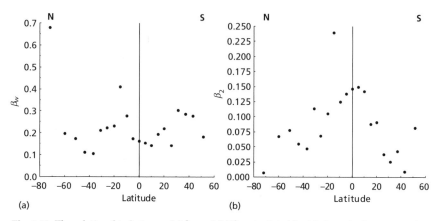

Fig. 3.13 The relationship between (a) β_w and (b) β_2 calculated for bird species between pairs of consecutive latitudinal bands across the New World, and the latitude (degrees) of the common edge of those bands. The Equator is indicated by a vertical line. From data sources in Blackburn and Gaston (1996a).

northernmost two bands clearly demonstrates the influence of species loss on the former. However, like β_w, β_2 also produces an outlying point, identifying particularly high spatial turnover at about 15–20°N. This is a reassuring result. The 15°N line of latitude coincides approximately with the southern tip of Mexico, where the borders of Belize, Guatemala, Honduras and Nicaragua all cluster. Central America marks a zone of floristic and faunistic change (e.g. Leith & Werger 1989; Groombridge 1992; Gauld & Gaston 1995; Ortega & Arita 1998). The humid, lowland rainforest typifying much of equatorial South America reaches its northern limit around southern Mexico. Likewise, northern forests reach their southern range limit around this point. Northern conifers, for example, extend no further south than Honduras and western Nicaragua. Many bird species show similar range limits. The main tract of tropical rainforest in northern Central America (the Olancho forest) extends through central Honduras to its Caribbean north coast. Howell and Webb (1995) exclude from their guide to the birds of Mexico and northern Central America 50 species which have their northern range limit in the Olancho. They also note that most temperate zone species reach their southern range limits in the north of the highlands of Honduras and western Nicaragua. The boundary is not impermeable, as many species are found both to the north and south of it (e.g. American kestrel). Nevertheless, β_2 identifies the relatively high level of spatial turnover in the avifauna of the New World at this latitude.

Most other studies of large-scale patterns in spatial turnover have addressed the question of how it may affect observed patterns in species richness (Chapter 2). Many studies of latitudinal richness gradients (Section 2.5) compare the number of species inhabiting different latitudinal bands. However, if spatial turnover was lower at higher latitudes, then a higher proportion of the total regional richness could occupy separate quadrats within those bands at high latitudes. This could conceivably cause the absence, or even a reversal, of the latitudinal richness gradient when examined by quadrat (McCoy & Connor 1980; Willig & Selcer 1989; Willig & Sandlin 1992; Blackburn & Gaston 1996b; Willig & Gannon 1997).

Willig and Sandlin (1992) demonstrated for New World bats that spatial turnover across areas at a given latitude decreased with latitude. However, they used a measure of β which may confound turnover and species loss (see above). Blackburn and Gaston (1996d) examined equivalent relationships for New World birds. For most measures of β employed, including that used by Willig and Sandlin (1992), spatial turnover in birds within latitudes was highest at low latitudes, mirroring the pattern in bats. However, no significant trend in turnover with latitude was revealed using β_2. Thus, low latitudes in the New World do not have higher avian turnover. Rather, they are the latitudes with the steepest longitudinal diversity gradients, a consequence of the extremely high avian diversity of the Andes (Haffer 1988; Fjeldså 1994; Blackburn & Gaston 1996b). No latitudinal trend in turnover was detected for marsupials in the New World (Willig & Gannon 1997). The results of all of these studies suggest that latitudinal richness gradients will not be consistently affected by longitudinal spatial turnover, as it shows no consistent latitudinal variation. Nevertheless, three studies is rather few on which to base firm conclusions.

β-Diversity may interact not only with latitudinal richness gradients, but also with local–regional richness patterns (Section 2.4). Caley and Schluter (1997) compared the observed slopes of relationships between local and regional species richness with those predicted from a random sampling model. The real slopes were lower, showing that the richness of local sites grew more slowly than expected relative to the regional species pool. Caley and Schluter hypothesized that this was a consequence of β-diversity within regions. Adding species to the regional pool does not add to the richness of all local sites, because species identities turn over between local sites. Whether or not this argument is correct, consideration of β-diversity once again highlights the relationship between species incidence and species richness (Section 3.1).

In summary, while all faunas so far examined show some degree of true spatial turnover, albeit it is often low, the only significant pattern so far demonstrated is that spatial turnover is higher for birds in the neotropics than at higher latitudes in the New World, and peaks there at around 15–20°N (Fig. 3.13b). Why such a pattern pertains should be clearer once we have considered another manifestation of spatial turnover in the occurrences of species, that of the systematic pattern of spatial variation in the average range sizes of species.

3.4.3   *Rapoport's rule*

The most widely discussed systematic pattern of spatial variation in the average range sizes of species is Rapoport's rule, which Stevens (1989) defined as the circumstance in which: 'when the latitudinal extent of the geographical range of organisms occurring at a given latitude is plotted against latitude, a simple positive correlation is found'. In other words, the rule defines the tendency for species living at higher latitudes to have larger range sizes. Although usually attributed to Rapoport (1982) and Stevens (1989), this rule has a longer history, having been noted early in the 20th century (Lutz 1921). However, allowing for a variety of potentially significant complications to the interpretation of empirical studies, evidence that it is a general pattern is, at the very least, equivocal (Pagel *et al.* 1991a; France 1992; Rohde *et al.* 1993; Letcher & Harvey 1994; Roy *et al.* 1994; Ruggiero 1994; Taylor & Gotelli 1994; Blackburn & Gaston 1996a; Gaston & Blackburn 1996c; Hughes *et al.* 1996; Rohde 1996; Rohde & Heap 1996; Stevens 1996; Cowlishaw & Hacker 1997; Lyons & Willig 1997; Mourelle & Ezcurra 1997; Price *et al.* 1997; Blackburn *et al.* 1998a; Ruggiero & Lawton 1998; Hecnar 1999; for a review, see Gaston *et al.* 1998b).

There are two basic methods for analysing latitudinal variation in range sizes (Gaston *et al.* 1998b). One can compare, for different latitudes, either the average range size of all species residing at each latitude (Stevens' method; Stevens 1989), or the range sizes of species whose distributions are centred at different latitudes (the mid-point method; Rohde *et al.* 1993). Stevens' method suffers from the problem that a single species can contribute to the mean range size at more than one latitude, so that latitudinal means are not statistically independent. For the mid-point method, each species is used only once in calculating the means, and values for neighbouring rows are therefore independent (Rohde *et al.* 1993). However, Stevens' method incorporates information on the range sizes of all species occurring at each latitude, whereas the mid-point method shows only the mean range size of those species whose ranges are centred on a latitude. The two methods therefore provide different information, and neither is a priori the correct one to use. The significance of latitudinal patterns detected using Stevens' method needs to be interpreted with caution, though, because of the problem of non-independence.

Whichever method is used, there is a documented tendency for the geographical ranges of bird species to decrease from the north towards the Equator (e.g. Stevens 1989; Brown 1995; Blackburn & Gaston 1996a; Price *et al.* 1997). The geographical ranges of British birds, for example, tend to be larger the further north the species resides in Europe as a whole (Gregory & Blackburn 1998). This result is dependent on removal of 'boundedness' effects. Species whose ranges abut the northern or southern edges of land masses may have the capacity to range much more widely, but have their ranges bounded by lands end. This may lead to artificially small ranges in these latitudinally

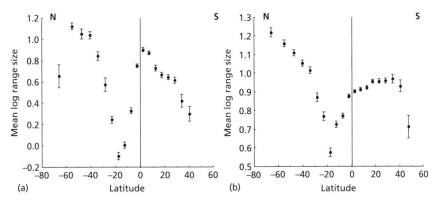

Fig. 3.14 The relationships between the mean (± SE) $\log_{10}$ breeding range size of bird species at different latitudes across the New World and latitude (degrees) using (a) the mid-point method and (b) Stevens' method (see text for details). The Equator is indicated by a vertical line. From Blackburn and Gaston (1996a).

restricted species, and hence to bias in latitudinal patterns in the range sizes of species (Colwell & Hurtt 1994). To examine these latitudinal patterns properly, it is standard practice that latitudinally restricted species should be excluded. Only once they are excluded does the relationship between range size and latitude become significant for British birds (Gregory & Blackburn 1998).

The existence of Rapoport's rule more widely in the Palaearctic seems to depend on the application of further restrictions to the set of species analysed. Stevens (1989) found that the range sizes of Russian birds were fairly uniform across most latitudes except the northernmost, where they were considerably smaller. However, when migratory species were excluded, the pattern of increasing range size with latitude re-emerged (Stevens 1989). Nearctic birds also show Rapoport's rule (Brown 1995; Blackburn & Gaston 1996a).

Outside the Holarctic, examples of Rapoport's rule in birds are scarce. In the New World, range sizes on average decrease with latitude only as far as about the southern border of Mexico (Blackburn & Gaston 1996a). Here, they start to increase again, and south of Mexico reach a maximum around the Equator (Fig. 3.14). South of the Equator, they are roughly constant or decrease slightly with latitude, depending on whether the mid-point or Stevens' method is used to examine their variation (Blackburn & Gaston 1996a). Ruggiero and Lawton (1998) found the same pattern for bird distributions across South America alone using the mid-point method. However, while bird geographical ranges decrease in size in absolute terms towards the southern tip of South America, they do not if measured as a proportion of the land area potentially available to species living at different latitudes (Fig. 3.15; Blackburn & Gaston 1996a; Ruggiero & Lawton 1998).

Global patterns of geographical range size variation have been examined for wildfowl (Gaston & Blackburn 1996c) and woodpeckers (Blackburn *et al.*

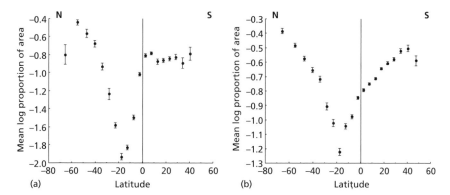

Fig. 3.15 The relationships between the mean (± SE) $\log_{10}$ proportion of the total land area available to a bird species at different latitudes across the New World that the species actually occupies and latitude (degrees) using (a) the mid-point method and (b) Stevens' method (see text for details). The Equator is indicated by a vertical line. From Blackburn and Gaston (1996a).

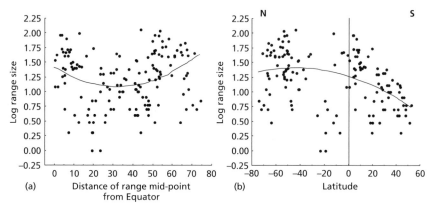

Fig. 3.16 The relationships between the global breeding geographical range size of wildfowl species and (a) the distance of the mid-point of the latitudinal spread of a species breeding range from the Equator (in degrees of latitude), irrespective of whether the mid-point is in the northern or southern hemisphere, and (b) the latitudinal mid-point of the breeding range. Solid lines are second-order polynomial regressions through the data. The Equator is indicated by a vertical line. From Gaston and Blackburn (1996c).

1998a). In wildfowl, range size is a U-shaped function of the distance a range lies from the Equator, peaking at low and high latitudes (Fig. 3.16a). However, when the hemisphere in which ranges lie is taken into account, range size shows a weak tendency to decrease from north to south across the globe (Fig. 3.16b): in other words, range sizes are larger further from the Equator for northern hemisphere species, but are larger closer to the Equator for southern hemisphere species. Unfortunately, interpretation of this global pattern is hampered by the lack of wildfowl whose ranges are centred at low latitudes in

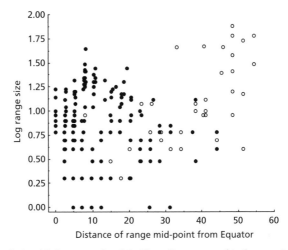

Fig. 3.17 The relationship between the global breeding geographical range size of woodpecker species and the distance (degrees) of the mid-point of the latitudinal spread of a species breeding range from the Equator, irrespective of whether the mid-point is in the northern or southern hemisphere. Species endemic to the Holarctic are distinguished by open circles. Species found in both the Holarctic and other geographical regions are excluded. From data sources in Blackburn *et al.* (1998b).

the northern hemisphere. Figure 3.16b suggests that range size might decrease from north to south in both hemispheres separately. Similar patterns are found for woodpeckers. Range size decreases towards the Equator, as Rapoport's rule predicts, but the pattern is largely due to variation in the range sizes of Nearctic and Palaearctic species (Fig. 3.17). Rapoport's rule is not shown by species endemic to other continents, and indeed woodpecker ranges decrease in size from north to south across the entire globe.

Latitudinal patterns of range size variation in other taxa largely mirror those in birds: Rapoport's rule seems to apply to taxa distributed across the Holarctic, but evidence for outside this region is scarce. The general conclusion is that the rule is a local effect, limited to the northern half of the northern hemisphere (Rohde *et al.* 1993; Roy *et al.* 1994; Rohde 1996; Gaston *et al.* 1998b).

Attempts to explain Rapoport's rule must account not only for the pattern, but also for its localized distribution. This has hampered acceptance of the most intuitively appealing hypothesis, proposed by Stevens (1989), which is that latitudinal patterns in range size arise as a consequence of climatic variability. Stevens argued that seasonal environmental variation, most especially in temperature, sets the minimum breadth of tolerances required by individual organisms that are found to reside at a given site. The greater climatic variation at higher latitudes requires the species living there to have broader tolerances, which in turn enables these species to exist in a wider range of climatic regimes than low-latitude species. Thus, they can attain larger range

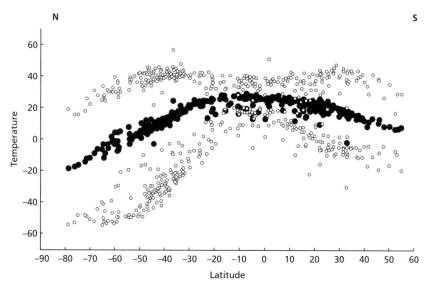

Fig. 3.18 Latitudinal variation in absolute maximum and minimum temperatures (open circles) and annual mean daily temperatures (filled circles) for mainland locations in the New World. Latitude is in degrees, temperature in degrees Celsius. Locations vary in altitude, the numbers of years of record and the period which those years embrace. From Gaston and Chown (1999b).

sizes. This mechanism can explain why species living at high latitudes have large geographical ranges, but cannot explain why tropical species often have similarly large ranges (e.g. Rohde *et al.* 1993; Roy *et al.* 1994; Ruggiero 1994; Blackburn & Gaston 1996a; Gaston & Blackburn 1996c; Lyons & Willig 1997).

A climatic tolerance hypothesis can, however, explain the observed latitudinal patterns in geographical range sizes if Stevens' original formulation is modified somewhat (Gaston & Chown 1999b). Stevens (1989) only considered the effects of variance in climatic conditions on range sizes. However, the mean is also important (Gaston & Chown 1999b). Mean environmental temperatures increase from the Poles until about 25° north or south, at which point they level out (Fig. 3.18; Terborgh 1973). Thus, environmental conditions are reasonably constant across the tropics. Treeline and snowline, for instance, fall at an altitude that is approximately stable throughout this region (Körner 1998). Any bird species that can endure tropical conditions can potentially therefore occupy all latitudes from 25°N to 25°S. The range size of such species is then only limited by the amount of land between these latitudes. In fact, Rosenzweig (1992, 1995) has shown that the land area of the tropics is vastly greater than that of any other biome (e.g. Fig. 3.19). Thus, tropical species can have large ranges because the area of the tropics is great (see also Section 2.5.2), whereas high-latitude species can have broad ranges because of their ability to withstand a broad range of conditions (Gaston & Chown 1999b); evidence for

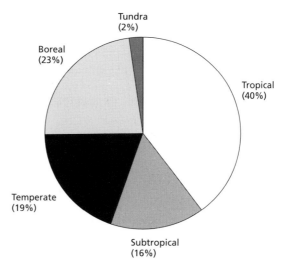

Fig. 3.19 The land areas of different biomes in the New World. The definition of biomes follows Rosenzweig (1992), with slight modification (see Blackburn & Gaston 1997a).

latitudinal variation in physiological tolerance is reviewed by Spicer and Gaston (1999).

If range sizes were constrained by the climatic tolerances of species in the way described, we would expect to see a marked spatial turnover of species coincident with regions where environmental conditions also change markedly. Because environmental conditions are reasonably constant across the tropics, but start to change at the boundaries, we might expect to see elevated levels of spatial turnover at the limits of the tropics, as species adapted to tropical conditions will be unable to extend their ranges outside this region. As observed above, this is certainly true in the northern hemisphere for New World birds, which show a peak in spatial turnover at about 15–20°N. This turnover is also reflected in avian range sizes reaching a minimum at about these latitudes (Fig. 3.14). Hard boundaries, such as that potentially imposed on tropical species by the edge of the tropical environment, can cause mean range size to decline towards them for reasons analogous to those discussed above for the effects of latitudinal extremes of land masses. A full discussion of how this arises is given by Colwell and Hurtt (1994; see also Sections 2.5.1 and 2.7). Nevertheless, that the boundaries of biogeographical regions may impose limits on the range sizes of species has been suggested by previous authors (Roy *et al.* 1994; Blackburn & Gaston 1996a). If those boundaries are also characterized by steep environmental gradients, the modified climatic tolerance hypothesis provides an explanation of why.

More puzzling is why no evidence of spatial turnover in New World birds is observed around the southern border of the tropics. How general this absence may be is difficult to assess, as this is the only terrestrial assemblage for which geographical range size variation has been examined across the full range of available latitudes. South American mammals do show such turnover, but the study of ranges that showed this effect (Ruggiero *et al.* 1998) did not extend far enough to test their pattern of turnover around the northern border of the tropics. There are, nevertheless, a number of possible reasons why the southern border of the neotropics may form less of a barrier to species distributions than the north. In particular, climatic variability seems to increase less steeply with latitude south of the Equator than to the north (Fig. 3.18; Gaston & Chown 1999b). The southern edge of the neotropics may therefore present a softer boundary to species distributions, allowing more species to cross it, and so producing a lower rate of spatial turnover. The more temperate climate in southern South America may explain why avian range sizes do not vary so dramatically with latitude there (e.g. Figs 3.14 & 3.15). Alternatively, the lack of obvious spatial turnover in the southern neotropics may be because this border is longer than the equivalent in the north, and spans a wider range of latitudes. The spine of the Andean mountains exerts a major influence on South American species, allowing many birds with southern distributions to extend north. The southern extension of tropical forest on the continent is less in the west (close to the Andes) than in the east. Spatial turnover at the southern edge of the neotropics may still occur in birds, as it does in mammals (Ruggiero *et al.* 1998), but over a wide enough range of latitudes that an abrupt boundary is not detected by an analysis based on broad latitudinal bands. Tellingly, the study of South American mammals by Ruggiero *et al.* (1998) was based on a grid of squares, and at a much finer spatial scale.

At least four explanations other than that of climatic variability have been proposed to explain Rapoport's rule (Gaston *et al.* 1998b), based on patterns in land area, differential extinction, competition and biogeographical boundaries (Table 3.3). This last explanation (Roy *et al.* 1994) proposes that range sizes may be determined by the extent of biogeographical provinces if species can expand their distributions more easily within than across province boundaries. This would explain the coincidence between the inflection point in range sizes at about 15–20°N for New World birds (Fig. 3.14) and the high level of faunal turnover at this latitude (Fig. 3.13b). However, it does raise the question of what determines biogeographical boundaries, and the spectre of circularity if biogeographical regions are defined on the basis of faunal turnover. Moreover, if it could be shown that biogeographical boundaries are areas where changes in environmental conditions limit species distributions, this explanation could largely be subsumed into the modified climatic variability hypothesis, as described above.

Of the other hypotheses for Rapoport's rule, that based on differential extinction (Brown 1995) has the potential to explain why variation in range sizes should differ north and south of the Equator. The hypothesis proposes that species at higher latitudes may have larger range sizes because those

Table 3.3 Mechanisms postulated as determining or contributing to latitudinal gradients in range size (Rapoport's rule).

Mechanism	Process
Climatic variability	Greater climatic variation at higher latitudes selects for broader environmental tolerances of individuals, which enables species to become more widespread
Land area	Greater land area at higher latitudes results in larger ranges
Differential extinction	Species at higher latitudes with restricted ranges underwent extinction due to glaciation and climate change
Competition	Species at higher latitudes have larger ranges because of lower levels of competition resulting from lower species richness
Biogeographical boundaries	Easier expansion of ranges within than between biogeographical provinces

species in these areas which had narrow tolerances, and hence restricted occurrence, underwent differential extinction due to glaciation and climate change (Brown 1995). A link between glaciation and Rapoport's rule has also been made by Price *et al.* (1997), who examined the geographical range sizes of Palaearctic taxa in the genus *Phylloscopus* (leaf warblers). They suggested that the pattern arises from the different colonization abilities of species invading areas following the retreat of glaciers. These ideas essentially propose the action of similar effects to those invoked by the ecological time hypothesis for latitudinal richness gradients discussed in the previous chapter (Section 2.5.4). Glaciation hypotheses can explain the hemispheric asymmetry in range size variation, because the effects of glaciation were far more severe in the northern than in the southern hemisphere (Markgraf *et al.* 1995). However, it is difficult to see how this explanation can account for the large average ranges of tropical species in many taxa (e.g. Rohde *et al.* 1993; Roy *et al.* 1994; Ruggiero 1994; Blackburn & Gaston 1996a; Gaston & Blackburn 1996c; Lyons & Willig 1997). The competition hypothesis, which suggests that species at higher latitudes may have larger range sizes because of lower levels of competition resulting from the lower species richness (Pianka 1989; Stevens 1996; Stevens & Enquist 1997), falls at the same hurdle. Finally, hypotheses based on variation in land area can be ruled out, as analyses controlling for variation in land area still reveal latitudinal variation in range sizes (e.g. Fig. 3.15; Rapoport 1982; Pagel *et al.* 1991a; Blackburn & Gaston 1996a). In sum, only the modified climatic variability hypothesis presently stands as an adequate explanation for latitudinal patterns in range size variation.

3.4.4   *Implications of patterns in range overlap for Eastern Wood*

Our consideration of patterns in the overlap of species geographical ranges has taken us a long way from Eastern Wood. This has been a journey borne of necessity, because the relatively small number of studies that have considered such issues at the large spatial scales of concern have largely dealt with faunas in other biogeographical regions. Nevertheless, the general conclusions that can be drawn from those studies do have relevance to the understanding of the composition of the fauna of Eastern Wood.

The large-scale distribution of species across the landscape affects how species are sampled by local sites. Species ranges affect the composition of local avifaunas, such as the assemblage at Eastern Wood. Pattern in range sizes is one important factor determining the outcome of this sampling process. Pattern in range overlap is another.

The latitudinal position of Eastern Wood in the northern hemisphere suggests that most of the species likely to occur in the wood should have relatively large geographical ranges. The introductory chapter showed that the species occurring in the Eastern Wood bird assemblage were those that are more widespread across Britain (Fig. 1.6). The ranges of British birds tend to be larger the more widespread the species is across Europe, and the further north it resides (Gregory & Blackburn 1998). Were Eastern Wood positioned a lot further south, its bird assemblage would be likely to include more species with restricted geographical ranges. Coupled with the latitudinal decline in species richness, Rapoport's rule implies that the avifauna of Eastern Wood should consist of relatively few, widely distributed species—those able to cope with the vagaries of the British climate. Thus, it should be rather similar in composition to most other British avifaunas occupying a similar habitat. Experience tells us that this does indeed seem to be the case.

Further evidence for the general homogeneity of the British avifauna as a whole is provided by studies of spatial turnover. Although there is some species loss as one moves north across Britain (as indicated by β_w), there is no significant pattern of true turnover, and spatial turnover is always very low (Fig. 3.12; Harrison *et al.* 1992). Thus, as is inevitable, some components of local bird assemblages do change across Britain, but mostly the same species are seen in Scottish deciduous woods as in Eastern Wood [compare, for example, Williamson (1974) and Beven (1976)], albeit that the latter would typically be more species rich. The pattern is mirrored more widely by the European bird assemblage, for which turnover is low at all except the highest and lowest latitudes (Gregory *et al.* 1998), where high values of β_w are caused by species loss and gain, respectively. Data from the New World suggest that, more widely, true spatial turnover across latitudes is only high at tropical latitudes (Fig. 3.13). This reflects a genuine tendency towards faunal change at such latitudes, where in addition many species with small geographical ranges reside.

Of course, the generality of this pattern is impossible to conclude, based as it is on only a single taxonomic group in one geographical region.

The low spatial turnover of bird species across Britain in turn contributes to this fauna showing a nested structure. The species inhabiting woodland sites more depauperate than Eastern Wood will tend to be a subset of the fauna of this site. Nestedness within a habitat type is likely to arise because some species in the landscape are more likely to colonize sites, or less likely to go extinct from them, or both. These are the widespread species, occupying most sites. Species with poorer colonization abilities or higher extinction likelihoods can only survive at the more hospitable localities. More widely, nestedness may also arise because richer areas are those with greater habitat diversity. Nestedness is most likely to be a feature of homogeneous regions. It is probably for this reason that species in tetrads in a 10×10-km square in Berkshire (Fig. 3.10) are more nested than those in the transect of 50×50-km squares running across Britain (Fig. 3.11). If the regional fauna shows a high rate of spatial turnover, the faunas of different sites are unlikely to be nested within each other. However, nestedness may still be a feature of such landscapes, albeit a multimodal pattern.

In sum, patterns in the overlap of species range sizes, and the processes generating them, help in the understanding of why the avifauna of Eastern Wood consists mainly of species that are widespread, and encountered at site after site across Britain.

3.5 Summary

Chapter 2 showed that the number of bird species recorded at a local site depends on how the site samples the regional avifauna, which depends on features of the site (e.g. area, isolation, habitat) and features of the region (e.g. area, isolation, latitude). However, because not all species that can potentially occur at a site actually do so, which species are sampled depends on features of those species. One of the principal determinants of the likelihood that a species will occur at any given suitable site is the extent of its spatial distribution, which we here term its range size. Widespread species are more likely to encounter and occupy suitable sites than are species with small ranges. Thus, the composition of local assemblages, such as the avifauna of Eastern Wood, will be influenced by patterns in the range sizes of species in the regional pool.

Most species in any given region tend to have small range sizes relative to the maximum attained by the most widespread species. However, logarithmic transformation reveals a distribution of range sizes that tends to be somewhat left-skewed, so that very small range sizes are also relatively uncommon.

The absolute range sizes attained by species depend on the latitude they inhabit, with the largest ranges on average in the northern temperate and tropical regions. Range sizes are large in the former region, probably because species need to be able to tolerate a wide range of environmental conditions to survive at these latitudes, and so can tolerate the conditions pertaining across a wide range of areas. Range sizes are large in the tropics, probably because of the comparative uniformity of environmental conditions across this large biome: species that can tolerate conditions somewhere in the tropics can tolerate them anywhere there. The tendency for significant faunal turnover at some tropical boundaries, as indicated by high values of β_2, may be because of a concomitant inability of tropical species to tolerate conditions outside the tropics, while biotic factors such as competition may prevent extratropical species from crossing the same boundary in the opposite direction. Ultimately, ranges are likely to be limited by a combination of biotic and abiotic factors, with the importance of the latter perhaps increasing with latitude.

At the largest spatial scales, the distribution of range sizes must ultimately be determined by patterns of speciation and extinction, and the dynamics of range expansions and contractions between these two endpoints in the life of a species. However, uncertainties surrounding these long-term dynamics mean that presently it is difficult to add much to this statement. Large-scale range size variation has been suggested to reflect variation in a species realized niche. Species with large ranges may be either those able to utilize a wide range of resources or conditions (those with large niche breadth), or those utilizing resources or conditions that are themselves more widespread (those with typical niche position). What evidence there is points to the latter as the more likely explanation, a conclusion that also concurs with the most likely explanation for the presence of large ranges in the tropics. Moreover, while large niche breadth appears to be the explanation for large ranges in the north temperate zone, variation in range sizes within this region may be a consequence of variation in niche position.

At smaller spatial scales, untransformed species–range size distributions tend to become bimodal, with species occupying either most possible sites, or very few. Development of the additional mode may be in part a consequence of the increasing uniformity of sites at smaller spatial scales, in turn a consequence of spatial autocorrelation in the environment. Range sizes nonetheless tend to be correlated across scales, so that regionally widespread species are also those found at most local sites.

The distribution of range sizes at smaller scales may be influenced by the occurrence of vagrants, which will increase the size of the left-hand mode, and by the position of sites relative to the geographical ranges of the species present. Species present at few sites may be on the edge of their geographical range in the area of interest, while the sites may fall close to the centres of the ranges of species present at most. However, this hypothesis begs the questions why some species penetrate further into a region than others, and why occupancy becomes patchier towards range limits. These questions must be answered with recourse to other explanations. One of these alternatives is that range sizes at smaller scales may be mediated by the colonization and extinction of species across sites in accordance with metapopulation dynamic theory. Depending on the specific extinction and colonization parameters, these models can predict species–range size distributions of a range of forms, including bimodality of the sort most often observed in real assemblages. Patterns of colonization and extinction also seem the most likely to explain the observation that species ranges frequently show a nested structure. Repeated colonization and extinction is certainly a feature of real sites, such as Eastern Wood, although for such mobile species as birds it is unclear whether the populations of even the most isolated sites in a landscape have independent dynamics, as is the case for true metapopulations.

4 Abundance

It has often been a matter of comment by ecologists that one or two species are extraordinarily abundant at a particular time and place: all others seem rare in comparison. [Preston 1948]

4.1 Introduction

At the start of the previous chapter, it was noted that the species richness of a set of sites and the range sizes of the species occupying them are equivalent to the row and column totals of an incidence matrix plotting the presence and absence of species across sites. Patterns in species richness and in range size are therefore necessarily associated. This association is of particular importance for the purposes of this book because the evident relevance of macroecological factors in determining the species richness of local sites like Eastern Wood makes it easier to accept that such factors similarly influence which species occupy that site, or put another way, include that site within their range. From this starting point, we went on to examine the interaction between large- and small-scale patterns in species ranges, to show how the presence or absence of species in the local avifauna of Eastern Wood was affected by range size patterns in the avifauna of the region in which the wood is embedded.

The species richness of sites and the range sizes of the species occupying them share a common currency in that both are quantified by the presence or absence of species at sites. Yet, studying the distribution of species purely in terms of whether or not they are present ignores a fundamental difference among those occurring at any given site that will be immediately obvious to any birdwatcher or ecologist visiting it. Simply, some species are present in greater numbers than others. Whatever the mechanisms generating the spatial distribution of birds, they result in an uneven distribution of individuals across sites. It is variation in the number of individuals of different species present at different sites—species abundance—that we explore in this chapter.

Variation in the abundance of species at a site is well illustrated by the avifauna of Eastern Wood. Figure 4.1 reproduces the frequency distribution of breeding species numbers plotted in the first chapter. It shows that most species that bred in the period 1949–79 were represented, on average, by rela-

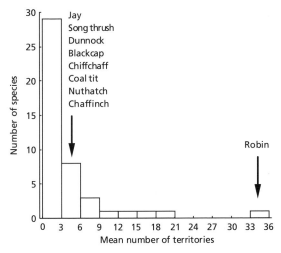

Fig. 4.1 The frequency distribution of the mean annual number of territories held by breeding bird species when present in Eastern Wood in the period 1949–79.

tively few pairs. Thus, while an average year found 33 pairs of robins and 16 pairs of blue tits nesting in the 16 hectares of the wood, there were never more than single pairs of sparrowhawk, woodcock, stock dove, tawny owl, lesser whitethroat, spotted flycatcher, greenfinch and hawfinch. Commonness is the exception, whereas rarity is the rule.

Over 5000 bird territories were recorded over the years 1949–79. Of these, more than 50% belonged just to the commonest five species, and over 80% to the commonest 15 species, despite the fact that 45 species bred in the wood in this period (Fig. 4.2). Thus, while most species breeding in Eastern Wood are rare within the wood, most individuals belong to the common species. It is individuals we encounter when we visit the wood. Whether we perceive commonness or rarity to be the norm depends on whether or not we weight species according to their abundances.

The birds of Eastern Wood can be viewed as a sample of the wider avifauna of the region in which the wood is embedded. As seen in the previous two chapters, factors that affect how the wood samples that wider avifauna, and patterns in the range size of the wider avifauna that cause some species to be sampled more readily than others, help determine the size and composition of the bird assemblage of Eastern Wood. Range size and abundance are both features of the distribution of individuals across the landscape. Hence, just as any feature of a species that is present in a region that affects the likelihood that it will be sampled by a local site may contribute to the likelihood that the species will occur at the site, so any feature that affects the likelihood that a local site will sample it may contribute to the abundance of a species at the site. Clearly, more abundant species regionally are more likely to be sampled

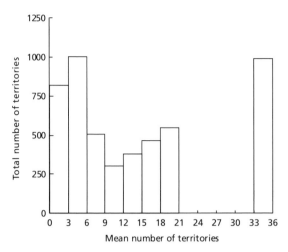

Fig. 4.2 The total number of territories held in Eastern Wood over the period 1949–79 by bird species of different mean annual abundance (mean number of territories per year when present). Thus, a total of 822 territories were held by those species which had three or fewer territories, on average, in any one year that they were present in the wood. The number of species contributing to each column is shown in Fig. 4.1.

locally. Thus, although most species present at Eastern Wood occur at low numbers, the assemblage should consist predominantly of those species that are most likely to be sampled from the region, which in turn are those that are regionally abundant.

This is indeed the case. At the scale of Britain, the population sizes of the 45 species recorded breeding at Eastern Wood in the period 1949–79 averaged significantly higher than species not recorded breeding there (two-sample t-test: $t = 7.92$, $n = 217$, $P < 0.0001$). The same is true if we restrict the species pool to those typical of deciduous woodland ($t = 6.8$, $n = 80$, $P < 0.0001$).

However, if the probability that Eastern Wood samples species from the wider environment depends on their abundances, abundances need to be accounted for when testing whether or not the Eastern Wood species are a random subset of the British avifauna. We selected 45 British bird species at random from this pool, with the probability that a species was selected weighted by its breeding population size (Appendix III). The range of geometric mean population sizes from 1000 such simulations was 728 853–1 372 208 individuals, compared to 472 402 for the real assemblage. Restricting the species pool just to those British bird species breeding in woodland, the range of simulated geometric mean population sizes was 583 875–946 694. Similar results were obtained for arithmetic means. Thus, although the Eastern Wood avifauna mainly comprises regionally abundant species, it includes more rare species than expected from a random selection on the basis of population sizes.

The tendency for the Eastern Wood avifauna to comprise those species that are regionally abundant matches the tendency for species in this fauna also to be regionally widespread (Section 3.2.2). This implies two corollaries. First, it suggests that large-scale patterns in species abundance affect the composition of local avifaunas. Although the breeding avifauna of Eastern Wood includes more rare species than expected on the basis of a random sample of the British bird assemblage weighted by population size, the fact that the species in this avifauna nonetheless tend to be those abundant in Britain suggests that many species that could potentially breed at a site actually do not because of their pattern of abundance. For example, Hinsley *et al.* (1996) showed that when the regional population sizes of most woodland bird species are low, whether or not they are found breeding in any given wood depends largely on its size. Woods that could be occupied by these species, and indeed are occupied when their regional population sizes are high, go unused. We argued in the previous chapter for a similar effect of range size.

Second, the tendency for species in the Eastern Wood avifauna to be both regionally abundant and widespread suggests that these specific features ought to be linked. This is not really surprising. One would expect those species that are represented in a region by more individuals either to occupy more local sites, to be more abundant at the sites that they occupy, or both. Therefore, some sort of positive relationship between abundance and range size should pertain. In fact, the association between these two variables is one of the most general patterns in macroecology, perhaps rivalling the species–area relationship in this regard (Gaston 1996a). It is with consideration of this association that we begin our examination of large-scale patterns in species abundance.

4.2   Abundance–range size relationships

Positive interspecific relationships between local abundance (averaged across occupied sites) and range size (measured as area of occupancy, or more exceptionally as extent of occurrence) are an almost universal feature of animal assemblages (Järvinen & Sammalisto 1976; Hanski 1982a; Brown 1984; Hanski *et al.* 1993; Gaston 1996a; Gaston *et al.* 1997a). They have been documented for groups as diverse as plants (Gotelli & Simberloff 1987; Collins & Glenn 1990, 1997; Boeken & Shachak 1998), helminth parasites (Poulin 1999), spiders (Pettersson 1997), insects (Gaston & Lawton 1988a; Owen & Gilbert 1989; Kemp 1992; Obeso 1992; Inkinen 1994; Kozár 1995; Collins & Glenn 1997), frogs (Murray *et al.* 1998b), birds (Brown 1984; Bock 1987; Gaston & Lawton 1990; Maurer 1990; Solonen 1994; Gaston & Blackburn 1996c; Collins & Glenn 1997; Poulsen & Krabbe 1997; Tellería & Santos 1999) and mammals (Brown 1984; Blackburn *et al.* 1997b; Collins & Glenn 1997; Johnson 1998b).

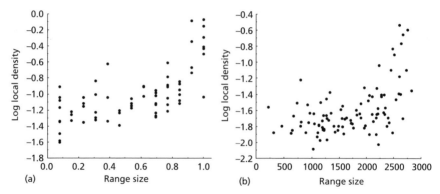

Fig. 4.3 The relationships between (a) local density (territories per hectare when present) and range size (proportion of sites contributing to the mean local density, out of a maximum of 13) for bird species breeding on woodland Common Birds Census (CBC) plots in the period 1968–72, and (b) local density (territories per hectare when present) and range size (number of 10 × 10-km squares from which a species was recorded in the first BTO atlas (Sharrock 1976), out of a maximum of 2830) for bird species breeding on farmland CBC plots in Britain in the period 1988–91. From Gaston *et al.* (1998c).

Figure 4.3 illustrates interspecific abundance–range size relationships for birds in Britain. Such relationships have been demonstrated for this assemblage in a number of separate analyses (Fuller 1982; Hengeveld & Haeck 1982; O'Connor & Shrubb 1986; O'Connor 1987; Gaston & Lawton 1990; Sutherland & Baillie 1993; Gregory 1995; Blackburn *et al.* 1997b; Gaston *et al.* 1997b,c, 1998c). Indeed, they have been more comprehensively explored for the British avifauna than for any other group of species.

Exceptions to the positive interspecific abundance–range size relationship do exist, but are scarce. Gaston (1996a) found that about 5% of the relationships in the literature were significantly negative. Often, the exceptions concern unusual sets of circumstances (Gaston & Lawton 1990; Gaston 1996a; Johnson 1998a). For example, negative relationships can arise when abundance is measured in an area of the environment which is highly atypical of the region over which distribution is assessed. Thus, Gaston and Lawton (1990) found a negative relationship for British birds when abundance was assessed as the number of breeding pairs on the small Scottish island of Handa, on the northwest edge of Britain (and indeed Europe), and range size was measured across the whole of Britain.

Although positive abundance–range size relationships are very widespread, abundance typically explains only a moderate proportion of the variation in range size (Gaston 1996a). For British birds this proportion is typically in the range of 20–30% for estimates of density, rising to about 80% if abundance is estimated as total population size (e.g. Fig. 4.4). The relationship tends to become poorer with cruder measures of range size. For British birds, the finest

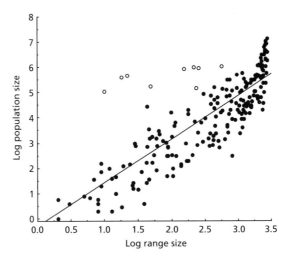

Fig. 4.4 The relationship between population size (number of individuals) and range size (number of 10×10-km squares on the British National Grid occupied) for breeding bird species in Britain ($y = 1.73x - 0.30$; $r^2 = 0.69$, $n = 217$, $P < 0.0001$). Open circles are seabirds; with these excluded, the equation for the relationship changes to $y = 1.87x - 0.78$ ($r^2 = 0.80$, $n = 208$, $P < 0.0001$). The slopes of both relationships are steeper than 1, as expected from the tendency for density to increase with range size across species (e.g. Fig. 4.3). From data in Appendix III.

scale at which distributions across the whole country are available is that of the 10×10-km square on the British National Grid. These fine scale data can be used to derive other, cruder measures of species occurrence, based on latitudinal or longitudinal extents, or minimum convex polygons (Quinn *et al.* 1996). The strengths of the resultant abundance–range size relationships, plotted using different range size measures, are essentially inversely related to the coarseness of this measure (Gaston *et al.*, in press).

If a causal link is present, the predominant direction of causality seems likely to run from local abundance to range size, rather than vice versa. As such, the abundance–range size relationship seems likely to be of limited value in understanding the determinants of species–abundance distributions (see below), but may be of relevance to the determinants of species–range size distributions (Section 3.3).

4.2.1   *The structure of abundance–range size relationships*

A number of features of the anatomy of interspecific abundance–range size relationships have become apparent, particularly for relationships in British birds. Most of these details derive from analyses of data from the BTO's Common Birds Census, or CBC (see Marchant *et al.* 1990). This is the main scheme by which the breeding populations of common British birds have been

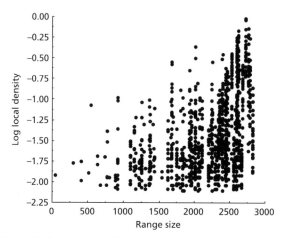

Fig. 4.5 The relationship between mean local density (territories per hectare) and range size (number of 10 × 10-km squares occupied in the second BTO atlas; Gibbons *et al.* 1993) for bird species breeding on 25 farmland CBC plots in Britain in the period 1968–72. Each data point represents one species on one site, with density averaged across years when present. From Gaston *et al.* (1998c).

monitored, where 'common' refers to species with breeding populations of more than about 100 000 pairs. It has been running since 1961, and details of methodology are given by Marchant *et al.* (1990; some of which were piloted by Beven's work at Eastern Wood). The important point for our purposes is that the CBC provides high-quality estimates of the abundances of birds in local communities on a range of woodland and farmland sites across Britain. It is an invaluable source of information for large-scale studies on the abundances of British birds, and has enabled several anatomical details of abundance–range size relationships to be elucidated.

1 For British birds on farmland and woodland CBC sites, the abundance–range size relationship appears to be driven by increases in the maximum abundance which a species can attain at occupied sites. Minimum abundances are, not surprisingly, not related to range sizes (Gaston *et al.* 1998c). Thus, all species are rare at some sites at which they occur, but only widespread species attain high densities at others. This pattern results in a triangular relationship between abundance and range size when the abundance at every site occupied by each species is plotted on the same axes (Gaston *et al.* 1998c; Fig. 4.5). Thus, mean density varies with range size because maximum density does. This pattern of variation in mean, minimum and maximum densities has at least three additional consequences.

First, the relationship between local abundance and range size is often triangular when abundances are taken from a single site. All of the narrowly distributed species present at the site are likely to be rare there, whereas some of the widely distributed species will be abundant at the site while others are not.

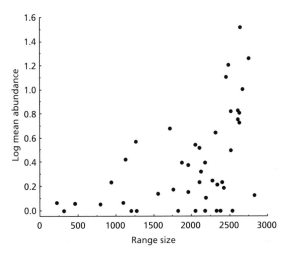

Fig. 4.6 The relationship for bird species between local density in Eastern Wood in the period 1949–79 (mean number of territories, averaged across years when present) and range size (number of 10 × 10-km squares occupied). From data in Appendices II and III.

This is true for Eastern Wood (Fig. 4.6), as well as many other local sites (Gaston *et al.* 1998c). Thus, while widespread and abundant at many other sites, both the carrion crow and house sparrow breed in Eastern Wood in low numbers. By contrast, the narrowly distributed lesser spotted woodpecker and hawfinch breed at low densities at Eastern Wood and everywhere else they occur in Britain too.

Second, species of British birds with low maximum densities occur at a greater proportion of sites with low absolute densities than do species with high maximum (or mean) densities (Gaston *et al.* 1998c). That is, in absolute terms, species with low local abundances are rare at a greater proportion of sites at which they occur than are species with high local abundances (see also Mehlman 1994).

Third, there is a relationship between the maximum density a species attains and the density it attains at other sites it occupies (K.J. Gaston, T.M. Blackburn & R.D. Gregory, unpublished analyses). Counter-intuitively, British bird species that attain higher maximum local densities on average occupy a higher proportion of sites at densities which represent a low percentage of this maximum; common species have relatively low density at most sites and relatively high density at a few, whereas rare species have relatively high (although absolutely low; see above) densities at all sites. These patterns should, however, be interpreted cautiously because the census technique may limit the range of abundances of the rarer species. The minimum density that can be detected on a CBC site is $1/A$, where A is the site area. This may overestimate true minimum density more for rare than for common species.

2 The relationship is consistent between different habitats. Thus, using data from the CBC, Blackburn *et al.* (1998b) found that the regression slopes of the interspecific relationships for species censused on farmland and woodland sites were very similar, and that year to year variation in these values was positively correlated between these habitats. In other words, in years when abundant species occupy a higher proportion of woodland CBC sites, they also occupy a higher proportion of farmland sites. This is presumably a simple consequence of whatever it is that constitutes good years for common or rare species (e.g. favourable weather conditions) being the same in both habitat types, and perhaps density-dependent habitat selection causing the spill-over of individuals from one to the other at high densities. Certainly, most birds tend to show some intraspecific synchrony, albeit weak, in population fluctuations across CBC sites (Paradis *et al.*, in press). The consistency in the abundance–range size relationship between habitats arises despite consistently lower densities being recorded on farmland than woodland sites. A bird species recorded on a single woodland CBC site in any given year will have a density three to four times higher, on average, than a species recorded on only a single farmland site (Blackburn *et al.* 1998b), this difference being a consequence of much of the area of most farmland CBC sites being relatively barren for birds (e.g. crop land).

3 The relationship remains stable from one season to another. Thus, the resident species of the breeding and wintering bird assemblages in Britain (i.e. those species they have in common) show a common interspecific population size–range size relationship, such that neither the slope nor the intercept differ significantly between the two seasons (Gaston *et al.* 1998c). However, resident birds do have significantly larger British population sizes and range sizes in winter than in summer. Given the similarity in the summer and winter population size–range size relationships, these results imply that bird species 'slide up' the interspecific relationship from summer to winter, and back down again from winter to summer. Presumably, this dynamic simply represents recruitment at the end of the breeding season followed by the death of many individuals over the course of the winter, as 'summer' population size estimates generally refer to the start of the breeding season.

More interesting regularities arise if migrant species are considered (Gaston *et al.*, in press). On average, summer migrants have smaller breeding population sizes and range sizes in Britain than do residents (Greenwood *et al.* 1996; Gregory & Blackburn 1998), yet summer migrants and residents do not differ in either the slopes or the intercepts of the relationships between population size and range size in the breeding season. Likewise, winter migrants to Britain have smaller breeding population sizes and range sizes than do residents, yet winter migrants and residents do not differ in either the slopes or the intercepts of their wintering population size–range size relationships. When all species are included, the slopes of the interspecific population size–range size relationships differ between summer and winter (Gaston *et al.* 1998c), but residents and migrants lie on the same line in both seasons. Because migrant and resident populations are subject to different environmental pressures for around half the year, yet lie on the same interspecific relationship when in Britain, this implies a commonality in the forces setting this relationship for both groups. What causes this commonality is presently unclear. In Section 4.2.2 we consider some of the possibilities.

4 The relationship remains reasonably stable from one year to the next. Thus, Blackburn *et al.* (1998b) found relatively little interannual variation in the slope, intercept and coefficient of determination for abundance–range size relationships for farmland birds in Britain, and similarly little variation for relationships across woodland birds. In major part, this occurs because abundant and widespread species remain abundant and widespread from year to year, while rare and restricted species remain rare and restricted. However, assessment of the amount of interannual variation in these interspecific relationships is necessarily subjective, as there is no objective baseline for what constitutes a large amount.

5 In the one example where this has been tested explicitly, the relationship is similar for two different groups of organisms in the same area. Thus, using estimates of abundance and range size which were broadly comparable, Blackburn *et al.* (1997b) found that the slopes of the relationships between overall population size and range size were not significantly different for birds and mammals breeding on mainland Britain. However, intercepts were different, and for a given range size mammals had abundances that averaged 30 times higher than those of birds (see also Greenwood *et al.* 1996), an observation which surprises many birdwatchers!

6 It is very difficult to account for the (substantial) variation around abundance–range size relationships as a product of variation in life history traits. One obvious feature of species that may cause variation around abundance–range size relationships is their body size. Species of different body mass require predictably different amounts of energy for their metabolism (Section 5.5.3). Consider two species attaining the same range size. If similar amounts of energy were available to both, then the larger bodied of the two, the individuals of which require absolutely more energy to survive, would be unable to attain a population size as large as that of the smaller-bodied species. The same total amount of energy can be divided either into many small or fewer large individuals. We might therefore expect larger-bodied species generally to fall below the regression line describing the abundance–range size relationship, and smaller-bodied species to lie above it.

In fact, there is no evidence of any consistent effect of body size on where species lie with respect to abundance–range size relationships (e.g. Blackburn *et al.* 1997a; Quinn *et al.* 1997a). For British birds, body mass explains an insignificant ($< 1\%$) amount of the variation in population size not explained

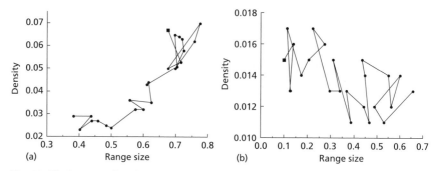

Fig. 4.7 The intraspecific relationship between abundance (territories per hectare when present) and range size (proportion of sites occupied) for (a) tree sparrow ($r = 0.898$, $n = 24$, $P < 0.001$) and (b) sparrowhawk ($r = -0.516$, $n = 24$, $P < 0.01$) on farmland CBC sites in the period 1968–91. The points are joined in temporal sequence, with 1968 indicated by a square point. In years when they occurred on a higher proportion of sites, tree sparrows were more abundant on those sites than in years when they occurred on fewer sites, and sparrowhawks were less abundant. From Gaston *et al.* (1999a).

by range size (Blackburn *et al.* 1997a). Similar conclusions can be drawn if population density is substituted for population size, or if metabolic rate, a more direct measure of energy requirements, is substituted for body mass. For British moths, the proportion of variation explained by body mass is 2% (Quinn *et al.* 1997a). One way these results might be accounted for would be if species of different body mass used consistently different amounts of energy (see also Section 5.5.3).

7 Positive intraspecific abundance–range size relationships might also be expected given that positive interspecific relationships occur. Most interspecific relationships are too weak, and the variance in abundances between species too much larger than that within species, for this inevitably to follow. However, such intraspecific relationships are also widespread. They have been documented most frequently for birds, although this probably better reflects the availability of suitable data than any necessary taxonomic bias in the occurrence of the pattern (Gibbons *et al.* 1993; Smith *et al.* 1993; Ambrose 1994; Tucker & Heath 1994; Fuller *et al.* 1995; Hinsley *et al.* 1996; Cade & Woods 1997; Gaston *et al.* 1997a, 1998d; Newton 1997; Blackburn *et al.* 1998b; Díaz *et al.* 1998; Donald & Fuller 1998; Gaston & Curnutt 1998; Venier & Fahrig 1998; Tellería & Santos 1999).

For birds in Britain, the frequency of positive intraspecific relationships has been determined using data derived from the CBC over a 24-year period. Of 75 species occurring on farmland sites, 69% exhibited positive relationships and 40% relationships that were both positive and statistically significant (e.g. Fig. 4.7a; Blackburn *et al.* 1998b). Of 56 species occurring on woodland sites, these proportions were 55% and 18%, respectively. The corollary, of course, is that 31% of farmland species and 45% of woodland species showed negative intraspecific abundance–range size relationships and, overall, around 6% of intraspecific relationships were significantly negative (e.g. Fig. 4.7b). Donald and Fuller (1998) reported that 18 out of 57 British bird species showed range and population size changes in opposite directions over this same time period, although their measure of range change in each case was based on comparison of two values only [species distributions given by Sharrock (1976) and Gibbons *et al.* (1993)]. A related analysis was performed by Venier and Fahrig (1998) for 20 species of forest songbirds censused in north-western Ontario, Canada. All these relationships were positive, and 16 (80%) were significant.

Although published studies reveal that significant, positive intraspecific relationships are far from universal, failures to find significant abundance–range size relationships may in part reflect the comparatively narrow range of variation in either variable which has been observed. Local abundances and geographical distributions may be quite stable over long periods, as illustrated by the high temporal concordance in abundances exhibited by assemblages of birds breeding on CBC sites in the period 1971–92 (Bengtsson *et al.* 1997), and by the generally small percentage changes in range sizes of British breeding birds from the early 1970s to the early 1990s (Donald & Fuller 1998). In that regard, it is interesting that Gaston *et al.* (1998d) found that significant intra-specific abundance–range size relationships were more likely to be observed in species undergoing simultaneous increases or decreases in both abundance and range. This is the case for the two species in Fig. 4.7, as the British popula-tion of the tree sparrow is undergoing a precipitous decline (Summers-Smith 1989; Marchant *et al.* 1990; Gibbons *et al.* 1993; Greenwood *et al.* 1994; Marchant & Gregory 1994; Fuller *et al.* 1995; Siriwardena *et al.* 1998), while that of the sparrowhawk is recovering from such a decline earlier in the century (Newton & Haas 1984; Gibbons *et al.* 1993). Presumably, the association between the two variables is strengthened because these temporal trends reduce the likelihood of random population fluctuations.

4.2.2 *What generates abundance–range size relationships?*

Nine separate mechanisms can be postulated as possible determinants of pos-itive interspecific relationships between abundance and range size (Table 4.1; Gaston *et al.* 1997a). These can broadly be classified as statistical, range posi-tion, resource-based and population dynamics explanations. This need not imply that the mechanisms are mutually exclusive or even necessarily funda-mentally different; in some cases, they are perhaps better regarded as provid-ing different perspectives on a common issue.

Statistics
The first, and simplest, explanation for positive interspecific abundance–range size relationships and, as we have argued before, the first explanation that the

Table 4.1 Mechanisms postulated as determining or contributing to positive interspecific abundance–range size relationships.

Mechanism	Process
Sampling artefact	Systematic underestimation of the range sizes of species with lower local abundances
Abundance distribution	Random or aggregated patterns of distribution of individuals of species across landscape
Phylogenetic non-independence	Non-independence of species as data points for statistical analysis
Range position	Species closer to edges of their geographic ranges have lower abundances in, and occupy a smaller proportion of, study area
Niche breadth	Attainment of higher local abundances and wider distributions by species with greater niche breadths
Niche position	Attainment of higher local abundances and wider distributions by species with lower niche positions
Habitat selection	Density-dependent habitat selection
Metapopulation dynamics	Metapopulation structures
Vital rates	Similar spatial patterns of density-dependent intrinsic growth rates across sites for different species, but species with lower density-independent death rates able to attain higher abundances and occupy more sites

macroecologist should normally consider, is that they are artefacts of inadequate data. If levels of sampling effort are insufficient, then a species which occurs at low densities will tend to be recorded from fewer localities compared with a species which occurs at high densities, even if they are actually equally widely distributed. If census data were of sufficient quality, there would be no pattern to explain.

In many data sets it seems very likely that a substantial component of observed abundance–range size relationships does arise in this way. Levels of sampling are simply too low to provide reliable estimates of relative range size. However, as a general explanation, this one seems highly unlikely. One of the desirable features of using data on British birds is that the abundances and distributions of these species are so well known. Although the census of Eastern Wood co-ordinated by G. Beven is of exceptionally high quality, the census data collected by amateur observers and collated by the BTO are very reliable. It is improbable that so many spurious positive associations between these variables could arise in this assemblage.

A second statistical mechanism is that positive abundance–range size relationships arise through the random distribution of individuals on the landscape. If this distribution followed a Poisson process, then a positive relationship between abundance and range size would be expected by chance (Wright 1991). Hinsley *et al.* (1996) found no evidence that a Poisson model could account for the distribution of individuals of several bird species among small woods in an eastern English landscape. Venier and Fahrig (1998) explicitly tested the abundance data they derived for Canadian birds for departures from Poisson expectation. No species showed a pattern of abundances that could be adequately modelled by this distribution. This follows from the general observation that the abundances of species, except when they are particularly rare, are typically aggregated in space (Taylor *et al.* 1978, 1979; McArdle *et al.* 1990).

Spatially aggregated distributions also predict positive relationships between local abundance and range size (Wright 1991; Hartley 1998). However, the extent to which this is strictly a mechanism for abundance–range size relationships, rather than essentially a restatement of the relationship in another form, is questionable (Gaston *et al.* 1998e). If we accept that spatial aggregation determines abundance–range size relationships, then logically we must seek mechanistic explanations for patterns in the spatial aggregation of species (and hence, by implication, the abundance–range size correlation). Unfortunately, despite extensive study, the patterns and determinants of aggregation are generally poorly understood, particularly at the large (geographical) scales of relevance to most observed abundance–range size relationships (for discussion, see, for example, Perry 1988; McArdle & Gaston 1995). To argue that spatial aggregation explains abundance–range size relationships is simply to supplant one poorly understood pattern with another.

In addition, the likely form of the interspecific abundance–range size relationship is rather difficult to predict a priori on the basis of aggregation. Individual species exhibit different levels of aggregation with changing density, these changes may be species-specific, and the universe of sites which can be occupied varies from species to species (Gaston *et al.* 1998e). Gaston *et al.* (1998e) found that increasing interspecific variance in the value of the shape parameter (k) of the negative binomial distribution (where the distribution is derived by expansion of $[q - p]^{-k}$) made little difference to the shape of the abundance–range size relationship predicted for British birds (Fig. 4.8). They concluded that the most important factor affecting the extent to which a species deviates from the predicted relationship may be how the universe of sites actually available to it differs from that assumed for the purposes of the model. The model assumes all species can use all sites, but for many this is unlikely to be true.

A third arguably statistical mechanism for positive abundance–range size relationships is the phylogenetic non-independence hypothesis. An artefactual positive interspecific relationship between abundance and range size

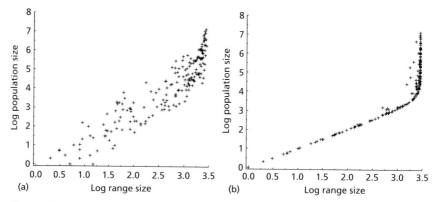

Fig. 4.8 The relationship between population size (number of individuals) and (a) actual and (b) simulated range size (number of 10×10-km squares occupied on the British National Grid) for 193 species of breeding bird in Britain. The simulated range sizes were generated using the negative binomial distribution with k values chosen randomly from the range 0.1 to 4. The mean population size averaged over all (2830) grid squares was used together with k to predict the number of grid squares unoccupied by a species, which was then used to calculate range size (number of occupied squares). The simulated relationship shown is just one of 100 generated, but all gave virtually identical results. Results were similar if k values in the more restricted range of 0.2–0.5 were used. From Gaston *et al.* (1998e).

could result from the shared common ancestries of species in an assemblage (Section 1.5). Because of their phylogenetic relatedness, species do not constitute independent data points for analysis, inflating the degrees of freedom available for testing statistical significance (Harvey & Pagel 1991; Harvey 1996). If sufficient, this inflation may falsely indicate relationships which in reality do not exist.

The phylogenetic non-independence hypothesis cannot, of course, explain positive intraspecific abundance–range size relationships, where they exist. In addition, a growing number of studies, including some for birds in Britain, have found that interspecific abundance–range size relationships persist when phylogenetic non-independence has been controlled for (Fig. 4.9; e.g. macrolepidoptera—Quinn *et al.* 1997a; frogs—Murray *et al.* 1998b; birds—Blackburn *et al.* 1997b; Gaston *et al.* 1997b; mammals—Blackburn *et al.* 1997b). Thus, the relatedness of species does not generate the positive interspecific relationship either.

Range position

Abundances and patch occupancy are widely held, on average, to decline towards the edge of the geographical range of a species (e.g. Shelford 1911; Kendeigh 1974; Hengeveld *et al.* 1979; Hengeveld & Haeck 1981; Brown 1984; Lawton 1993; Safriel *et al.* 1994; Maurer 1999). This being so, species whose range edges are close to or are overlapped by a study region will appear to

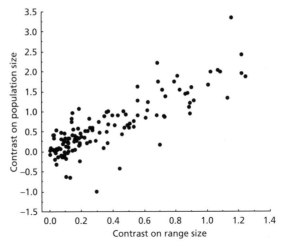

Fig. 4.9 The relationship between population size (number of individuals) and range size (number of 10 × 10-km squares occupied on the British National Grid) for breeding bird species in Britain, controlling for phylogenetic relatedness using the independent contrast method ($r^2 = 0.83$, n (number of independent contrasts) = 132, $P < 0.0001$). The method was implemented by the CAIC computer program (Purvis & Rambaut 1994, 1995), using the CRUNCH option and assuming equal branch lengths. Species were classified following Sibley and Monroe (1990, 1993). From data in Appendix III, which were $\log_{10}$-transformed before analysis.

Table 4.2 The $\log_{10}$-transformed mean number of territories in Eastern Wood of bird species in different range position categories. The means differ significantly across categories (ANOVA: $F_{2,41} = 3.85$, $P = 0.029$). Central species occur throughout Britain and their geographical ranges in Europe extend to both north and south of the region, subcentral species occur throughout Britain but this is on the edge of their European northerly or southerly limit, while a geographical range limit crosses Britain for submarginal species. See Gaston et al. (1997b) for details.

Group	n	Mean	Standard error
Central	28	0.52	0.08
Subcentral	5	0.24	0.12
Submarginal	11	0.17	0.06

have lower abundances and more restricted ranges than species close to the centre of their ranges in the region, even if there are no interspecific differences in the overall range size and average abundance of the species.

The effect can be illustrated with the birds of Eastern Wood. We can classify species in this avifauna according to how close Britain lies to the edge of their total European ranges (except for the mandarin, for which the introduced British population is greatly isolated from the species natural range). Table 4.2

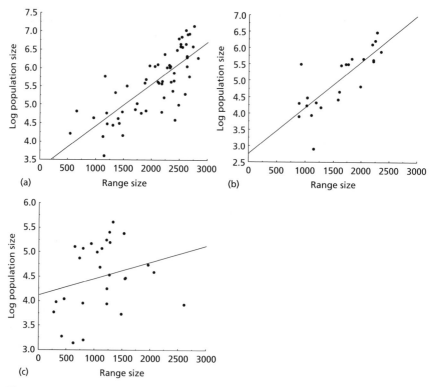

Fig. 4.10 The relationship between population size (number of individuals) and range size (number of 10×10-km squares occupied in the second BTO atlas; Gibbons *et al.* 1993) for those British bird species found on CBC plots and having ranges in Britain classified as (a) central ($r^2 = 0.59$, $n = 59$, $P < 0.0001$), (b) subcentral ($r^2 = 0.625$, $n = 23$, $P < 0.0001$) and (c) submarginal ($r^2 = 0.07$, $n = 28$, $P = 0.19$) to their European ranges. See Table 4.2 for more details. From Gaston *et al.* (1997b).

shows how the mean abundance of species in Eastern Wood is related to the species range position. Birds for which Britain lies close to their range edge have lower abundances in Eastern Wood than species for which Britain lies towards the range centre.

Controlling for the effects of range position reveals that it generally cannot explain interspecific abundance–range size relationships. For example, Gaston *et al.* (1997b) tested this mechanism as an explanation for the relationship in British birds. Dividing species into the same range position categories as listed in Table 4.2, they showed that positive abundance–range size relationships persisted within the categories. Only those species on the edge of their range in Britain upheld the predictions of the range position hypothesis by showing no significant abundance–range size relationship (Fig. 4.10). For the abundances of bird species in Eastern Wood, a significant positive relationship persists for central species ($r^2 = 0.24$, $n = 28$, $P = 0.0085$), but not subcentral ($r^2 = 0.25$,

$n = 11, P = 0.45$) or submarginal ($r^2 = 0.11, n = 5, P = 0.58$) species. However, the sample sizes in the last two groups are rather low.

This is not to say that the range position effect does not contribute to many documented patterns, especially in partial analyses ('partial' analyses are performed over areas which embrace the entire geographical ranges of none or only a small proportion of the species concerned, 'comprehensive' ones are performed over areas which embrace all or a very large proportion of the extents of the geographical ranges of the species concerned; Gaston & Blackburn 1996a). The extent to which it does contribute will depend on the proportion of between-species variation in abundances which can be accounted for simply by the position of the area in which those abundances were measured with respect to the centres of the geographical ranges of the different species.

Other information also fails to support the range position hypothesis. In particular, while the majority of British bird species show significant positive intraspecific abundance–range size relationships, only a small minority are undergoing changes in distribution sufficient to alter the position of Britain with respect to their wider geographical range. The range position hypothesis also unnecessarily begs the question of why abundance and range size decline towards range limits. In fact, it is not at all clear that this is a general feature of species distributions: while some studies claim to have shown this pattern (McClure & Price 1976; Hengeveld & Haeck 1981, 1982; Brown 1984; Bart & Klosiewski 1989; Svensson 1992; Tellería & Santos 1993; Maurer 1994; Brown *et al.* 1995; Curnutt *et al.* 1996), others have failed to find such evidence (Rapoport 1982; Brussard 1984; Carter & Prince 1985; Wiens 1989). Of particular relevance here is that Blackburn *et al.* (1999a) failed to find evidence that abundance declines towards the range edge in Britain for most of the 32 species of passerine bird they examined. There are several methodological reasons why no such relationships might be forthcoming in these data (Blackburn *et al.* 1999a), but their absence is equally likely to be genuine; a glance at the most recent BTO atlas of breeding birds in Britain encourages belief that this is often so (Gibbons *et al.* 1993; note that the 'abundance' maps presented actually show variation in tetrad occupancy in different 10-km squares, but that maps illustrating variation in numbers of individuals in point counts are presented on pp. 457–461; two are reproduced here as Fig. 4.11). That is not to say, however, that these species do not become rarer in other senses of the word towards their range edges in Britain. They may, for example, be more patchily distributed close to edges, but attain average densities in those patches that they do occupy (see also Carter & Prince 1985, 1988). Nevertheless, if densities do not, on average, decline towards range limits in British birds, the range position mechanism is unlikely to produce a positive density–range size relationship.

Finally, the range position mechanism cannot explain interspecific abundance–range size relationships based on data for the entire geographical

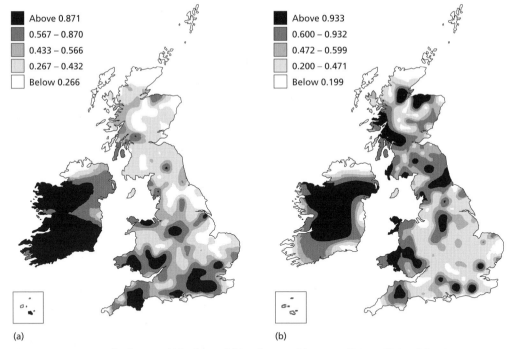

Fig. 4.11 Variation in the density of (a) robin and (b) willow warbler across Britain. Units of density are mean number of individuals recorded in timed point counts in designated 10×10-km squares. From Gibbons *et al.* (1993), with permission from Academic Press.

ranges of species (e.g. Brown & Maurer 1987; Gaston & Blackburn 1996c; Murray *et al.* 1998b).

Resources

A positive interspecific abundance–range size relationship could arise if there were interspecific variation in the realized niche breadth of a species, and if local abundance and regional range size were both functions of niche breadth; this is commonly known as Brown's hypothesis (Brown 1984). Then, species with wide niche breadths would be more abundant and widespread than species with narrow niche breadths.

Although the niche breadth hypothesis is intuitively appealing, as discussed in the context of range sizes (Section 3.3.5), it is impossible to conduct an unequivocal test because of the *n*-dimensional nature of the niche. The problems can be illustrated with reference to the avifauna of Eastern Wood. For example, the blue tit has twice the British range size (measured as number of occupied 10×10-km squares on the British National Grid) of the marsh tit (Appendix III), and on average maintains six times as many territories in Eastern Wood (Appendix II). Yet, reference to standard works (e.g. Cramp & Perrins 1993) indicates that the biology of these species scarcely differs. They are of similar size, both feed mainly on arthropods in summer and seeds and fruit in winter, and both are monogamous hole nesters, laying large clutches incubated by the female for similar amounts of time. They show similar latitudinal ranges across Europe. The blue tit is found in more habitat types, but that could be a function of its greater abundance: we expect to see common species in more habitats, whether or not rare species can use fewer (for discussion, see Gaston 1994a; Gaston *et al.* 1997a and references therein). Why should the British range sizes and relative abundance in Eastern Wood of these two species differ so much? It is difficult to suggest a plausible reason based on niche differences.

Despite the problems of unequivocally identifying the correct axes for comparison, as with interspecific relationships between range size and resource usage (Section 3.3.5), there have nonetheless been a host of studies of relationships between local abundance and resource usage. Unfortunately, these generally suffer from innumerable further difficulties, mostly associated with the statistical difficulty of dissociating sample size effects from genuine differences between the resource usage of common and rare species. The evidence for the hypothesis as an explanation for positive interspecific abundance–range size relationships appears to us to be largely unconvincing.

A further problem in successfully testing the niche breadth hypothesis lies in making an effective distinction from a second resource-based mechanism, the resource availability hypothesis. A positive interspecific abundance–range size relationship could arise if there were variation in the abundance and distribution of the resources on which different species depend (Hanski *et al.* 1993; Gaston *et al.* 1997a). Species utilizing widespread and abundant resources would themselves be widespread and abundant, and vice versa, regardless of how broad a spectrum of resources the species could use.

The distribution and abundance of some species are obviously affected by resource abundance. Synchronous population cycles of specialist predators (or those of restricted diet) and their prey, for example, clearly demonstrate how the availability of a resource can affect the population of a species dependent on it. Unfortunately, however, we know of no case where an abundance–range size relationship has been plotted for a set of species showing such resource-driven population dynamics.

Resource use in 85 British breeding bird species has been examined by Gregory and Gaston (2000; see also Section 3.3.5). Across all species, none of five measures of abundance and distribution chosen were correlated with niche breadth, whereas four out of five of these measures were correlated negatively with niche position. Repeating the analyses using a method designed to control for phylogenetic non-independence confirmed these general patterns. Birds that tend to utilize resources that are more atypical of the British environment tend to be rarer and more thinly distributed, while those utilizing typical resources are common and widely distributed (see also Fuller 1982). These results refute the resource breadth hypothesis, but support the resource availability hypothesis for British birds (but cf. Quinn *et al.* 1997b, 1998; Arneberg *et al.* 1998).

A third, and largely ignored, resource-based explanation for abundance–range size relationships derives from the ideas of habitat selection. It has long been known that some species exhibit density-dependent habitat selection (driven through intraspecific competition), occupying more habitats when densities are high and less when they are low (Kluyver & Tinbergen 1953; Fretwell & Lucas 1970; for a review, see Rosenzweig 1991). For example, as the British population of the wren recovered from the severe winter of 1962/63, numbers rose sequentially in different habitats. Only once its favoured woodland and streamside habitats had started to become saturated did populations in gardens and hedges begin to increase (Williamson 1969; Fuller 1982; O'Connor 1987). Density-dependent habitat selection has also been recorded for other birds in Britain (e.g. O'Connor 1981, 1982, 1987). This can give rise to positive intraspecific abundance–range size relationships, as explored, with particular regard to marine fish, in MacCall's (1990) 'basin model' (see also Newton 1997; Gregory 1998a). Assuming some broad commonality across species in this dynamic, then locally abundant species will tend to occupy more habitats and to be more widespread, without having to postulate niche differences.

The role of habitat selection is difficult to judge at present. It could be important for assemblages of largely generalist species, such as the birds of Britain. As mentioned earlier, the only obvious way in which the blue and marsh tits differ is in habitat use. In other taxa, however, examples are known of assemblages for which species do not show broader habitat use at higher densities (e.g. Rogovin & Shenbrot 1995). Indeed, habitat selection seems unlikely to be of major importance in determining interspecific abundance–range size relationships for many taxa, but may perhaps play a role for some.

At present, while resource issues seem likely to play a potentially significant role in the determination of interspecific abundance–range size relationships, the dearth of unequivocal tests means that the manner in which they do so remains unclear.

Population dynamics

Metapopulation dynamics Further explanations for abundance–range size relationships have been framed in terms of population dynamics constructs. In particular, a positive relationship is a prediction of metapopulation structures which incorporate a rescue effect (Hanski 1991a,b; Gyllenberg & Hanski 1992; Hanski & Gyllenberg 1993; Hanski *et al.* 1993). This assumes that immigration decreases the probability of a local population going extinct (the rescue effect), and that the rate of immigration per patch increases as the proportion of patches which are occupied increases. Here, for many parameter values, a positive relationship between local abundance and number of occupied patches can result.

The significance of such a mechanism rests fundamentally on whether or not in practice species exhibit metapopulation dynamics of an appropriate form. In the case of our exemplar assemblage, this would necessitate the avifauna of Eastern Wood being an isolated independent population whose dynamics are affected by immigration of individuals from other such populations. Certainly, the pattern of woodland habitat fragmentation in the landscape in which this wood sits fits well with the metapopulation ideal of a series of isolated patches (Fig. 4.12). Eastern Wood, which is part of the larger area of woodland covering Bookham Common, is just one of many small fragments of the forest that would originally have covered most of this landscape. Thus, the question becomes whether the separate woodland patches contain separate avian populations. As discussed in Section 3.3.3, the distances across which birds are able to disperse, at least in north temperate landscapes such as this, are likely to be large relative to between-patch distances, reducing the probability that this is true. Average natal and breeding dispersal distances for British woodland birds are usually of the order of 5 km (Paradis *et al.* 1998), which would take individuals reared or breeding in Eastern Wood to the edges of the map in Fig. 4.12. Maximum dispersal distances are much greater (Paradis *et al.* 1998). Moreover, abundance–range size relationships observed at much larger scales, such as that shown by birds across the whole of Britain (Fig. 4.4), are unlikely to be generated by a mechanism based on interpatch dispersal.

A metapopulation-based mechanism seems unlikely to be the most frequent explanation for positive abundance–range size relationships given the present understanding of the proportions of species exhibiting metapopulations (Harrison 1994; Gaston *et al.* 1997a). However, there is some evidence for some of the predictions of the rescue effect hypothesis (e.g. Hanski *et al.* 1995; Gonzalez *et al.* 1998), and there may be certain taxa for which metapopulation structure is the rule, rather than the exception. Some of this evidence arises from studies of systems which can be manipulated experimentally, because although large for the organisms concerned the spatial scales are small to the human observer.

For example, Gonzalez *et al.* (1998) experimentally fragmented a natural moss microecosystem. When dispersal between the moss fragments was prevented, most species in the microarthropod assemblage inhabiting the fragments declined in both abundance and range size, and many became extinct. These declines caused the collapse of the positive interspecific abundance–range size relationship. However, when the patches were connected by habitat corridors, allowing dispersal of animals between them, the declines in both abundance and range size were arrested, and their positive relationship was maintained. Control patches linked by broken corridors exhibit similar patterns

Fig. 4.12 The Surrey landscape in which Eastern Wood is located. The wood is part of the larger area of woodland covering Bookham Common, unsurprisingly falling to the east of this area. Solid lines represent roads, dotted lines are railways. On the bottom map, light shading denotes woodland and dark shading represents urban areas.

to isolated patches, showing that the effect of corridors is not due simply to the extra area they provide. Nor did manipulation adversely affect microclimatic conditions within the patches. Thus, the positive abundance–range size relationship in this natural microecosystem is maintained by dispersal in a connected landscape, as predicted by the metapopulation dynamics hypothesis.

Other microcosm experiments, however, are less supportive. Warren and Gaston (1997) used laboratory communities of protists to test the effects of different dispersal regimes on the abundance–range size relationship. Positive relationships were present in all treatments, even in those communities where there was no dispersal between patches. This contradicts the metapopulation dynamics hypothesis. Warren and Gaston suggested that range size may be driven by local abundance, combined with a very general (not specifically metapopulation structured) set of extinction and colonization processes.

The experimental design in the study by Gonzalez *et al.* (1998) rules out the action of other mechanisms, so that metapopulation dynamics must be generating the relationship in this system, at least. However, the authors point out that the scale of fragmentation and the dispersal distances of the organisms in their system are likely to be appropriate for observing metapopulation dynamics, and they do not necessarily expect the metapopulation mechanism to pertain at biogeographical scales, where interpatch dispersal is unlikely to be important.

Vital rates If interpatch dispersal seems unlikely to cause the birds of Eastern Wood to behave as part of a metapopulation, but this wood is part of a habitat fragmented into a series of discrete patches, a better model for the population dynamics of the system may be described by the vital rates hypothesis (Holt *et al.* 1997). The idea that habitat is divided into discrete patches is also central to this mechanism for the abundance–range size relationship. However, it differs from the previous one in that it is population dynamics within each individual site, rather than interactions between sites, that are hypothesized to forge the link between abundance and range size.

Consider a set of species distributed at a scale large enough that regular immigration is sufficient to permit colonization of suitable sites, while not a dominant determinant of local abundance. It follows that species will only persist at sites at which their birth rate, b, exceeds their density-independent death rate, d. In other words, since $b - d = r$, the intrinsic rate of increase of a species at a site, a site is occupied when $r > 0$. Species with high average r will occupy more sites. In addition, population dynamics theory predicts that the abundance attained by a species at a site will be proportional to r, assuming that the effect of density dependence is uniform across all sites. Therefore, both the number of sites occupied and mean abundance at those sites will depend on the value of r for a species, and hence on the relationship between birth and death rates across sites. It follows that interspecific variation in birth and/or

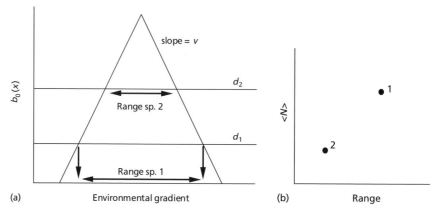

Fig. 4.13 Graphical representation of the vital rates model. (a) Birth rates (b_0) at low density at each site (x) along an environmental gradient decline linearly away from the optimum site, with the same gradient v for all species. Species 1 and 2 have different density-independent death rates (d_1 and d_2), giving rise to (b) a positive relationship between mean abundance ($<N>$) and range size. From Holt *et al.* (1997).

death rates will lead to variation in both the abundance attained by species at occupied sites, and the number of sites occupied. A positive abundance–range size relationship will result (Fig. 4.13; Holt *et al.* 1997).

The relative youth of the vital rates hypothesis means that, at present, there are too few data with which to evaluate it in any detail. We can, however, speculate about some of its assumptions. For example, the assumption that all sites that can be occupied are so is probably not valid for British birds. The tendency for turnover at sites over time, with species regularly colonizing and going extinct, suggests that species are frequently absent from sites that are suitable for them. The data on Eastern Wood in Appendix II provide several examples. Also, for the vital rates mechanism to generate a positive abundance–range size relationship, there has to be some degree of commonality in how b and d vary across the environment across species. More specifically, there should be little interspecific variation in v, defined in the model depiction in Fig. 4.13. If v varies markedly, then non-significant or negative abundance–range size relationships can result (Holt *et al.* 1997). The consistency of v in real assemblages is unknown. Certainly, there will be substantial variation in b and d. For example, the annual adult survival rates of the sparrowhawk and blue tit are 0.58 and 0.48, respectively (Sæther 1989), while the reproductive output of the latter is concomitantly higher. However, this variation does not necessarily translate into variation in v. Even if it does, this variation may provide an explanation for the often considerable scatter around abundance–range size relationships, and for relationships that are not positive. More generally, whether violations of these assumptions by real assemblages are sufficient to invalidate the vital rates model remains to be seen.

Table 4.3 A qualitative assessment (on a scale of 0–4) of the available empirical evidence in support of the role of the mechanisms postulated as determining or contributing to the form of positive interspecific abundance–range size relationships.

Mechanism	Assessment
Sampling artefact	++
Abundance distribution	++
Phylogenetic non-independence	–
Range position	++
Niche breadth	+
Niche position	++
Habitat selection	+
Metapopulation dynamics	+
Vital rates	–

4.2.3 Synthesis

At present there is no unequivocal answer to what determines the existence of positive interspecific abundance–range size relationships, despite their near ubiquity (Table 4.3). They are undoubtedly real, rather than artefactual, and are predicted from most realistic statistical models of the distribution of individuals of species across landscapes. Indeed, this last observation may provide an important clue as to why no single ecological mechanism has yet emerged as the most significant, and why there is some evidence in support of each of several such mechanisms. If abundance–range size relationships follow from a diverse set of aggregated and non-aggregated distributions, then these distributions may be generated by a whole host of processes, which may vary not simply from species to species, but from one region to another. This argument would fit well with the recognition that multiple mechanistic models, from often divergent roots, have been found to be able to predict the same patterns of aggregation of a species (e.g. Anderson *et al.* 1982; Perry 1988; Routledge & Swartz 1991). It does, however, run counter to the argument (e.g. MacArthur & Connell 1966) that for any ecological pattern which occurs widely there must be a single general explanation.

Whatever their cause may be, the existence of general positive relationships between abundance and range size has a number of implications for species assemblages. A wide-ranging discussion of these is given by Gaston (1999; see also, Gaston *et al.*, in press). Here, attention is drawn to four.

First, the existence of a positive interspecific abundance–range size relationship implies that species face 'double jeopardy' in terms of extinction risk (Lawton 1993, 1995, 1996b). Both the magnitude of local abundance and the extent of spatial distribution of a species may contribute independently to its risk of global extinction. Low local abundance increases the likelihood of stochastic extinction (although the dynamics of local density and population size may be very different; Gaston & McArdle 1993), while a narrow geographical range increases the likelihood that all populations will simultaneously be subject to adverse abiotic and/or biotic environmental conditions (Section 3.3.6; Hanski 1982a; Diamond 1984; Pimm *et al.* 1988; Tracy & George 1992; Gaston 1994a; Lawton 1995; Gaston & Blackburn 1996c). There has been little empirical work to discriminate the relative importance of abundances and range sizes to extinction risk (but see Gaston & Blackburn 1996c; Johnson 1998a), and opinions seem divided. Whatever it is, though, the fact that species occupying few sites will tend to do so at low densities does nothing to ameliorate the extinction risk they face.

Second, the general positive relationship between local abundance and range size also suggests a dual effect of the distribution of individual birds across the landscape on the likelihood that they will be sampled by local sites like Eastern Wood. We saw in Chapter 3 that patterns in the frequency distribution of avian range sizes at all scales reveal that many species in assemblages occur at only a very limited proportion of all possible sites, so that many sites are unoccupied not because the sites have inappropriate habitat, but instead because by chance the individuals that are present in the landscape do not coincide with them. The positive abundance–range size relationship additionally reveals that narrowly distributed species are less abundant than widely distributed species at sites that they do occupy, further reducing the probability that they will be present at occupiable sites.

Third, positive interspecific abundance–range size relationships suggest that latitudinal variation in range sizes (Section 3.4.3) might be paralleled by a similar pattern for abundances. In this context, it is interesting that there is a small but growing body of evidence that species abundances do indeed vary spatially in general ways. However, whether this variation follows a consistent pattern is currently unclear. Johnson (1998b) found that population density on average increased with latitude across species for the mammalian fauna of eastern Australia. Currie and Fritz (1993) showed similar patterns for a variety of animal groups. However, Currie and Fritz's analysis is accompanied by a plot showing body size decreasing with latitude for the same groups of species for which they plotted latitudinal variation in abundance. As this pattern runs counter to Bergmann's rule (Section 5.4), for which there is evidence from much more extensive and comprehensive compilations than those used by Currie and Fritz (for birds, for example, see Blackburn & Gaston 1996e), the reliability of their analyses must be questioned.

For birds, Gaston and Blackburn (1996c) showed how abundance varied with latitude in wildfowl. Population size does indeed tend to increase with the distance from the Equator at which wildfowl species reside (Fig. 4.14a). However, the relationship shows distinct curvilinearity, such that the least

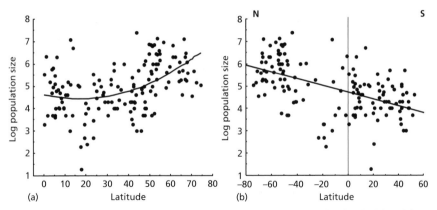

Fig. 4.14 The relationships between global population size (number of individuals) and the latitudinal mid-point (degrees) of the geographical range for the wildfowl species of the world ($n = 152$): (a) irrespective of hemisphere; (b) with hemisphere taken into account. Solid lines are second-order polynomial and linear regressions, respectively, through the data. The Equator is indicated in (b) by a vertical line. From Gaston and Blackburn (1996c).

abundant species reside at around 20° of latitude. This is a consequence of hemispheric differences in the pattern of abundance variation. Wildfowl population sizes decrease slightly with latitude in the southern hemisphere, resulting in an overall decline from north to south across the globe (Fig. 4.14b). In these regards, population size variation matches range size variation in this group (Fig. 3.16; Gaston & Blackburn 1996c). Moreover, the pattern of variation is not dissimilar from the general tendency of range sizes to decrease from north to south across the Holarctic, to be relatively large in the tropics, and to decrease again south of the tropics (e.g. Fig. 3.14; Section 3.4.3).

Other evidence for systematic patterns of variation in abundance with latitude for birds is more piecemeal and indirect. For example, Thiollay (1990) reports a comparison of the structure of bird assemblages in temperate forest in France and tropical forest in French Guiana, based on data obtained from multiple 0.25-hectare sample plots (515 in France, 440 in Guiana). In Guiana, the number of species (260), number of individuals (3047) and estimated mass of birds (221.6 kg) are all greater than in France (number of species = 53, number of individuals = 1998, mass = 76.6 kg), despite the somewhat larger number of sample plots for the latter. The number of species is five times greater in the tropical forest than in the temperate one, while the number of individuals and the mass of birds differ by considerably smaller multiples, suggesting that the average density and biomass of species must be lower in the tropics than in the temperate region. Similarly, Terborgh *et al.* (1990) contrast the structure of the avian assemblage of their 97-hectare study plot of floodplain forest in Amazonian Peru with one of the best studied bird communities in North America, that of the secondary northern hardwood forest at Hubbard Brook. The 245 resident

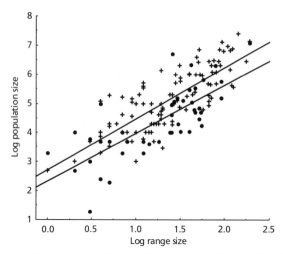

Fig. 4.15 The relationship between global population size (number of individuals) and geographical range size (number of equal-area squares occupied on the WORLDMAP grid; Williams 1992, 1993) for the wildfowl of the world, distinguishing between tropical (filled circles) and extratropical (crosses) species. A species was defined as tropical if the latitudinal mid-point of its geographical range was within 23° of the Equator. Regressions of population size on geographical range size for the two groups do not differ significantly in slope ($F_{1,149} = 0.23$, $P = 0.64$), but the intercept for extratropical species is significantly higher ($F_{1,149} = 19.6$, $P < 0.0001$). From data sources in Gaston and Blackburn (1996c).

species found to hold territory on the floodplain forest plot comprised about 955 pairs of nesting birds, with a conservative biomass estimate of 190 kg/km². Unsurprisingly, even allowing for area effects, the species richness of the hardwood forest is much lower, with 24 species being present on a 10-hectare plot in an average year. However, the density of birds is similar to the floodplain forest, at 1000 breeding pairs per 100 hectare, and their biomass is lower, at about 40 kg/km². Thus, once again, individuals, and probably biomass, are more finely divided in the tropical forest, resulting in lower densities per species. Such partitioning is reflected in a tendency for species–abundance distributions for bird assemblages to be more equable (we use the term 'equable', which is a synonym of 'evenness', rather than 'equitable' which is widely used but means 'reasonable'; Cotgreave & Harvey 1994b) towards lower latitudes (Short 1979; Nee *et al.* 1992a; for a similar pattern in marine invertebrates, see Rex *et al.* 1993). Finally, wildfowl species with ranges centred outside the tropics have larger population sizes, for a given range size, than do species with ranges centred within the tropics (Fig. 4.15). This suggests that densities are greater for wildfowl species living at higher latitudes, all else being equal.

Overall, it seems likely that, for comparable vegetation types and areas, the numbers of species, biomass and equability of avian assemblages, and perhaps the overall numbers of individuals (although these may equally remain roughly constant), increase towards lower latitudes, while the mean density, the mean biomass and the mean body mass (Section 5.4.2), expressed *per species*, decrease. However, the explicit analyses with which to substantiate these assertions largely remain wanting. A similar set of patterns may also occur across elevations (e.g. Sabo 1980), emphasizing the similarity of variation in community structure across the two kinds of gradients.

Finally, the correlation between abundance and range size suggests that the frequency distribution of species abundances ought to take a similar form to that for range sizes (Section 3.2). It is thus to a consideration of the shape of species–abundance distributions that we next turn.

4.3 Species–abundance distributions

4.3.1 *Data*

As already seen (Fig. 4.1), the species–abundance distribution for the avifauna of Eastern Wood is strongly right-skewed on arithmetic axes: most species are present in the wood at low numbers. This pattern is typical of many such distributions at local scales (Preston 1948; Williams 1964; May 1975; Hughes 1986; Gray 1987; Magurran 1988; Wilson 1991; Gaston 1994a). There is no evidence that local species–abundance distributions mirror the bimodal pattern for range size, probably as a consequence of the broad spread of abundances exhibited by widely distributed species at small scales (e.g. Fig. 4.6).

As observed in the previous chapter, logarithmic transformation of right-skewed distributions often results in distributions that are approximately normal. However, that is certainly not the case for the abundances of the birds of Eastern Wood, which remain highly right-skewed on logarithmic axes (Fig. 4.16). Figure 4.16 illustrates the log average abundance of species when they are actually present at the site. A distribution closer to a log-normal is obtained if mean abundance is calculated as total territory number divided by the number of years of the census (Fig. 4.17), rather than the number of years that each species was present. This distribution is actually skewed to the left, but the skew is not significant (skew $[g_1] = -0.22$; $t = 0.62$, $P > 0.05$ using the test given by Sokal & Rohlf 1995), and the distribution is not significantly different from log-normal (Lilliefors test, $P = 0.46$).

At a larger spatial scale, the frequency distribution of the population sizes of British breeding birds is, like the Eastern Wood assemblage, highly right-skewed (Fig. 4.18a). The distribution becomes significantly skewed to the left when the population sizes are logarithmically transformed ($t = 2.6$, $P < 0.05$), and is significantly different from log-normal (Fig. 4.18b; Lilliefors test, $P = 0.002$). Those species–abundance distributions that have been plotted for other large-scale assemblages seem more often than not also to be left-skewed (e.g. Järvinen & Ulfstrand 1980; Nee *et al.* 1991b; Gregory 1994, 2000; Gaston

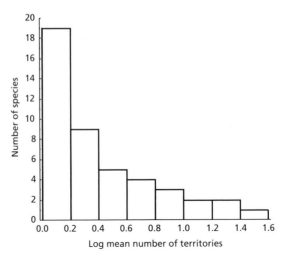

Fig. 4.16 The frequency distribution of the $\log_{10}$-transformed mean number of territories held per year by breeding bird species when present in Eastern Wood in the period 1949–79.

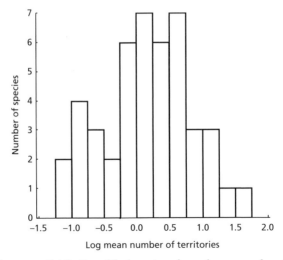

Fig. 4.17 The frequency distribution of the $\log_{10}$-transformed mean number of territories held in Eastern Wood by breeding bird species in the period 1949–79. The mean for each species is calculated across all years censused, not just those in which the species was present.

& Blackburn 1996c; Murray *et al.* 1998b; Maurer 1999; but see Osborne & Tigar 1992), but even when present the skew is often weak. Figure 4.19 shows two examples, for the birds of Slovakia and the Czech Republic. The former differs from log-normal (Lilliefors test, $P = 0.005$), and is left-skewed, but not significantly so ($t = 0.9, P > 0.05$). The latter is neither different from log-normal (Lilliefors test, $P = 0.16$) nor significantly skewed ($t = 0.25, P > 0.05$).

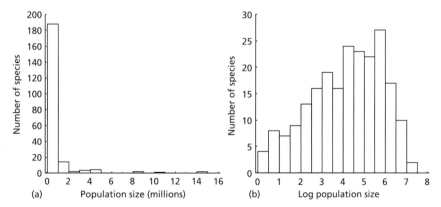

Fig. 4.18 The frequency distributions of (a) the untransformed and (b) the $\log_{10}$-transformed breeding population size (number of individuals) of bird species in Britain ($n = 217$). From data in Appendix III.

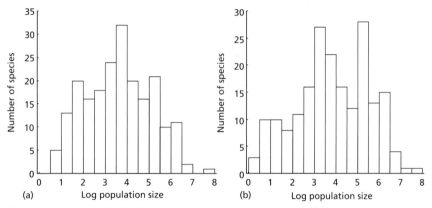

Fig. 4.19 Frequency distributions of $\log_{10}$-transformed population size (number of individuals) for breeding birds in (a) Slovakia (mean ± SD = 3.64 ± 1.53, skew = 0.04, $n = 209$), and (b) the Czech Republic (mean ± SD = 3.85 ± 1.70, skew = −0.16, $n = 197$). From data in Trnka (1997) and Hudec *et al.* (1995).

The high degree of right-skew for the untransformed species–abundance distributions for the birds of Eastern Wood and Britain means that most species in the assemblages are much less abundant than the most abundant species, but also that there are relatively few very abundant species. Interpretation of the left-skew in the logarithmically transformed distribution is more equivocal. It could be taken to mean that there are relatively few very rare species in these assemblages, in comparison to a log-normal distribution covering the same range of abundance values. Alternatively, the left-skew could be interpreted as a consequence of more very rare species in these assemblages, in comparison to a log-normal distribution with the same mode and maximum. The latter has been the more usual interpretation (e.g. Gregory

1994; Harte *et al.* 1999). Whichever, the shape typically assumed by large-scale species–abundance distributions is very similar to that typically taken by large-scale species–range size distributions, as expected from the general positive correlation between these two variables.

To what extent is the distribution of species abundances seen in Eastern Wood a consequence of the broader species–abundance distribution for British birds? We examined this question using the, by now familiar, technique of random draws to ask what the avifauna of Eastern Wood would look like if it was the result of the random selection of individuals from the entire British avifauna. The answer is that it would look somewhat different.

The simulation model developed to examine abundances is more complicated than those used so far in this book. The reason is that the average abundance of species is usually calculated as abundance when present. If a species is absent from a census, then that census is not included when calculating the species abundance. This practice has the consequence that we need to know in how many censuses a species is present. This means that our simulation of the Eastern Wood assemblage has to model each year separately. Therefore, we proceeded as follows. We started by summing the total number of territories occupied in Eastern Wood in each of the 30 years in the period 1949–79 that the wood was censused (we included the incomplete census of 1949 as territories of all species except blue tit and dunnock were counted in this year). For each year, we then selected this many pairs at random from the total British avifauna. This gives a simulated species list for each year, and a simulated abundance for each species. This approach produces a table very much like that in Appendix II, from which measures of abundance, and the descriptive statistics associated with them, can be calculated. We produced and analysed 1000 of these tables. Each has the same total number of territories that were recorded from Eastern Wood in the period 1949–79, and the same number in each year. No other features were constrained.

The species–abundance distributions of the random assemblages drawn from the entire British avifauna share several qualitative features with the avifauna of Eastern Wood (Table 4.4). In particular, they tend to be highly right-skewed, only becoming left-skewed when abundances are averaged over all years (and not just years when present) and are log-transformed. Mean abundance is low in all cases, and less than the standard deviation. The commonest species is always much more abundant than the average. Quantitatively, however, only in the degree of skew do the random assemblages match the Eastern Wood assemblage. The average and maximum abundances attained by species in Eastern Wood are significantly higher. Thinking back to Section 2.2.1, this is hardly surprising. There, we showed that random analogues of Eastern Wood are far more speciose than the real assemblage. Thus, the same total number of territories is being divided between more than twice as many species.

Table 4.4 Results of models simulating the abundances of bird species in Eastern Wood as random draws of individuals from three different species pools: all breeding bird species in Britain, breeding deciduous woodland bird species in Britain, and bird species known to have bred in Eastern Wood in the period 1949–97. Abundance here is number of territories in the wood. Simulated values derive from 1000 iterations of the model. Where real values do not differ significantly from the results of a given simulation, these latter are italicized. See text for further details.

Statistic	Real value	Simulated range based on species pool consisting of:		
		all breeding birds	woodland breeding birds	birds that have bred in Eastern Wood
Mean abundance when present	4.15	2.02–2.30	3.26–3.90	*4.09–4.75*
Standard deviation	5.89	2.80–3.49	4.58–5.42	*5.26–6.05*
Skew	3.25	1.71–3.24	1.49–2.41	*1.65–2.50*
Highest mean abundance when present	33.03	16.77–21.47	20.70–25.70	*22.33–27.8*
Mean log abundance when present	0.39	0.18–0.22	0.30–0.39	*0.36–0.46*
Standard deviation	0.40	0.31–0.37	0.41–0.49	*0.42–0.47*
Skew	1.02	*0.69–1.42*	0.18–0.93	0.32–1.08
Mean abundance over all years	3.73	1.30–1.63	2.66–3.36	*3.50–4.30*
Standard deviation	6.03	3.01–3.57	4.86–5.73	*5.55–6.36*
Skew	3.17	2.18–3.12	1.43–2.29	1.48–2.34
Mean log abundance over all years	0.14	−0.568−−0.331	−0.23–−0.05	*−0.12−0.17*
Standard deviation	0.68	0.84–1.27	0.79–1.26	*0.80–1.02*
Skew	−0.23	*−0.57−0.30*	−0.87−−0.07	−0.85−−0.06

Of course, once again the entire British avifauna is not a sensible species pool from which to draw random assemblages to compare with Eastern Wood. If we make the species pool for the random draw model more realistic by restricting it to those species typical of British deciduous woodland, the simulated values for mean, maximum and variation in abundance are indeed all closer to the real values (Table 4.4). However, in no case do the simulated values encompass the real ones: the model still does not accurately predict the abundances observed in the wood. Moreover, the skew values predicted are less accurate than before. The species richness that this model predicts for the wood is also too high (mean $\pm$ SD = 56.55 ± 1.83, range 50–63, n (number of null assemblages) = 1000). Territories are still being randomly assigned too equably between too many species.

Are the differences between the random and real species–abundance distributions consequences of the difference in species richness alone? We can go some way toward answering this question by assuming that the species pool for Eastern Wood comprises only species that have been observed breeding there. We thus set the pool to be the 45 species that have been observed breeding in the wood in the period 1949–79, plus three species that have been recorded breeding there since (grey heron, collared dove and wood warbler). We add these three other species to ensure that the species richness difference between the random and real assemblages is removed. If we randomly select individuals only from the British populations of the 45 species breeding in the period 1949–79, the richness of simulated assemblages is less than 45 species in the vast majority of cases (99.3%). Thus, there could still be a species richness effect on differences in the abundance distributions of real and simulated assemblages. By contrast, if we set the species pool to 48 species, the mean richness of random assemblages becomes 44.23 species (range 39–48), and the richness of Eastern Wood does not differ significantly from the random expectation ($P = 0.86$). As the three extra species did breed in Eastern Wood, albeit outside the period of interest to us, they can legitimately be viewed as members of its species pool.

With the species richness difference between Eastern Wood and its random analogues removed, we find more similarities between them. In particular, when individuals of the 48 species observed breeding in Eastern Wood are plucked at random from their British populations, the resulting species–abundance distributions have means not significantly different from those observed for the wood for three out of the four methods of calculating the mean (Table 4.4). Nevertheless, most other features of the real species–abundance distributions do differ from this random expectation. Real distributions are usually more skewed to the right, and the maximum abundance is always greater than its random analogue. Although the means of real and random distributions are usually very similar, abundances are usually more equably distributed in the simulations. The species–abundance distributions of real and random assemblages differ for reasons other than their species richness.

To make sense of the patterns observed in the distributions of species abundance, it is necessary now to consider what it is that generates these distributions.

4.3.2   Descriptive models

The first real effort to make sense of the distribution of abundances was made by Fisher *et al.* (1943). They modelled the abundances of species in samples as a logarithmic series, such that if N species were represented in the sample by a single individual, then approximately $N/2$ species would be represented by two individuals, approximately $N/3$ species by three individuals, and so on. This model gave a reasonable fit to some data on the abundances of insect species (for discussions of the technicalities of fitting models to species–abundance data, see Dennis & Patil 1988; Wilson 1991; note, differentiating between particular models is problematical when species numbers or sample sizes are small—e.g. Wilson 1993; Wilson *et al.* 1998).

Despite the work of Fisher *et al.* (1943), interest in species–abundance distributions has largely been stimulated by the classic work of Preston (1948, 1962; see also May 1975). His innovation was to model the distribution of abundances in a sample with species grouped into logarithmic abundance classes. He found that, so grouped, the number of species attaining different abundances approximately followed a normal distribution. Since Preston's first paper on the subject, many species–abundance distributions of this form have been plotted, and with an important caveat (see below), most have a shape that is not wildly inconsistent with the log-normal. We have already presented several examples.

Later, Preston (1962) significantly extended his earlier work by additionally considering the relationship of the species–abundance curve to the number of individuals' curve. He noted that if the number of individuals of every species in each logarithmic abundance class was summed, then the peak of this curve falls approximately at the right-hand limit of the species curve—that is, the highest abundance class contains more individuals than any other. Figure 4.20 shows that this relationship is approximated by British breeding birds. This is interesting because the parameters of these 'canonical' log-normal curves are fixed within narrow limits.

Normally, log-normal curves require three parameters for complete characterization. One defines the location of the curve along the abscissa, a second defines its variance and a third defines its height on the ordinate (or the actual area under the curve). Typically, these parameters are given by the mean, variance and sample size, respectively. (Note that mean and variance alone are

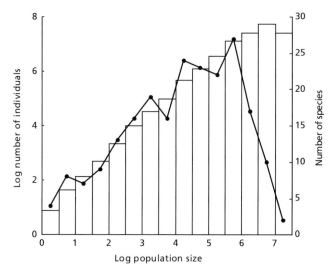

Fig. 4.20 The summed number of individuals (bars) for species of bird breeding in Britain at different population sizes. Thus, the number of individuals of species with a population size of fewer than antilog 0.5 (i.e. < 3 individuals) totalled 8. The number of species in each population size class is indicated by the filled circles.

sufficient to define a normal probability density function, where the total area under the curve sums to 1, and hence the height of the curve is relative. However, when we are dealing with real entities, such as number of species, then we additionally require information on sample size to tell us how high the curve is in absolute terms.) Canonical curves, by contrast, require one less parameter. Here, the standard deviation is a function of the sample size (number of species in this case). This means that knowledge of the total number of species in the sample specifies the shape, as well as the absolute height, of a canonical log-normal species–abundance distribution (Preston 1962; May 1975). Additional information on the number of individuals is still required to fix the absolute position of the curve along the abscissa. However, May (1975) notes that for most ecological purposes, this information can be circumvented by use of the dimensionless ratio $J = N_T/m$, where N_T is the total number of individuals, and m is the number of individuals of the least abundant species. The value of J is also specified by the total number of species in the sample (May 1975).

The conclusion that most species–abundance distributions are approximately log-normal does carry one significant caveat, however. The distributions plotted are usually based on a small sample of the entire underlying universe of individual animals. Thus, depending on the size of the sample, very rare species—those forming the left-hand tail of the log-normal curve—will theoretically be represented in the sample by only a fraction of an individual. These species will most likely be missed from the sample. Thus, the species–

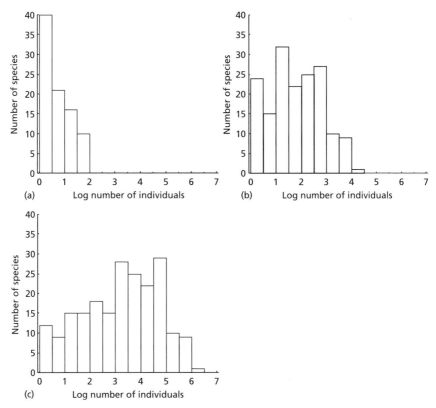

Fig. 4.21 The effect of sample size on the species–abundance distribution. Frequency distributions of the number of species with different population sizes were produced by sampling individual birds at random from the breeding avifauna of Britain. As the sample size increases from (a) 1000 to (b) 100 000 to (c) 10 000 000 individuals of the approximately 125 000 000 breeding birds in Britain, the distribution becomes increasingly unveiled, and more similar to the true distribution (Fig. 4.18b). From data for the breeding avifauna of Britain in Appendix III.

abundance distribution based on the sample will be 'veiled'; that is, part of the left hand of the distribution will be absent (Preston 1948). If the sample is very small, even the mode of the distribution may not be apparent. As the sample size increases, the position of this veil line shifts further left, until the sample size equals the total number of individuals in the sample universe, at which point the distribution is completely unveiled. An example of this unveiling process is shown in Fig. 4.21 for the breeding birds of Britain.

This means that we can only know the true species–abundance distribution when we know the entire sampling universe. For large-scale assemblages, such data are scarce. However, some attempts to estimate the entire sampling universe do exist. The birds of Britain provide one example (Fig. 4.18; Nee *et al.* 1991b; Gregory 1994), and other such exercises have been conducted for the national avifaunas of Sweden (Järvinen & Ulfstrand 1980), the Czech Republic and Slovakia (Fig. 4.19), all European countries (Gregory 2000), and the assemblage of wildfowl of the world (Gaston & Blackburn 1996c). As already seen, there is evidence that the species–abundance distributions of many of these assemblages are log left-skewed. Thus, left-skew seems to be an important feature of species–abundance distributions, as it is posited also to be such for species–range size distributions (Section 3.2). If so, the Prestonian view of these distributions would need some adjustment.

Elegant as Preston's work was, it was nevertheless purely empirical (May 1975): Preston described the relationship between numbers of individuals and species, but proposed no mechanistic basis for it. The same is true of the earlier treatment by Fisher *et al.* (1943), and of many subsequent attempts to fit models of species–abundance distributions. These approaches view such distributions in terms of statistical processes. For example, since log-normal distributions can be explained in terms of the action of multiplicative factors and the central limit theorem, log-normal species–abundance distributions are interpreted as a consequence of species responding independently to factors influencing exponential population growth (MacArthur 1960; May 1975; Gotelli & Graves 1996).

Although statistical interpretations of species–abundance distributions are attractive in their simplicity and generality, they are ultimately unsatisfactory for two reasons. First, the interpretation of interspecific log-normal species–abundance distributions is flawed (Pielou 1975). The theory on which it is based assumes that the action of multiplicative factors and the central limit theorem apply to repeated and independent samples of the same variate. Thus, while it is reasonable to expect that populations of the same species will be random variates from a log-normal distribution, the same does not follow for populations of different species (Pielou 1975).

Second, although the proposed models may fit observed distributions well, they say little about any underlying biological forces that might generate them. Indeed, it can be assumed that because a log-normal abundance distribution can be generated by a set of factors acting independently on population growth, it does not matter what those factors are. Yet, different taxonomic assemblages exhibit different distributions. For example, while British birds and mammals both inhabit the same environment, their abundances are not sampled from the same underlying distribution (using data from Appendix III for birds and Harris *et al.* (1995) for mammals, with mammal species included following criteria defined by Greenwood *et al.* (1996); ANOVA: $F_{1,265} = 18.7$, $P < 0.0001$). This suggests a difference in the response of different taxonomic groups to the environment but a consistent response within the taxa, and hence that explanations for patterns in species–abundance distributions should be sought in the biologies of the species involved (Sugihara 1980). An alternative approach therefore has been to model species–abundance distributions in terms of factors that can be equated to real biological processes.

4.3.3   *Mechanistic models based on niche partitioning*

Probably the most influential mechanistic model for species–abundance distributions was presented by Sugihara (1980), based on models of niche apportionment by MacArthur (1957, 1960) and Pielou (1975).

MacArthur suggested that the resources available to a community could be thought of as a stick. He then modelled the resources obtained by n different species in the community as the fragments derived from simultaneous breakage of this stick into n different pieces. The frequency distribution of stick lengths will give the frequency distribution of resources, and hence of abundances if it is assumed that resources and abundances can be directly equated. Unfortunately, the distribution of stick fragment sizes so derived is more equable than it should be to model a log-normal.

Pielou (1975) suggested that a log-normal distribution of stick fragments would be obtained if breakage was a sequential process. Here, the first two species in an assemblage divide up the resource stick between them. The next species to invade the community can be imagined to pre-empt some of the resources utilized by one or other of the two original species, which results in the fragment of the stick belonging to this original species being subdivided. This sequential process continues as more species invade the community. If breakage is random, the larger resultant fragment will constitute 50–100% of the original. Sugihara (1980) modelled this by assuming that the larger fragment would lie at the mid-point of this range, and so constitute 75% of the original piece (see also Nee *et al.* 1991b; Tokeshi 1996). An important point about sequential breakage is that the probability that a stick is chosen for subdivision with each invasion is independent of stick size. This is a key difference from MacArthur's model, which can also be derived from a sequential breakage process, but one in which the probability that a fragment is chosen for subsequent breakage is proportional to its length (Tokeshi 1990).

Sugihara's 'sequential broken stick' model generated random assemblages with log-normal distributions of resource fragments, for which the relationship between the number of species in the distribution and its standard deviation is in reasonable agreement with that predicted by Preston's canonical hypothesis and with data from real animal and plant assemblages (Sugihara 1980). Moreover, the model is not necessarily falsified by the unexpected tendency of completely unveiled distributions to be left-skewed. Nee *et al.* (1991b) showed that the degree of left-skew in the British bird species–abundance distribution was not significantly different from that observed in 1000 random assemblages of equivalent sample size generated using the sequential broken stick model. Although the distribution of skew values from their model assemblages embraced the log-normal expectation (i.e. zero skew), log left-skew was twice as common as log right-skew (Nee *et al.* 1991b). If the sequential broken stick model was correct, we would expect left-skew to be more common in real assemblages, as appears to be the case.

Despite its quantitative fit to real data, Sugihara's model has been criticized from a number of perspectives. One fundamental issue concerns the relationship between resources and individuals (Harvey & Godfray 1987; Sugihara 1989; Pagel *et al.* 1991b; Taper & Marquet 1996). The sequential broken stick model generates a canonical log-normal distribution of resources, whereas the distribution it is modelling is for abundances. Sugihara's model assumes correspondence between these distributions. However, Harvey and Godfray (1987) argued that such correspondence is probably unlikely. In particular, per capita resource use and species abundance are unlikely to be independent, and both should be related to species body mass. Large-bodied species are likely to be rarer, on average, than small-bodied species, but to have higher per capita resource requirements (Chapter 5). Therefore, species in the left-hand tail of the species–abundance distribution should use more energy, and hence resources, in total than expected were per capita resource use independent of abundance, while species in the right-hand tail of the distribution should use less. If so, population energy use would be more equably divided among species than abundances, and a model generating canonicity in the distribution of the former would not result in canonicity in the distribution of the latter.

Whether this argument is correct depends on two key factors. The first is the precise nature of the relationships between abundance, per capita resource use and body mass. The relationship between the latter two variables is reasonably well established, but, as we will see in the next chapter, the relationship between abundance and body mass is a matter for considerable debate. Depending on the specific slope of this second relationship, resource use may be more or less equably divided among species than abundances, or equally so (Pagel *et al.* 1991b).

The second key factor is the form of the causal model by which abundance, body mass and per capita energy use combine to generate population energy use (Taper & Marquet 1996). Although the statistical relationships between the variables can be measured, the causal links that generate them are unknown. As is frequently noted in textbooks, correlation of itself tells us nothing about causation. The relationship between per capita resource use, abundance and body mass could potentially develop along a number of different pathways (Fig. 4.22). For example, body mass may determine per capita energy use and abundance, the product of which is population energy use. Taper and Marquet (1996) call this the *m*-causal model. Alternatively, per capita energy use may determine body mass, which then determines abundance (*p*-causal model). Taper and Marquet point out that previous treatments of the question (e.g. Pagel *et al.* 1991b) implicitly assume a third model, the *n*-causal model: abundance determines body mass, which then determines per capita energy use and hence population energy use. This point is important because the estimated

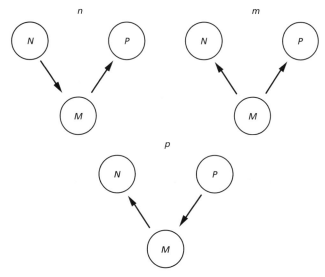

Fig. 4.22 Graphic representation of Taper and Marquet's (1996) *n*-, *m*- and *p*-causal models for the relationship between per capita resource use (*P*), abundance (*N*) and body mass (*M*). The arrows indicate the direction of causality. Modified from Taper and Marquet (1996), with permission from University of Chicago Press.

variance in the distribution of population energy use depends on how body mass, per capita energy use and abundance are combined in the causal model. Different causal models result in different variance estimates. Because the canonical hypothesis assumes that variance in population energy (resource) use and abundance should be equal, whether or not the hypothesis is validated by comparison of these variances clearly will depend on how the former is calculated. Taper and Marquet find that variance estimates for population energy use in 41 bird assemblages match those for abundance if the *m*- or *p*-causal models, but not the *n*-causal model, are followed (Fig. 4.23). They also note that only the *n*-causal model is sensitive to variance in the slope of the abundance–body mass relationship. Contrary to the previous paragraph, the form of that relationship may be largely irrelevant to this whole issue.

In sum, arguments about the relationship between the distribution of resources and individuals among species do not necessarily refute Sugihara's sequential broken stick model. However, firmer statements will have to wait for more evidence as to which causal model provides the true description of the real world. Taper and Marquet (1996) suggest that the *m*-causal model seems most appropriate. We agree. Interestingly, though, estimates of variance in population energy use for the 41 avian assemblages generated by the *m*-causal model are almost always (39/41) greater than those for abundance (Fig. 4.23). This implies that population energy use is consistently less equably divided among species than is abundance. If the *m*-causal model is correct, Sugihara's model may well be wrong.

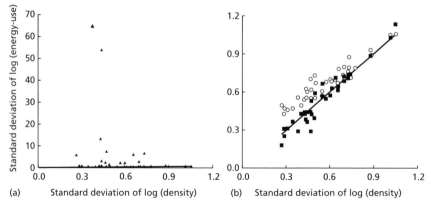

Fig. 4.23 The relationship between the standard deviation of logarithmically transformed energy use and the standard deviation of logarithmically transformed population density for 41 local bird communities. (a) Filled triangles are estimates of standard deviation in population energy use calculated under the n-causal model; (b) circles are estimates of standard deviation in population energy use calculated under the m-causal model, and filled squares are estimates of standard deviation in population energy use calculated under the p-causal model. From Taper and Marquet (1996, with permission from University of Chicago Press).

Aside from the question of whether the distribution of individuals equates to that of resources, the sequential broken stick model has mainly been criticized for the model of resource division on which it is based. Pielou (1977b) questioned whether the division of multidimensional niche space can actually be modelled by breakage of a one-dimensional resource 'stick'. Even if it can, there is then the issue of the process by which the resource 'stick' is fragmented. This process has two components, concerning which fragment is chosen for division, and the sizes of the fragments resulting from that division. Taking the second of these concerns first, Sugihara (1980) assumed that the larger fragments resulting from breakages would average 0.75 of the total fragment length prior to breakage, and that overall there would be a triangular probability distribution of larger fragment sizes that peaked at 0.75. However, he simulated this by assuming that breakages always result in two fragments that are 0.75 and 0.25 the length of the original fragment. As Tokeshi (1996) points out, this simulation does not give the same answer as the triangular probability distribution of larger fragment sizes. Moreover, it is not clear that the random division of stick fragments should result in larger fragment sizes that cluster around 0.75—a uniform distribution of larger fragment sizes between 0.5 and 1.0 seems more logical.

Before the outcome of an individual fragmentation is decided, though, it is necessary to select one of the existing pieces of stick to fragment. The selection process may follow a number of paths. The sequential broken stick model assumes that fragments are chosen independently of their length. Thus, species entering a community are as likely to pre-empt the resources of a rare species as a common one. Arguably, however, species are more likely to invade a more abundant resource base, and so pre-empt resources of abundant species with higher probability. An analogous situation may relate to the probability that species of different abundances or distributions will speciate (Tokeshi 1996), with this perhaps less likely for rarer species (Section 3.3.6).

Other niche apportionment models allow for inequality in the likelihood that a species niche will be invaded. MacArthur's simultaneous stick breakage model is equivalent to a sequential model where the probability that a fragment is chosen for breakage is proportional to its length (Tokeshi 1990). Tokeshi (1996) proposes a third model where selection probability is proportional to stick length raised to an exponent k. Here, $k = 0$ corresponds to Sugihara's model (but with a uniform, not triangular, distribution of larger fragment sizes), while $k = 1$ corresponds to MacArthur's model. Tokeshi (1996) found that the species–abundance distribution of most real communities can be modelled by k in the range 0–0.2. Distributions produced by simulations with k in this range exhibit realistic levels of left-skew. This 'power fraction' model can be thought of as a niche apportionment model where the probability that the niche of an abundant species gets invaded is slightly higher than for a rare species, which seems reasonable. Varying k allows virtually any species–abundance distribution to be fitted by the model. However, this flexibility is as much a weakness as a strength: the generality of the model is somewhat compromised.

Finally, two additional points can be noted in relation to niche apportionment models. First, the various models discussed up to this point all assume that the resource stick is completely allocated. The invasion of an assemblage by an additional species must consequently result in a decline in the abundance of one of the species already present. Thus, the abundances of species are mutually related. This is not necessarily the case in nature. In particular, there is a considerable body of evidence suggesting that communities often are not saturated with species (Section 2.4, e.g. Lawton 1982; Cornell 1985a; Ricklefs 1987; Lawton *et al.* 1993; Cornell & Karlson 1996; Pärtel *et al.* 1996; Willson & Comet 1996; Caley & Schluter 1997; Griffiths 1997; Hugueny *et al.* 1997; Srivastava 1999), that the abundances of some members of a community can increase without apparent negative effects on the abundances of others (e.g. von Haartmann 1971; Perrins 1979; Lawton 1982; East & Perrins 1988; Gustafsson 1988) and that all communities are more or less equally invasible (when propagule pressure and the availability of suitable species are accounted for; Williamson 1996). This situation was modelled by Tokeshi (1990), who assumed that the abundance of the commonest species in the community was 1, and that the abundances of all other species took random values constrained to be less than those of all species of higher abundance rank. This model provided a good fit to data on the abundances of species in communities of chironomids (Tokeshi 1990). However, if, as Nee *et al.* (1991b) claim, it generally predicts a log-uniform distribution of abundances, it seems unlikely to be of general application.

Second, as for abundance–range size relationships, it has proven difficult to provide convincing evidence that the abundances of species are proportional to a quantitative definition of their niches (Section 4.2.2). Certainly, species with broader niches are not necessarily more abundant. For British birds, there is evidence that more abundant species are those utilizing resources more typical of the environment (Gregory & Gaston 2000). Although this distinction has not been explicitly discussed in treatments of niche apportionment, if such processes are determining the form of species–abundance distributions, they are presumably acting through variation in niche position, rather than niche breadth.

4.3.4   *Other mechanistic approaches*

Thus far, we have only considered models of the species–abundance distribution based on niche apportionment. This bias merely reflects that of the literature, where these models have dominated debate about the causes of such distributions. They are not the only models that exist. Among the alternatives, recent interesting suggestions are that species–abundance distributions derive from self-similarity (Harte & Kinzig 1997; Harte *et al.* 1999) and from the incorporation of speciation into island biogeography theory (Hubbell 1997).

The idea that species–abundance distributions could be modelled as a self-similar process was first considered by Williamson and Lawton (1991). They noted that the division of niche space among species in an assemblage was conceptually identical to a fractal tiling process. Some shapes of tile, such as squares or hexagons, will cover an area completely leaving no gaps (tessellate). Other shapes, such as pentagons, do not tessellate. If an area is covered as far as possible with a shape of this type, the gaps left are covered as far as possible with smaller such shapes, and so on *ad infinitum*: the result is fractal tiling. The proportion of total space covered by tiles of each size can be equated to the proportion of niche space pre-empted by each species in an assemblage. Williamson and Lawton (1991) showed that species rank–abundance curves were quantitatively similar to curves produced by fractal tiling processes, and concluded that it may be possible to model resource division by species as a fractal process.

The approach taken by Harte *et al.* (Harte & Kinzig 1997; Harte *et al.* 1999) differs from that of Williamson and Lawton (1991) by modelling self-similarity in the probability distribution of species across ecosystems. The self-similarity criterion can be illustrated with reference to a golden rectangle (length to width ratio = $\sqrt{2}$) of area, A_0. This can be divided in half perpendicular to the long dimension to form two equal areas, A_1, the same shape but half the size of the

original. (This shape-preserving characteristic of golden rectangles is important for certain predictions of this model, but will be of no further concern
here.) Each A_1 area can be similarly divided to give two A_2 areas, and so on.
The area of the rectangles formed by the ith division is $A_0/2^i$. If area A_0 has an
associated number of species, S_0, then some fraction of those species will be
found in each A_1 after division. It is this fraction that the model assumes to
show self-similarity. The probability that any species present in A_i will be present in at least one A_{i+1} is given by the constant a. This constant is independent
of i, and of which A_{i+1} is chosen. a must fall between 0.5 and 1, because these
values assume that no or all species, respectively, occupy more than one of the
A_{i+1}. From these simple relationships, it follows that the number of species S_i
found in any particular rectangle A_i must equal $a^i S_0$.

To produce a species–abundance distribution from the self-similarity criterion, some further quantities need to be derived. First, the probability that a
species is found in only one of two A_{i+1} rectangles is $1 - a$, or 1 minus the probability that it exists in at least the other A_{i+1} rectangle. Then, the probability that
a species is found in both halves is 1 minus the probability that a species is only
found in the left-hand rectangle, minus the probability that a species is only
found in the right-hand rectangle: $(1 - 2(1 - a))$, or $2a - 1$. Finally, the smallest
rectangle possible, A_m, is defined as that containing a single individual. Since,
from above, $A_m = A_0/2^m$, there must be 2^m individuals in A_0. Using these quantities, it is possible to derive the species–abundance distribution, or the proportion of species in a patch A_0 that contain n individuals, $P_0(n)$, for each value of
n. This can be done recursively using the $P_i(n)$ for $0 < i \leq m$. Harte et al. (1999)
demonstrate this process using the example of $P_4(3)$, the probability that a
species is present in A_4 with three individuals. This is the sum of the following
probabilities: the probability that it is present in either the left-hand or right-
hand A_5 only ($1 - a$ in each case), and the probability that it is present in both
left- and right-hand A_5 areas, with either two individuals in the left-hand area
or two in the right ($2a - 1$ in each case). Thus:

$$P_4(3) = 2(1 - a) P_5(3) + (2a - 1) [P_5(2)P_5(1) + P_5(1)P_5(2)] \qquad \text{(Eqn 4.1)}$$

The values for $P_5(n)$ can be obtained from $P_6(n)$, and so on. Denoting $2(1 - a)$ as
x, the equation above generalizes to:

$$P_i(n) = xP_{i+1}(n) + (1 + x) \sum_{k=1}^{n-1} P_{i+1}(n - k)P_{i+1}(k) \qquad \text{(Eqn 4.2)}$$

Harte et al. (1999) show numerically that this equation produces species–
abundance distributions that have more rare species than expected from a
log-normal. In other words, the theoretical distribution is log left-skewed.
As we saw above, left-skew is a feature of the species–abundance distribution of British birds, and also of most other well-characterized large-scale

assemblages (Section 4.3.1). That this self-similarity model predicts left-skew is not, however, its only quality. There are three others that are of relevance here (see also Harte & Kinzig 1997; Harte *et al.* 1999).

First, the self-similarity model also predicts the power form of the species–area relationship (Harte & Kinzig 1997; Section 2.2). Thus, it provides a firm link between the species–abundance distribution and the species–area relationship. The species–area exponent z is related simply to x in the species–abundance distribution as $z = -\ln_2(1 - (x/2))$. A connection between these two relationships was first suggested by Preston (1962), who claimed that the power form of the species–area relationship was a consequence of the log-normal species–abundance distribution. However, this association has been brought into question by recent attempts to simulate the species–area relationship from a log-normal distribution of abundances, which showed that the latter does not generate the correct shape of the former (Leitner & Rosenzweig 1997). In fact, the self-similarity model as outlined above demonstrates that a log-normal species–abundance distribution cannot generate the power form of the species–area relationship, because this form of species–area relationship is formally associated with a log left-skewed abundance distribution.

Second, the self-similarity model makes explicit predictions about patterns of species spatial turnover. In particular, it predicts that the fraction of species common to two areas is inversely proportional to the square of the distance between them, raised to the power z (Harte & Kinzig 1997). By contrast, models of species distributions across regions based on the log-normal make no predictions of faunal turnover. As we noted in Chapter 3, studies of large-scale patterns of turnover are somewhat scarce. Birds, at least, do show increased β-diversity, and hence reduced faunal similarity, on both longitudinal and latitudinal transects across Britain (Harrison *et al.* 1992). The form of this turnover is such that the fraction of British bird species common to two areas does indeed increase as the inverse of the square of the distance between them raised to the power z (Fig. 4.24). However, the relationship is curvilinear for these data, not linear as predicted.

Third, the self-similarity criterion allows prediction of the slope of the relationship between abundance and range size in British birds (J. Harte & T.M. Blackburn, in preparation). The prediction is in close agreement with the slope derived from the data for breeding birds in Appendix III.

Thus, the self-similarity model is interesting not only because it predicts log left-skewed species–abundance distributions, but also because it provides a unified mechanism generating several large-scale ecological patterns. Moreover, a recent study has provided evidence that birds in part of the Iberian peninsula do indeed show an approximately self-similar distribution, at least over scales in the range 0.2–10 000 km² (Finlayson 1999). Nonetheless, the model cannot be accepted uncritically. The self-similarity assumption implies that the probability that a species occurs in one half of an area is the

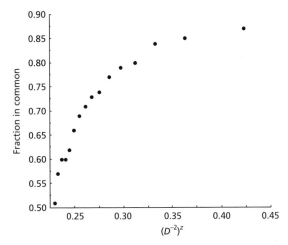

Fig. 4.24 The relationship between the fraction of bird species in common to pairs of consecutive 50 × 50-km quadrats on a transect south–north across Britain, and the inverse of the square of the distance (D) between the centres of the squares raised to the power z. z is the exponent of the power form of the species–area relationship, estimated to be 0.11 using data for bird species breeding in Britain (Fig. 2.13). The fraction of species in common is calculated as the average value for all pairs of quadrats in the same distance class; therefore, sample size increases from 1 for the left-hand data point to 16 for the right-hand point. The relationship is significantly curvilinear (second-order polynomial regression: $r^2 = 0.96$, $n = 16$, $P(x) < 0.0001$, $P(x^2) < 0.0001$). From data also used in Fig. 3.11.

same as the probability that it occurs in each half of that half. Clearly, even if self-similarity applies generally in the first place, the range of scales over which it does so will be limited (e.g. Rosenzweig 1995; Kunin 1998). In particular, it is likely to break down at very small scales, such as that of individual territories in Eastern Wood. Nevertheless, the manner of this breakdown has predictable consequences for the precise shape of the species–abundance distribution under the model (Harte *et al.* 1999). If the breakdown is influenced by intraspecific attraction at small spatial scales (e.g. coloniality), decreased left-skew in the log-transformed distribution will result. Intraspecific repulsion (e.g. strong territoriality) will lead to the reverse. This suggests that territoriality in the birds of Eastern Wood will influence the degree of left-skew in their species–abundance distribution.

All ecological models caricature nature to a greater or lesser degree. The self-similarity model makes quantitative predictions. It is on the outcome of tests of these that it will stand or fall. If it stands, the next step will be to determine what it is that generates self-similarity.

A different, but equally interesting, approach to the question of species abundances is that taken by Hubbell (1997, in press). Hubbell's 'neutral' model distinguishes between processes acting at two distinct scales: those of the local community, and those of the metacommunity of which this is a part.

Thus, it makes a similar separation to the one we have pursued throughout this book between the avifauna of Eastern Wood and the wider regional bird assemblage.

The neutral model first assumes that local communities are saturated with individuals, such that an individual must die before a new individual can enter the community. For each species in the metacommunity, it is then possible to calculate the probability that that species increases, decreases or maintains its abundance in the local community, based on its local abundance, its metacommunity abundance and its probability of immigrating from the surrounding metacommunity. The mean local abundance of a species depends on its abundance in the source area: regionally abundant species tend also to be abundant locally. At high immigration rates, the local abundance of a species is unimodally distributed about this mean, but as immigration rate declines, its probability density function of abundance becomes U-shaped, such that the species tends either to be locally monodominant or locally extinct.

At the metacommunity level, processes of speciation and extinction generate an equilibrium species richness and species–abundance distribution. Both these patterns depend on a dimensionless parameter θ, which is the product of the total metacommunity size and the speciation rate. When θ is high (e.g. high speciation rate for a given metacommunity size), metacommunities have low dominance and many rare species. Samples from this equilibrium metacommunity give local species–abundance curves that fit with observed data, once the effect of dispersal is taken into account.

The shape of the metacommunity species–abundance distribution depends on the model of speciation (Hubbell, in press). With point speciation (species arise as one mutant individual), a log-series distribution results. With a fission model (extant species split into two more or less equally abundant daughter species), a more log left-skewed distribution is obtained. With no limitation on dispersal, local community species–abundance distributions reflect that of the metacommunity. However, the less likely species are to immigrate to local communities from the surrounding metacommunity, the more log left-skewed local community species–abundance distributions become, regardless of the metacommunity distribution. This is because species becoming extinct in the local community are less likely to be restored by immigration, and so the local community contains a lower proportion of very rare species than the metacommunity. Because the model assumes saturation, the 'missing' individuals from rare species are compensated for by higher local abundance in the commoner species: common species will become more common because most new individuals in the local community are of local origin. Thus, dispersal limitation causes the species–abundance distribution of local assemblages to differ from that of the region, such that common species are more common locally than expected, rare species are too rare, and the variance in abundance in the local community is higher than expected. This is exactly the pattern we see when comparing the avifauna of Eastern Wood to random samples from the British avifauna (Table 4.4).

The neutral model also predicts the species–area relationship. Because the species richness of communities is determined by θ, which is the product of the total metacommunity size and speciation rate, then as the number of individuals is proportional to area, so too will be the number of species. The slope of the relationship will be greater at metacommunity than local community scales because dispersal limitation is more important at larger scales. Increasing area adds proportionally more species over scales greater than those across which species typically disperse. Thus, like the self-similarity model of Harte and coworkers (Harte & Kinzig 1997; Harte et al. 1999), Hubbell's neutral model links species–abundance distributions and species–area relationships.

The neutral model implies a dynamic coupling between local and metacommunities. Dispersal links local community structure to regional dynamics, but also stabilizes regional metacommunity composition. Dominance in abundance arises by chance, rather than through competition, and there are no assumptions about niche differences or division. These features all fit with a view of ecological systems where the structure of local assemblages depends on the regional assemblage in which they are embedded. A less appealing feature of the model from our perspective is the assumption that assemblages are saturated in terms of number of individuals. As we will see in the next chapter (Section 5.5.3), this is unlikely to apply to bird (and perhaps many other) communities. It is telling that the neutral model has so far largely been tested on data for tropical forests, where saturation of this sort is perhaps most likely to apply. Nevertheless, the model makes clear and reasonable predictions, in particular about the effects of dispersal limitation on the structure of local communities, and their similarity to the metacommunity of which they are a part. Tests of these should prove illuminating.

Finally, since Hubbell's neutral model and Harte's self-similarity model both make such clear predictions about the structure of assemblages, it begs the question of whether the two approaches may be complementary. Given the current fledgling state of theoretical development, that remains unclear. It seems to us to be an exciting avenue for exploration.

4.3.5   *Synthesis: abundance, range size and their distributions*

The positive interspecific relationship between abundance and range size suggests that the frequency distributions of the two variables should show a degree of commonality. This does seem to be the case at large spatial scales, where both distributions tend to be highly right-skewed when data are untransformed, but slightly left-skewed once the data have been logarithmically transformed. This commonality, together with the positive abundance–range size relationship, suggests that all three patterns might plausibly result from a common cause. Yet, consideration of the causes of each has largely developed along independent lines (compare Sections 4.3.3 and 4.3.4 with Table 3.1 and 4.1). Thus, metapopulation dynamics have been invoked to explain the species–range size distribution and the abundance–range size relationship, but not the species–abundance distribution. Self-similarity has been invoked to explain the species–abundance distribution and the abundance–range size relationship, but not (yet) the species–range size distribution. The principal attempt to understand species–range size distributions has concentrated upon patterns of speciation and extinction, yet these processes have been largely overlooked as determinants of species–abundance distributions and abundance–range size relationships (although see Tokeshi 1996; Hubbell 1997; Warren & Gaston 1997).

The only factor that has consistently been proposed as a cause for all three patterns is variation in the niches of species. Ironically, though, niche partitioning models of the sort explored in depth for species–abundance distributions have been dismissed as a cause of species–range size distributions, because species share space rather than apportioning it. In fact, whether this dismissal is correct is unclear. On the one hand, it is niche space, not real space, that is modelled by niche partition: species can share the latter without sharing the former. Species with higher abundances tend to use more space, and so it is plausible that those factors determining abundance might also help determine range size. On the other hand, this argument seems to imply that niche partitioning determines abundances, and that range sizes follow from abundances as a by-product of the distribution of individuals across the environment. Yet, it is clear from studies of abundance–range size relationships that range size is not a simple statistical consequence of abundance (Section 4.2.2), which suggests that an additional mechanism would be needed to translate species abundances generated by niche partition into realistic range sizes. This is clearly an issue that requires more detailed attention. Although the close linkage between abundance and range size suggests to us that a unified approach to investigating the causes of their association and their separate distributions would be instructive, it is unlikely that any one pattern will turn out to be a simple linear consequence of any other.

4.4 Summary

We opened this book by reflecting on the lack of excitement offered by the avifauna of Eastern Wood to the typical British birdwatcher. This observation was made not to deny the intrinsic beauty and fascination of birds, but rather to indicate that a morning spent watching birds at the site involves a good degree of repetition. Relative to many sites around the world, the avifauna of Eastern Wood consists of few species, and most individuals encountered are from a small subset of the few that do breed there. Indeed, the results of random draw models of this assemblage (Table 4.4) show that individuals are less equably divided among breeding species than expected given the overall abundances of species in the wider environment. Selecting individuals at random from a pool of those species known to have bred at Eastern Wood produces random assemblages that share many similarities with the real one, but even so the maximum abundance of no random species was ever as high as that actually observed for the robin.

A large-scale perspective helps us to place the abundance structure of the avifauna of Eastern Wood in a broader context. First, the skewed distribution of local abundances reflects to some degree a skewed distribution of regional abundances. Most species in the environment are rare, but a few are common. This distribution influences the probability that a species is sampled by a local assemblage in a manner directly analogous to that described for the effect of range size (Section 3.1): species rare in terms of either variable, which we have also seen will tend to be rare in terms of both, are less likely to occur at suitable sites. The effect of abundance is not simply a direct consequence of that for range size, however, because a species range size is not a simple statistical consequence of its abundance (Section 4.2.2). Instead, some additional factor must influence both abundance and range size to produce the association between them. Those that have been suggested encompass processes operating on a range of scales, from effects of resource use by individuals or species to dispersal among patches in a metapopulation and the juxtaposition of entire geographical ranges.

While providing a first step, the regional distribution of abundances alone is not sufficient to explain the distribution in Eastern Wood. However, links between the two are potentially provided by additional large-scale processes. For example, Hubbell's (1997) neutral model explains the differences observed between the species abundance of Eastern Wood and random samples from the wider regional pool as a consequence of dispersal limitation in the species involved. The shape of local species–abundance distributions will be influenced by the tendency of small populations at local sites to go extinct. If dispersal fails to re-establish the species, then the resources not used by it are more likely to be pre-empted by individuals of species already established at the site, and the commonest of those species in particular. This leads to local assemblages with common species of higher than expected abundance (e.g. Table 4.4), and rare species that relatively are too rare. The magnitude of this effect depends on the extent to which dispersal is limited across the whole landscape. Gradients in the distribution of bird species across the British landscape mean that, even in this vagile group, not all species are likely to colonize all possible sites, potentially generating the mismatch between the random and Eastern Wood assemblages. This scenario also fits with the closer matches obtained when the regional pool is successively limited to species more and more likely to be able to colonize Eastern Wood. Whether the tendency for the species–abundance distribution of the Eastern Wood avifauna to be less left-skewed than expected fits with this model is less clear, however.

A link between local and regional species–abundance distributions may also be provided by Harte's self-similarity model (Harte & Kinzig 1997; Harte *et al.* 1999). If self-similarity breaks down at scales equivalent to local sites such as Eastern Wood, and this breakdown is a consequence of intraspecific attraction (e.g. clumping of territories), then distributions more right-skewed than expected can result. This would imply that the skew observed in the species–abundance distribution of the wood is a consequence of intraspecific attraction: species are more likely to occupy areas in the wood close to conspecifics. If there were any systematic habitat variation within the wood, this would not be an unreasonable expectation.

Variation in species abundance is linked intimately with variation in the size of the range of a species, and both affect the composition of the avifauna of Eastern Wood. Species breeding there tend to be more abundant and more widespread than British bird or woodland bird species in general. Presumably, as argued in the previous chapter, regional abundance affects the probability that a species is sampled by a local site. The abundances attained by many species in Eastern Wood are nonetheless much lower than expected on the basis of their distribution, reflecting the skewed intraspecific distribution of abundances—the tendency for most species to be relatively rare at a high proportion of sites that they occupy, even when their average and maximum densities are high (Section 4.2.1). What causes this pattern is still unknown.

What is the direction of causality in the relationship between abundance and range size? Although abundance is usually plotted as the dependent variable (e.g. Figs 4.3, 4.4 & 4.6), it seems much more likely to us that local abundance primarily influences range size. As the local abundance of a species increases, its individuals are likely to disperse away from occupied sites, to colonize new sites and increase the species range. More widely distributed, abundant species are then more likely to be sampled by other local sites. If this characterization is correct, the abundance–range size relationship suggests a feedback loop from local to regional effects, reinforcing the interdependence of regional faunas and local sites like Eastern Wood.

5 Body Size

It seems not to be too much to say, too, that knowledge of a law of interspecific [body size] variation will, by itself, be of theoretical significance; and that, in divers, perhaps partly hitherto unsuspected ways, it is likely to be of directory significance in evolutionary speculations. [Hemmingsen 1934]

5.1 Introduction

The previous chapter addressed large-scale patterns in the abundances of species. There is considerable variation in abundances in a region, which relates to the extent of the ranges species attain. There is also some evidence that abundances tend, on average, to increase with the latitude at which species occur (Currie & Fritz 1993; Gaston & Blackburn 1996c; Johnson 1998b). If this last pattern were a general one, it would run counter to that normally exhibited by species number. As is generally appreciated, the tropics tend to hold more species in most major taxonomic groups than extratropical regions (Section 2.5). The opposition of these gradients has potentially interesting consequences for patterns in the distribution of life across the planet.

The absolute amount of life cannot exceed that which can be supported by the harnessing of all energy arriving from the sun (Section 2.5.3). Partly for this reason, the amount of solar radiation, or its surrogates, has frequently been cited as a primary determinant of species richness (e.g. Wright 1983). Deriving species numbers directly from energy availability, however, ignores an intermediate stage. Effectively, energy is first converted to biomass, or to some number of individuals. It is the way in which this biomass is apportioned into species that then determines the level of species richness (Section 2.5.3; Blackburn & Gaston 1996c).

If the richness of a region is determined first by how much living material the area can support, and then by how that material is apportioned into individuals and species, two features of species attain fundamental importance. The first is the amount of material appropriated by each individual. This depends in large part on how individuals use energy. It has been widely reported that the metabolic rate of species increases with their body mass according to a power relationship of exponent 0.75 (e.g. Kleiber 1962; Peters

1983; Calder 1984; Reiss 1989). Thus, a large-bodied species requires more energy in total than a small-bodied species to fuel its metabolism, although it requires less energy per gram. In short, the amount of energy appropriated by a species depends on its body size.

The second feature of species that is of importance, if species richness is determined by how living material is apportioned into individuals and species, is the number of individuals belonging to each species, or the species abundance. In a world of finite resources, it follows that there will be an interaction between species richness, the abundances of those species and their body sizes. We have already considered large-scale patterns in species richness and abundance. This chapter turns to the third of these interacting variables, and considers large-scale patterns in species body sizes.

We start by returning once again to Eastern Wood and its bird assemblage (Appendix II). One of the most striking features of that assemblage as we encountered it on our morning in the wood was that most of the species were rather small bodied (Chapter 1). This observation is interesting for two quite different reasons. On the one hand, it reveals something about our perceptions. For whatever reason, the size of an object is one of the principal ways by which we classify and judge it. Animal assemblages and their constituents are no exceptions. As Nee and Lawton (1996) observe, ecologists record body sizes just as journalists report people's ages.

On the other hand, our observation reveals something about the species that make up the Eastern Wood bird assemblage. Most life history traits of animal species are strongly correlated with their body size (Peters 1983; Calder 1984; Harvey & Pagel 1991). Figure 5.1 shows several examples for British birds. While relationships between the ecology of a species and its size tend to be weaker, in many cases body size is still the best predictor we have of variation in ecological parameters across large numbers of species (Peters 1983; Calder 1984). Moreover, unlike many of the traits of species with which it is correlated, body size has the significant advantage that it is comparatively simple to measure reliably. Thus, while theoretical advances mean that body size is no longer considered to be the cause of many evolutionary and ecological patterns (Harvey & Pagel 1991; Kozłowski & Weiner 1997), it is a convenient surrogate for variables that are so considered, but which are harder to estimate. The distribution of body sizes of the species in an assemblage is a useful first indication of the likely characteristics of that community. In consequence, body size is probably the single attribute of animal species most studied in the ecological and evolutionary literatures. For all these reasons, and despite frequent protestations to the contrary, size *is* important, at least to biologists.

Against this background, an early step towards elucidating the factors that structure animal assemblages may be to understand how the body sizes of their component species are distributed (Hemmingsen 1934; Blackburn & Gaston 1994c). It is not surprising, then, that the interspecific frequency distribution of

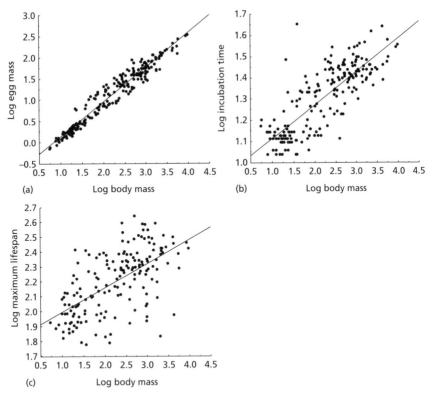

Fig. 5.1 The relationships between (a) egg mass (g; $r^2 = 0.95$, $n = 217$, $P < 0.0001$), (b) incubation time (days; $r^2 = 0.66$, $n = 207$, $P < 0.0001$) and (c) maximum recorded lifespan (months; $r^2 = 0.38$, $n = 181$, $P < 0.0001$) and body mass (g) for breeding bird species in Britain. From data in Appendix III and sources in Blackburn *et al.* (1996).

animal body sizes has long been a subject of interest, and that distributions have been reported for many animal taxa. Indeed, Loder (1997) found over 300 such distributions during a survey of the biological literature on body size, dating back to those presented by Boycott (1919) for Lepidoptera and Coleoptera. The distribution for the bird species breeding in Eastern Wood, which is shown again here as Fig. 5.2a, is not atypical.

The distribution of body masses for the Eastern Wood bird assemblage can be summarized in a number of ways (Table 5.1). Care needs to be taken when interpreting some summary statistics (see, for example, Sections 5.2.3 and 5.3.2) which can vary as a simple consequence of the class interval chosen for the plot (Loder *et al.* 1997). Of the statistics independent of the arbitrary details of the frequency distribution, the most obvious descriptor is the mean body mass. For the entire Eastern Wood breeding assemblage (Fig. 5.2a), the arithmetic mean (± SE) is 123 ± 27.9 g. This relatively high figure is illustrative of the weakness of the arithmetic mean as an estimator of the central tendency of

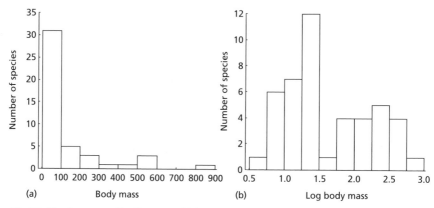

Fig. 5.2 The frequency distributions of (a) untransformed and (b) log₁₀-transformed body masses (g) of bird species recorded breeding in Eastern Wood in the period 1949–79.

Table 5.1 Statistics of the frequency distributions of untransformed and $\log_{10}$-transformed body masses (g) for the breeding bird species of Eastern Wood and Britain, and for a large sample ($n = 6214$) of the bird species of the world. Values have been back-transformed for the log-transformed distributions, but the median and range are only given once for each assemblage because these values are the same as for the untransformed distribution (the back-transformed mean of the log-transformed distribution equals the geometric mean of the untransformed distribution), and the range of values encompassed by the standard deviation is given because it is asymmetric after back-transformation.

	Untransformed masses	Log-transformed masses
Eastern Wood		
Mean	123.1	43.5
Standard deviation	187.2	10.1–187.5
Median	23.4	
Range	5.6–850	
Skew	2.1	0.49
Britain		
Mean	538.6	130.6
Standard deviation	1183.5	21.7–787.0
Median	146.0	
Range	5.3–9700	
Skew	4.98	0.12
World		
Mean	301.8	53.2
Standard deviation	1583.6	10.4–272.3
Median	37.6	
Range	2–83 500	
Skew	30.9	0.79

skewed distributions. In an unskewed distribution, the arithmetic mean, geometric mean and median would all be expected to attain similar values. However, the distribution in Fig. 5.2a is significantly right-skewed (skew = 2.1, $t = 5.9$, $P < 0.001$). In consequence, the arithmetic mean is much higher than either the geometric mean or median (Table 5.1).

The difference between the median and the geometric mean (the back-transformed mean of the log-transformed body masses) indicates a further feature of the distribution. As already noted, highly right-skewed distributions such as that in Fig. 5.2a often will be approximately normalized by logarithmic transformation. In that case, the median and the geometric mean are expected to coincide. However, the geometric mean body mass for the birds of Eastern Wood is almost double the median. This implies that even the distribution of log-transformed body masses will be right-skewed, and indeed it is, albeit not significantly so (Fig. 5.2b; skew = 0.49, $t = 1.4$, $0.2 > P > 0.05$).

Although the body mass distribution of the Eastern Wood bird assemblage is still right-skewed when log-transformed, the modal size class is not the smallest. While the range of body masses of species breeding in Eastern Wood spans 5.6–850 g (goldcrest—ring-necked pheasant), most species (over 25%) fall in the 18- to 32-g range. Few are very large bodied, but few also are very small bodied. The log-transformed mass distribution also appears distinctly bimodal (Fig. 5.2b), with a secondary peak of species at masses an order of magnitude greater than the mode (180–320 g). The modal class includes species such as the robin, blackcap and chaffinch, while the sparrowhawk, green woodpecker and magpie contribute to the secondary peak.

While we can readily quantify the body sizes of the birds of Eastern Wood, and equally easily generate statistics that describe a range of features of their distribution, these bare numbers raise several fundamental questions. Principal among these is how typical of the wider avian community is the body size distribution of birds in Eastern Wood? As repeatedly emphasized throughout this book, Eastern Wood is not a closed system, but rather is set within a regional context that inevitably exerts an influence on the characteristics of its fauna and flora. That region is in turn set within the broader global context.

5.2   The distribution of body sizes

5.2.1   *Body size measures*

Before expanding the discussion of patterns in the body sizes of birds, a comment is needed on size metrics. There are a variety of ways in which body size can be quantified. Each captures a different aspect of an organism's size, and their properties differ. Which size measure is considered the most appropriate in any given case is likely to depend on the taxon or taxa in question, and the

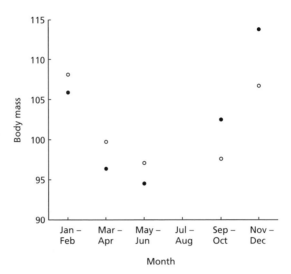

Fig. 5.3 Annual variation in the body mass (g) of adult female (open circles) and adult male (filled circles) blackbirds. From data in Cramp (1988).

purpose for which the estimate is required. For birds, common measures include body mass, body length, wingspan and tarsus length.

Body mass has the advantage that it is universally comparable. All species, whatever their taxon, have a body mass. This allows comparison among families and orders of birds, as well as between birds and mammals, fish, insects and so on. The disadvantage of mass is that it is a relatively variable trait. The mass of an individual will vary from month to month (Fig. 5.3), and depend on such factors as season, breeding condition, health and individual history. For comparative purposes, it is important to pay close attention to the equivalence of mass estimates.

Problems of intraindividual variation can be overcome by using size measures that are independent of the season, or of the condition of the individual. Tarsus length is such a metric widely used in avian research. The disadvantage of these measures is that they are of limited comparative value. They may be used for intraspecific comparisons, or for comparing closely related species, but their utility declines as the taxonomic distance between the objects of comparison increases. Thus, the great northern diver is considerably more massive than the grey heron, but a comparison of tarsus lengths would significantly favour the latter. Yet, their energetic requirements, and hence their relative impact on fish populations, for example, will be proportional to their masses, not the lengths of their legs. In the extreme, comparison may be required with a species which does not even possess a tarsus: estimating the likely comparative impact on prey fish of populations of herons and pike (*Esox lucius*) is a case in point.

As discussed in the opening chapter, macroecology essentially involves comparisons of the distribution of traits among species, or of patterns in variables measured for different communities or in different regions, with the aim of identifying causes of variation in those traits or variables. The primary need for the macroecologist interested in body size variation is a metric that allows the greatest possible comparability. For this reason, body mass is the metric of choice in macroecology, and whenever possible this measure will be the focus of the remainder of this chapter.

5.2.2 Scale and the body mass distribution

Having settled on body mass as the measure of body size, it can be used to set the Eastern Wood body mass distribution in context. However, rather than beginning with the patterns in Eastern Wood and expanding our frame of reference, we do the reverse and focus down from the largest spatial scales. Most interest has been centred on body size distributions at large scales.

Continental and global scales

The existence of more small species than large would be an obvious observation to incorporate into any general description of the composition of life on Earth. An agreed understanding of the details of this pattern has, however, been remarkably slow to emerge.

For taxonomic assemblages at continental scales and higher, species–body size distributions are strongly right-skewed on untransformed axes. This is well illustrated by data for the world's birds (Fig. 5.4a). The tail of small numbers of large-bodied species is frequently very marked, and the modal body size class may or may not be the smallest. Logarithmic transformation of body sizes reduces the skew of these distributions but it typically remains positive,

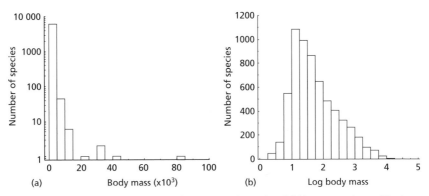

Fig. 5.4 The frequency distributions of (a) untransformed and (b) $\log_{10}$-transformed body masses (g) of a large sample of the world's bird species ($n = 6214$). Note the logarithmic scale on the ordinate in (a). From data sources in Blackburn and Gaston (1994a).

and frequently statistically significant (e.g. Van Valen 1973b; Maurer & Brown 1988; Brown & Nicoletto 1991; Maurer *et al.* 1992; Blackburn & Gaston 1994c; Maurer 1999). Early explicit recognition of this pattern in body size distributions, although not necessarily at this scale, was made by Schoener and Janzen (1968) and Stanley (1973).

Concerns have repeatedly been raised over how representative of the real patterns are documented species–body size distributions. This is because body size data are often unavailable for some species in an assemblage, either because their membership of the assemblage is unknown, or because although it is known their sizes are not (Blackburn & Gaston 1994c). That the real patterns are indeed typically right log-skewed has now been firmly established, using data for taxa for which body sizes are available or can reasonably be estimated for all known species, and for which additional species are unlikely to be discovered in such numbers that the overall form of the observed species–body size distribution would change markedly.

Blackburn and Gaston (1994a) demonstrated that the distribution of body masses of the two-thirds of the world's bird species for which data were available was right log-skewed, with declines in the numbers of species in size classes both above and below the mode (Fig. 5.4b). They argued that this pattern was probably close to the real one for the entire extant global avifauna, on the grounds:

1 that there was no evidence of a strong relationship between the proportion of species in a taxon for which data were not available and the average masses of the species in those taxa (based on species for which data were available); and

2 that there was no marked difference between the body size distribution based on species for which sizes were available and one based on these data and estimates of the masses of the species for which data were not available (derived from the mean sizes of species for which masses were available in the tribes to which they belong).

Right log-skewed distributions of body sizes for global taxonomic assemblages unbiased by missing species have also been demonstrated for a taxonomic subset of birds, the wildfowl (Fig. 5.5; Gaston & Blackburn 1996c).

This is not to say, however, that missing species cannot have a profound effect on observed species–body size distributions, when they occur in sufficient numbers. While arguing that: '. . . there is a priori no reason to suspect any correlation between the body size and the absence of measurements in the records', Hemmingsen (1934) also warned that: 'in groups including particularly small forms, the smaller forms are more likely to escape the attention of the specialists than are the larger forms, and such groups therefore require an especially sharp look-out for a false frequency distribution of the species body sizes.' In fact, it is now known that such biases may be quite widespread (Gaston 1991a; Gaston & Blackburn 1994; Patterson 1994; Blackburn & Gaston

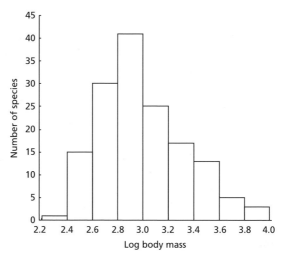

Fig. 5.5 The frequency distribution of $\log_{10}$-transformed body masses (g) of the wildfowl species of the world ($n = 150$). From data sources in Gaston and Blackburn (1996c).

1995; Gaston *et al.* 1995; Allsopp 1997). For example, those bird species which have been described in recent decades have tended to be smaller than those described earlier (Gaston & Blackburn 1994), and there are negative relationships between body size and year of taxonomic description for British beetles (Gaston 1991a) and South American birds (Blackburn & Gaston 1995). Blackburn and Gaston (1994b) showed, for several assemblages, that as progressively more species were described taxonomically, the mean of the body size distribution steadily declined, and the skew of the distribution steadily increased (Fig. 5.6). In the extreme, the distributions for some taxa changed from left log-skewed to log-normal to right log-skewed as the proportion of species described increased. It would be ironic if such trends explain why Hemmingsen (1934), although he expressed some reservations, proposed that body size distributions were approximately log-normal.

The form of the synoptic body size distribution for all extant terrestrial animal species has been addressed by May (1978, 1988). Based on 'crude approximations and outright guesses', he offered a first (and still, as far as we are aware, the only) attempt at what this might look like (Fig. 5.7). Again, it exhibits a right log-skew, with the modal size class being larger than the smallest. May suggested two significant caveats about the form of this distribution. First, the decline in numbers of species in size classes below 1 mm could be a real phenomenon, or a result of a breakdown in conventional taxonomy when applied to protists. Second, size classes below 10 mm could be underestimated by a factor of two or more because small invertebrate species are poorly known. These problems could potentially mean that the mode of the real distribution in fact lies in the smallest size class. However, Fenchel (1993) has

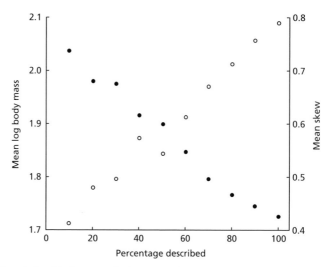

Fig. 5.6 The relationship between the $\log_{10}$-transformed geometric mean (filled circles) and the skew (open circles) of the body size frequency distribution of a large sample of bird species of the world ($n = 6199$) for successively increasing percentages of currently known species to have been formally described (starting with the earliest descriptions by Linnaeus and ending with descriptions published in the early 1990s). As more species have been described, the mean body mass of known species has declined, while the skew has increased. From Blackburn and Gaston (1994b).

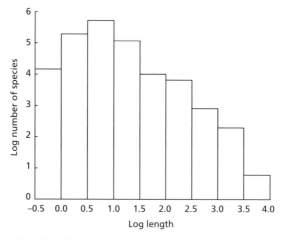

Fig. 5.7 An approximation of the frequency distribution of body lengths (mm) of all known terrestrial animal species. From May (1978).

argued that species numbers of terrestrial protozoa are inadequate for this to be so, and that differences in taxonomy are not likely to be a major contributor to the observed pattern. Fenchel also compiled a synoptic species–body size distribution, but this time for all named free-living aquatic species (Fig. 5.8). Although it is dissimilar to that for terrestrial animals with no evidence of

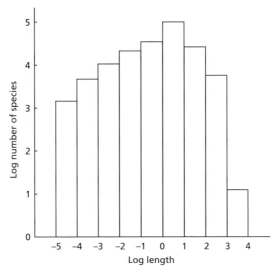

Fig. 5.8 The frequency distribution of body lengths (cm) of all named free-living aquatic animal species. From Fenchel (1993).

marked right-skew, the smallest size class is again not the most speciose. In sum, it remains reasonable to suppose that the species–body size distribution of all extant terrestrial animals is essentially much as May (1978) proposed, albeit composed of very many more species, while that for aquatic species is rather different in form.

There is a general tendency for species–body size distributions at large spatial scales to be right log-skewed, with a mode above the smallest size class. This general pattern in large-scale body size distributions does, however, admit of exceptions, in addition to that for aquatic animals. First, at the level of genera, the body mass distributions of terrestrial mammals are right-skewed under logarithmic transformation for assemblages from 'large' continents (e.g. Eurasia, Africa, South America, North America), but not significantly skewed for assemblages from 'small' ones (e.g. Australia, New Guinea, Madagascar; Maurer *et al.* 1992). This difference results from the absence of large-bodied genera from the small land masses. Although it has not been explored at the species level, it seems likely to pertain there also, as there is typically relatively little variation in body mass among species within genera (Elgar & Harvey 1987; Promislow & Harvey 1989; Read & Harvey 1989; Gaston & Blackburn 1997a; Webb *et al.* 2000). This said, among birds, species–body size distributions for both North America and Australia appear to be significantly right log-skewed (Maurer & Brown 1988; Maurer *et al.* 1991).

Second, when body size distributions are severely constrained taxonomically, or to trophic or ecological subsets of species, departures from a right log-skew may be observed. Thus, subdividing the overall right-skewed body size distribution for the world's birds by orders generates species–body size

Table 5.2 Values of skew of the distribution of $\log_{10}$ body masses of species for separate bird orders. Skewness = g_1 (see Sokal & Rohlf 1995), n = number of species in sample. From Maurer (1998a).

Taxon	Skewness	n
Anseriformes	0.66*	152
Apodiformes	0.40	68
Bucerotiformes	0.29	29
Ciconiiformes	0.19*†	777
Coliiformes	−0.52	6
Columbiformes	0.15†	167
Coraciiformes	−0.17†	107
Craciformes	0.50	31
Cuculiformes	0.29†	82
Galbuliformes	0.13†	36
Galliformes	0.01†	192
Gruiformes	0.26†	123
Musophagiformes	2.36*†	19
Passeriformes	0.86*†	3450
Piciformes	0.05†	292
Psittaciformes	0.25†	185
Strigiformes	0.84*	180
Struthioniformes	−0.73	6
Tinamiiformes	0.54	24
Trochiliformes	0.07†	247
Trogoniformes	0.61	27
Turniciformes	−0.52†	9
Upupiformes	−0.09	8

* Skew significantly ($P < 0.05$) different from zero, using the test given by Sokal and Rohlf (1995).
† Skew significantly ($P < 0.05$) different from a random sample from the distribution of body masses of a large sample of the bird species of the world.

distributions which exhibit wide variance in skew (Table 5.2; Maurer 1998a). These range from −0.73 to 2.36, compared to 0.79 for the complete assemblage (Table 5.1). The negative skew exhibited by the Turniciformes and Coraciiformes is more extreme than expected were the distributions for these orders random samples from the complete distribution (Maurer 1998a). In a similar vein, Gaston and Blackburn (1995a) demonstrated that if the global avifauna was divided into terrestrial and aquatic species, the log-skewed form of the overall distribution for all birds was also found for the terrestrial species, but not for the aquatic. The latter had an approximately log-normal body size distribution (Fig. 5.9).

Although data are not available for birds, available evidence from other taxa suggests that the general form of species–body size distributions exhibited by present-day assemblages has been a persistent feature across evolutionary time (e.g. Stanley 1973; Arnold *et al.* 1995; Jablonski & Raup 1995). Indeed,

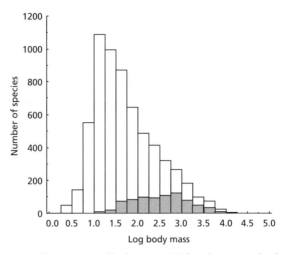

Fig. 5.9 The frequency distributions of body masses (g) for a large sample of terrestrial (unshaded bars; $n = 5423$) and aquatic (shaded bars; $n = 786$) bird species of the world. Aquatic species are those that are dependent on the aquatic environment at some stage in their life history either for feeding or protection (e.g. Procellariiformes, Alcidae, Cinclidae). All other species are classified as terrestrial. Unshaded bars should be read from the top of the shaded bars, not the abscissa. Mass classes 4.25–4.49, 4.5–4.749 and 4.75–4.99 contain two, two and one terrestrial species respectively, although the scale of the y-axis means that these classes appear empty. From Gaston and Blackburn (1995a).

Jablonski (1996b) documents a right-skewed distribution for a fossil mollusc assemblage (bivalves of the Gulf and Atlantic Coastal Plain Province from the Late Cretaceous) which is statistically very similar to that seen for the present-day Eastern Pacific fauna. What counter-examples do exist seem predominantly to concern relatively small groups of related species (e.g. McShea 1994), for which substantial variation in the form of body size distributions would be predicted from present-day assemblages also (e.g. see below and Table 5.2).

Smaller spatial scales

The numerous species–body size distributions which have been documented at micro- and meso-scales present a confused picture, which is not readily interpreted by attempting to order them in terms of spatial scale. On the one hand, a number of studies have reported species–body size distributions which are not significantly different from log-normal or, at least, do not appear to be so even if no formal test has been applied (e.g. Hutchinson & MacArthur 1959; Morse *et al.* 1988). Eadie *et al.* (1987) tested 41 species–body size distributions for small assemblages (mean number of species = 20) for departure from a normal, a log-normal and a log-uniform distribution using Kolmogorov–Smirnov tests. Thirty-five (85.4%) were not significantly different from a log-normal, and 19 (46.3%) were not significantly different from either a normal or a log-uniform. A log-normal fitted the observed distributions better than a normal in 38 (92.7%) cases, and better than a log-uniform in 31 (75.6%). They concluded that a log-normal distribution provides a reasonable fit to the body size distributions of a variety of animal assemblages.

By contrast, many other studies have reported species–body size distributions at local scales that are plainly not log-normal (e.g. Terborgh *et al.* 1990; Thiollay 1990, 1994; Basset & Kitching 1991; Brown & Nicoletto 1991; Damuth 1992; Blackburn & Gaston 1994c). Schoener and Janzen (1968) observed that a logarithmic transformation reduces the skew of species–body size distributions for sweep-net samples of insects from various localities, but that the distributions tend to remain positively skewed and depart significantly from normality. They employed a three-parameter model further to reduce this skew. Likewise, Holling (1992) found that the log-normal provided a poor fit to the four species–body size distributions of birds and mammals of boreal forest and prairie, while Fenchel (1993) documented size distributions for five different aquatic communities which were all apparently right log-skewed. The avifauna of Eastern Wood appears to fit most comfortably into this group. Although its body mass distribution is not significantly log-skewed, it does appear to have a distinctly bimodal form, and certainly differs significantly from a log-normal (Lilliefors test, $P = 0.001$).

These results may be closely associated with the observations made above regarding the effects of continent size and taxonomic inclusiveness on observed large-scale distributions. Both of these factors serve substantially to reduce the numbers of species in an analysis. Random sampling of a right-skewed distribution alone predicts that the body size distribution of smaller numbers of species belonging to a much larger assemblage may be highly variable (see, for example, figure 2 in Maurer 1998a), so that large- and small-scale distributions may differ regardless of whether or not they are structured by similar mechanisms. Body mass distributions for small-scale assemblages like Eastern Wood thus may simply be random samples from the larger-scale distributions of which they are a part. However, while a great many body size distributions have been documented, it is only recently that any attempt has been made to explore systematic differences between size distributions at different scales. Interest in this issue has chiefly been stimulated by a paper by Brown and Nicoletto (1991; see also Brown & Maurer 1989).

Using data for North American terrestrial mammals, Brown and Nicoletto demonstrated that while the logarithmically transformed body masses of the species assemblage for the entire continent are right-skewed, at the scale of individual biomes body sizes are distributed approximately normally, and at the scale of local patches of relatively homogeneous habitat are distributed approximately uniformly. While most frequency distributions were positively skewed (19/21 biomes, 16/24 habitat patches), the degree of skew declined from continent to biomes to local patches. The size distributions for assemblages at the scale of biomes all had medians significantly larger than those expected on the basis of a random draw of the appropriate numbers of species from the continental assemblage. Likewise, all but one of the assemblages at individual sites had distributions with medians which were larger than those expected on the basis of the size distributions of the biomes in which those sites lay, and the medians were significantly larger for nine of these assemblages (Brown & Nicoletto 1991).

The generality of Brown and Nicoletto's findings remains to be determined. Those studies that have explicitly replicated their analyses using different faunal assemblages have not in general been supportive. Thus, Marquet and Cofré (1999) found that most local and biome assemblages of mammals in South America did not differ significantly from random draws from appropriate species pools. Arita and Figueroa (1999) showed that body mass distributions for Mexican bat assemblages were right-skewed at all scales, and local assemblages were well modelled by random draws, although patterns for non-volant mammals tended to agree with those demonstrated by Brown and Nicoletto (1991). Loder (1997) tested the body size distributions of local- and biome-scale North American butterfly assemblages against random samples from the overall continental distribution. Statistics of the body size distributions of the real assemblages showed no tendency to differ in any consistent manner from the random samples. Similarly, Thiollay (1994) argued that the distribution of body masses of bird species recorded on a 100-hectare plot of primary rainforest in French Guiana was similar to that of all bird species in Guiana, although no formal test was provided.

Brown and Nicoletto's (1991) conclusions can be tested on the body mass data for the birds of Eastern Wood. Figure 5.10 shows the body mass distribution for the entire British breeding bird assemblage using both untransformed and $\log_{10}$-transformed data. The statistics for these distributions are given in Table 5.1. Compared to the avifauna of Eastern Wood, that of Britain as a whole shows higher arithmetic and geometric means, higher median body mass, and a greater range of body masses. Thus, the Eastern Wood avifauna seems not to be a simple random sample of the British avifauna with respect to body mass.

This can readily be examined using a similar randomization approach to that of Brown and Nicoletto (1991). The breeding fauna of Eastern Wood in the period 1949–79 comprised 45 species. We simulated this assemblage by selecting 45 species at random from the total British breeding avifauna (217 species), repeating the procedure 5000 times. Examining the statistics of these random assemblages revealed that their arithmetic and geometric mean masses were always much higher than those of the real Eastern Wood assemblage ($P < 0.0004$ in each case). This difference was not due to the absence of small-bodied species in the random samples—21.4% included a species smaller than the smallest present in the wood ($P = 0.43$). Instead, it resulted from the absence of

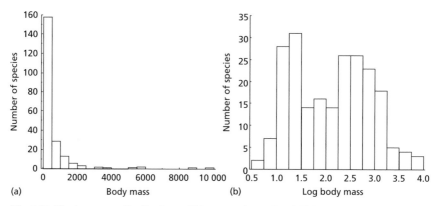

Fig. 5.10 The frequency distributions of (a) untransformed and (b) $\log_{10}$-transformed body masses (g) of breeding bird species in Britain ($n = 217$). From data in Appendix III.

large-bodied species in Eastern Wood: the largest species in all but three of the 5000 random samples was heavier than the largest species ever recorded breeding there ($P = 0.0012$). The skew of the Eastern Wood assemblage was high relative to most random assemblages, but not significantly so ($P = 0.072$).

One obvious reason for the difference between the body masses of the Eastern Wood avifauna and those expected given the overall body size distribution of the British avifauna is that the wood exhibits only limited habitat variation. Bird body masses are related to habitat type. For example, Gaston and Blackburn (1995a) showed that aquatic bird species tend to be larger bodied than terrestrial. Southwood *et al.* (1986) showed that the mean body size of bird species using sites at Silwood Park declines with the successional age of habitat (see also Helle & Mönkkönen 1990). Early successional stages (vegetation < 6 years old) have a high proportion of large-bodied ground-feeding herbivores, whereas the number of smaller-bodied arboreal-feeding passerines is high in the late succession plot (essentially a young woodland); similar patterns were shown by some insect groups at the same site. These and other studies suggest that the body masses of species sampled at random from the entire British avifauna may differ from those of the Eastern Wood species because many British species utilize habitats that are not available in the wood.

A more appropriate null hypothesis may be that the Eastern Wood avifauna is, with respect to body mass, a random selection of British woodland bird species. We repeated our simulations using only those species that breed in deciduous woodland. This time, the masses of the simulated and real assemblages were more similar. Nevertheless, the arithmetic mean ($P = 0.008$), geometric mean ($P = 0.031$) and maximum body mass ($P = 0.028$) of the samples were all greater than those of the Eastern Wood avifauna significantly more often than expected by chance. The skew of the simulated and real assemblages still did not differ significantly ($P = 0.61$). The avifauna of Eastern Wood

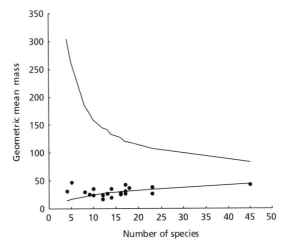

Fig. 5.11 Results of simulations of body mass frequency distributions of deciduous woodland local bird assemblages in Britain. For a given assemblage size (number of species), the pool of British deciduous woodland species ($n = 80$) was sampled at random without replacement, and the geometric mean body mass of the random assemblage calculated. This procedure was repeated 5000 times for each sample size. The lines bracket the range of geometric mean masses within which 95% of simulated assemblages fell. The filled circles indicate geometric mean masses from real British woodland bird assemblages; those falling outside the lines differ significantly from the random expectation. The right-hand point is for the bird assemblage of Eastern Wood, the rest are for woodlands censused by Ford (1987).

is composed of a slightly smaller set of species than expected given the composition of the fauna within which it is embedded, although they generate a body mass distribution of similar shape.

We tested the generality of this last set of findings using data from the 20 Oxfordshire woodland islands censused by Ford (1987; used also in Section 2.1). These islands contained between four and 23 breeding species, compared to the 45 recorded in Eastern Wood. This relatively low richness is a combination of their generally small area, and the fact that the data derive from a single census year. Small random samples from a species pool are likely to display a wider range of variation in their calculated statistics than are large samples (an unbiased coin tossed four times is not unlikely to come down heads every time, but such an outcome from 40 tosses would be much more surprising), making it harder to reject the null hypothesis that these statistics do not differ from random expectation. Nevertheless, nine of these 20 assemblages show geometric mean masses that are lighter than expected by chance. Moreover, Fig. 5.11 clearly shows that even the geometric mean masses of those woodland bird assemblages that did not differ from the random samples were close to the lower limit of random variation. All of the distributions were positively skewed, but only three assemblages were more skewed than expected by chance. Only one assemblage differed from expectation in maximum body mass, while none differed in minimum. Thus, we find that the body mass distributions of bird assemblages in British woodlands tend to have small means, but shapes and limits that would be expected given the woodland species pool from which they draw.

The tendency of the body masses of the Eastern Wood bird assemblage not to constitute a random sample of the British avifauna is mirrored at the larger scale in the relationship between the body masses of the British avifauna and those of bird species in general. The frequency distribution of $\log_{10}$-transformed body masses for a large sample of the world's birds is shown in Fig. 5.4, with the statistics of the distribution given in Table 5.1. The arithmetic mean body mass of species in the British bird assemblage is high relative to that for birds in general (Table 5.1), albeit not significantly so ($P = 0.084$). However, the geometric mean body mass of British birds is significantly higher than expected ($P = 0.0004$). This difference is a consequence of the lack of very small-bodied species in the British avifauna: the smallest-bodied breeding bird species is larger than expected from random samples of the world fauna ($P = 0.0056$), although the largest species does not differ from that expectation ($P = 0.65$).

The Eastern Wood and other British woodland bird data support Brown and Nicoletto's finding that local body mass distributions often deviate from random samples of the regional distribution in mean, but contradict them in that the means are always lower than expected, and by not differing in skew. The body size distribution for the Eastern Wood assemblage also differs from those of Brown and Nicoletto's local mammal assemblages by being significantly different from log-uniform (Kolmogorov–Smirnov one-sample test, $P = 0.002$). Taken together with the fact that the body mass distributions of 15/24 of the local mammal assemblages studied by Brown and Nicoletto (1991) did not differ significantly from random samples of the biome in which they were situated, and with the results from Loder's (1997) butterfly study, there seems little strong evidence that local assemblages differ in any consistent manner from what would be expected given the regional faunas of which they are part. Clearly, however, there is scope for more investigation here. It is an issue to which we later return.

5.2.3   *Discontinuities*

In considering the patterns of body size in species assemblages we have largely restricted concerns to the fit, or departure from, continuous distributions. However, the Eastern Wood assemblage and the British assemblage from which it draws both appear to exhibit a frequency distribution of log-transformed body masses that is bimodal (Figs 5.2b, 5.10b; see also Cousins 1980). This raises the question of the extent to which species–body size distributions are discontinuous. In fact, a number of studies have discussed the occurrence of multiple modes in body size distributions (e.g. Boulière 1975; Caughley & Krebs 1983; Caughley 1987; Maiorana 1990).

Support for polymodality in birds was provided by Griffiths (1986), who found evidence for it in 60% of the assemblages he analysed. He discussed a number of possible causes, including that body sizes may be associated with vegetation structure, and that suites of adaptive traits may be discretely, rather than continuously, distributed. Related analyses and explanations were provided by Holling (1992), based on analyses of the body masses of birds and mammals that breed in the boreal forest of North America east of the foothills of the Rocky Mountains, and of birds and mammals of the short-grass prairie of southern Alberta. He argued that these size distributions show distinct size ranges of clumps characterized by small body mass differences between neighbours, and that these are more frequent than would be expected by chance drawing of samples from a log-normal species–body size distribution. There appears to be little correlation between the taxonomic, guild or habitat affinities of species and these clumps. However, Holling argues that there is such a correlation between clumps and the scales at which different species '... make decisions to start, continue, or abandon activities associated with survival and opportunity' (p. 457). Moreover, he proposes that discontinuities in body size distributions 'are universal in landscapes associated with fixed physical structures' (p. 467), but are seldom noted because they are obscured by the expression of data as frequency histograms, or are regarded as noise about some underlying continuous distribution. Additional evidence for discontinuities in avian body mass distributions is provided by Restrepo *et al.* (1997).

Holling's (1992) arguments about the causes of clumps in body size distributions rest on the validity of his tests for detecting such discontinuities. These have been critically assessed by Manly (1996), who found the evidence they present for multiple modes in the body size data Holling analysed to be 'not compelling'. Indeed, three distributions out of four claimed by Holling to have at least six discontinuities were found by Manly to be consistent with unimodal or bimodal distributions. The explanation for this discrepancy is that clumps can easily occur by chance even in random samples from unimodal distributions (Manly 1996). However, Manly's approach has in turn been criticized by Restrepo *et al.* (1997), who argued that it is likely to be too conservative. Clearly, until clumps in body size distributions can be identified unequivocally, explanations for their presence will be premature.

5.3   What determines the shape of species–body size distributions?

The simplest explanation for the form of large-scale body mass distributions is that they are the result of random diversification (Maurer 1998a). In the absence of any other process, this would lead to a log-normal distribution. However, log right-skew can result if there is a lower limit to the body masses achievable by the species in a taxon that acts as a reflecting boundary, causing body masses to evolve away from this limit (McShea 1994). If this were the case, and begging the question of why it should be, subtaxa close to the lower body mass limit would be more likely to show log right-skewed body mass distributions than subtaxa further away, which are unaffected by the lower boundary constraint. Maurer (1998a) tested this for birds, finding no support for the idea of random diversification. Thus, the log right-skewed avian species–body mass distribution must be a consequence of non-random (with respect to body mass) cladogenetic or anagenetic processes.

Several models have been proposed to explain the observed patterns, and we examine these below. For smaller spatial scales, the key question is whether or not the body size distributions of species occupying local sites like Eastern Wood are random samples from the regional distribution. If they are, then the mechanisms determining regional patterns are sufficient also to explain local patterns. If they are not, then additional processes need to be invoked to generate the differences. As we have seen, the jury is still out on the question of the nature of the association between local and regional patterns in body sizes. Thus, for each of the models of large-scale body mass distributions that follow, we consider what additional processes might cause non-random small-scale patterns to be produced.

5.3.1   The ultimate explanation—speciation and extinction rates

Ultimately, the composition of faunas of biogeographic regions (i.e. areas with only a small fraction of their faunas derived from immigration) will be dependent on patterns of speciation and extinction within those regions. It follows that perhaps the most obvious, and general, mechanism for generating species–body size distributions is based simply on size-related speciation and extinction rates, ignoring what might cause those relations.

Dial and Marzluff (1988) presented a pictorial outline of how such a speciation–extinction mechanism might work, but pursued it no further (see also Fowler & MacMahon 1982). The first explicitly to simulate such a mechanism were Maurer *et al.* (1992), who used a modified random birth–death model. Speciation and extinction probabilities were either equal for species above and below the modal body size or a small bias in these probabilities was added to favour species either larger or smaller than the modal size. At the next step, the model took one of the following three forms.

1 Cladogenetic fixed jump—for each new species, body mass was either increased or decreased by a multiplicative factor relative to the species from which it originated, with the instantaneous jump being constant (= 1.1).

Table 5.3 Essential results of the Maurer *et al.* (1992) models, based on average skew of 500 body mass distributions generated for each combination of speciation and extinction probability for each of the three different models (see text for details). Distributions of logarithmic body sizes are normal (*N*), positively skewed (+) or negatively skewed (–).

(a) Cladogenetic—fixed jumps

		Likelihood of speciation		
		Small-biased	Unbiased	Large-biased
Likelihood of extinction	Small-biased	+	*N*	–
	Unbiased	+	*N*	–
	Large-biased	+	*N*	–

(b) Cladogenetic—random jumps

		Likelihood of speciation		
		Small-biased	Unbiased	Large-biased
Likelihood of extinction	Small-biased	–	–	–
	Unbiased	*N*	–	–
	Large-biased	+	+	+

(c) Anagenetic

		Likelihood of speciation		
		Small-biased	Unbiased	Large-biased
Likelihood of extinction	Small-biased	+	–	–
	Unbiased	+	*N*	–
	Large-biased	+	+	–

2 Cladogenetic random jump—as for the fixed-jump model, except that the multiplicative factor was a normal random variable with the same mean as that in the fixed-jump model (SD = 0.01).

3 Anagenetic—as for the fixed-jump cladogenetic model, except that species were allowed to change in body size between speciation events.

The direction of change was chosen at random, with the constraint that new species evolved in a direction opposite to that taken by their direct ancestor (i.e. sister taxa diverge in body mass).

The essential results of these models are given in Table 5.3, and allow a number of broad generalizations to be made. If the likelihoods of speciation and extinction are unbiased with respect to body size, then resultant species–body size distributions are approximately log-normal. Small-biased speciation rates give rise to positively skewed body size distributions (i.e. an excess of small-bodied species), while large-biased speciation rates give rise to negatively skewed distributions. Extinction rates have little effect in the fixed-jump cladogenetic model, but skew distributions generated by the other two models in the expected direction: large-biased extinction leads to positively skewed distributions. Speciation is the dominant effect in the fixed-jump cladogenetic model, extinction in the random-jump model, and both significantly modify the skew of body size distributions generated by anagenesis (Maurer *et al.* 1992).

Johst and Brandl (1997) also used a simulation approach to suggest that the effect of extinction alone is sufficient to produce positively skewed species–body size distributions. They assumed that extinctions are caused by environmental perturbations, which can be modelled by a first-order autoregressive process. This generates frequent small perturbations and rarer large ones. The effects depend on the birth and death rates of species. Small-bodied species, with high birth and death rates, are more susceptible than are large-bodied species (low birth and death rates) to relatively small environmental perturbations. Large environmental perturbations have similar effects on all species, but the lower birth and death rates of large-bodied species make them slower to recover. This increases the likelihood that they will be exterminated by subsequent smaller environmental perturbations. The sum of these two effects favours medium-sized species, which persist in higher numbers, so generating a hump-shaped body size distribution (although it is unclear how right-skewed it may be). This model extends that of Maurer *et al.* (1992) by identifying a process that can cause size-biased extinction. However, Maurer *et al.* showed that speciation could significantly modify patterns in body size distributions, to the extent that even small-biased extinction could cause right log-skew. The key questions begged by these models are therefore whether either speciation or extinction in practice shows any body size bias.

Views on size biases in speciation and extinction seem to be strongly dominated by perception or supposition, with empirical evidence being weak or inconsistent. Smaller-bodied species are generally held to speciate at a faster rate, on the grounds that they have shorter generation times and thus more opportunities for speciation. Such an argument is, however, potentially compromised by any tendency for likelihood of speciation to be influenced by differences in the population sizes (Chapter 4) or geographical range sizes (Chapter 3) of species. Empirical analyses of relationships between speciation rate (rather than cladogenesis) and body size remain largely wanting. One recent test found no significant difference between the origination rates of large- and small-bodied Cenozoic globerigerine planktonic foraminifera (Arnold *et al.* 1995).

Larger-bodied species are widely held to be at differentially greater risk of extinction. However, the theoretical grounds for such a belief are weak. Some factors can be envisaged which might make large-bodied species more prone to extinction (e.g. low abundance, slow rate of population increase), and others which might make small-bodied species so (e.g. susceptibility to environmental perturbation, short lifespans). It is difficult, if not impossible, to assess which are likely to be the more important, and to weigh them against the relative importance of extrinsic factors (e.g. catastrophes, climate change). Perceived relationships between body size and likelihood of extinction may also be strongly influenced by patterns of loss of the largest-bodied species. Even were extinctions to be random with respect to body size, most of the largest-bodied species would be expected to be lost relatively early, because there are very few of them. Empirical studies of the relationship between body size and extinction risk do not serve to clarify matters greatly. Those carried out on contemporary faunas report a variety of relationships (e.g. Brown 1971; Leck 1979; Järvinen & Ulfstrand 1980; Terborgh & Winter 1980; Karr 1982; Patterson 1984; Pimm *et al.* 1988; Soulé *et al.* 1988; Burbidge & McKenzie 1989; Laurila & Järvinen 1989; Gotelli & Graves 1990; Ceballos & Navarro 1991; Laurance 1991; Maurer *et al.* 1991; Kattan 1992; Rebelo 1992; Rosenzweig & Clark 1994; Angermeier 1995; Gaston & Blackburn 1995b, 1996c; Sieving & Karr 1997). Interpretation is complicated, however, because such studies typically concern local extinctions and not global ones. Where they are concerned with global extinctions they tend to be based on whether species are or are not included in listings of threatened species, which raises questions over the adequacy of identifying species under threat, and also ignore the possible influence of extinctions which resulted from human activity in earlier times. Prehistoric extinctions may have been size biased for other reasons, such as the disproportionate likelihood that large-bodied animals would be hunted.

The relationship between body size and global extinction risk has been assessed for birds by Gaston and Blackburn (1995b). They found that species presently considered to have a high risk of extinction were, on average, larger bodied than those which do not face such a risk. This difference was not due to body size differences between island endemic species and species with continental distributions; island endemics are well known to have a greater likelihood of being threatened (Johnson & Stattersfield 1990). Gaston and Blackburn considered four possible reasons for their result: (i) large-bodied species are genuinely more prone to extinction than small bodied; (ii) small-bodied species are actually generally more prone to extinction, and the current observed relation between body mass and extinction results from the differential extinction of small-bodied species; (iii) the perception of extinction threat is affected by body size; or (iv) there is better information on large-bodied species threatened with extinction, causing them to be disproportionately represented in lists of threatened birds. They concluded that, on balance, the evidence favoured a real association between high extinction risk and large body size (reason (i)), but it is difficult unequivocally to demonstrate that this is so.

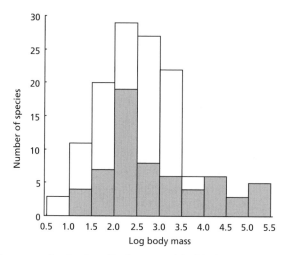

Fig. 5.12 The frequency distribution of body masses (g) of the New Zealand pre-human avifauna (*n* = 132). The stippled areas represent those species that went extinct from the three main islands (North, South and Stewart) after human arrival (*n* = 62), although some of these persist on small inshore islets (e.g. stitchbird) and offshore islands (e.g. common diving-petrel). Open bars should be read from the top of the stippled bars, not from the abscissa. From data in Holdaway (1999).

A good example of differences in body size distributions before and after recent extinctions is provided by the avifauna of New Zealand. Humans first colonized these islands within the last 1000 years, precipitating a wave of avian extinctions. Holdaway (1999) lists 132 species that were known to have been breeding on the three main New Zealand islands at that time, of which 62 have since been lost. Figure 5.12 gives the species–body size distribution of the original fauna, distinguishing between extinct and extant species. Species of a range of body masses have been driven extinct, but large-bodied species have clearly suffered disproportionately. The geometric mean body masses of extinct and extant species are 851 g and 250 g, respectively, and are significantly different (unpaired *t*-test: *t* = 3.19, *n* = 132, *P* = 0.002). However, this difference is due largely to the 11 species of moa, which are all large bodied and extinct. Excluding these, species that have gone extinct on the three main islands of New Zealand are not significantly larger than those that are still present (unpaired *t*-test: *t* = 0.89, *n* = 121, *P* = 0.38). Holdaway (1999) argues that the impact in New Zealand of predators of a range of different body sizes (from Pacific rat *Rattus exulans* to human) has resulted in extinctions right across the avian body size distribution. Although large-bodied species do tend to be lost from faunas, they do not go alone.

If the general relationship between body size and risk of extinction in modern faunas is unclear, empirical studies of palaeontological data are equally inconclusive (e.g. Van Valen 1975; Bakker 1977; Van Valen & Sloan 1977; Stanley 1979, 1986; Martin 1984; Norris 1991; Raup 1991, 1994; McLain 1993;

Arnold *et al*. 1995; Jablonski & Raup 1995; Jablonski 1996b). These suffer from the usual constraints of the fossil record, particularly uncertainty as to the extent to which observed patterns are products of body size-, abundance- or range size-biased patterns of preservation. Patterns may also differ between periods of background and mass extinction. Moreover, 'the issue is clouded by the lack of rigorously controlled statistical analysis' (Raup 1994, p. 6761).

If there are no repeatable relationships between the likelihoods of extinction or speciation and the body sizes of species, is there a relationship between the level of effective cladogenesis and body size? Once again, a simple answer is hampered by the inadequacy of the majority of studies. Negative relationships between the number of species in a taxon and the average body size of its constituent species (or representatives thereof) have been documented on several occasions (e.g. Van Valen 1973b; Dial & Marzluff 1988; Kochmer & Wagner 1988; Martin 1992; Gardezi & da Silva 1999). However, these may be severely compromised by treating taxa as independent data points. Nee *et al*. (1992b) illustrate this problem for birds. There is a highly significant negative correlation between body mass and the number of species in a bird family (see also Van Valen 1973b), but this is a result of taxonomic non-independence due to two monophyletic taxa, the species-poor large-bodied Eoaves (ostriches, rheas, cassowaries and emus, kiwis and tinamous) and the species-rich small-bodied Passeriformes. When sister taxa are compared, body mass does not predict which of the two sisters will be the more speciose. Thus, body size does not appear to be a general correlate of effective cladogenesis within and among bird families. In fact, a relationship between the level of effective cladogenesis and body size has only been convincingly demonstrated once, by Gittleman and Purvis (1998). They used sister taxon analyses to show that small-bodied lineages of primates and mammalian carnivores do indeed tend to be speciose. This tendency was never great, varied among clades, and may not apply more widely in mammals (Gardezi & da Silva 1999). Nevertheless, it stands as the best demonstration that body size may affect net speciation and extinction rates in an animal taxon.

Whatever their relative significance, the processes of speciation and extinction determine in large part the body size distribution of global and continental avian assemblages. If assemblages in smaller regions or at individual sites are random draws from these pools, then it is easy to see how these processes impinge directly on body size distributions at these smaller scales. Moreover, it implies that the rates of species colonization and extinction at the smaller scales must be related to body size in much the same way as are speciation and global extinction. If these assemblages are not random draws from the larger regional pools, then it implies that extinction and colonization at the smaller scales are related to body size in a different way. Since we have shown that the species inhabiting Eastern Wood were smaller bodied than expected from a random draw of the British total or woodland avifaunas, at this particular site this must ultimately be a consequence of small-biased colonization, large-biased extinction, or both, relative to the British avifauna.

Although body size distributions at large scales must develop primarily through the processes of speciation and extinction, and various simulation approaches prove that they can, this simply begs the question of why these processes favour small body size in birds (and other taxa). This question is especially pertinent given that it is possible, as shown above, to suggest reasons why both large- and small-bodied species might have higher speciation rates or lower extinction rates. It is to attempts to address the question of why small body size is apparently favoured by evolution that we now turn.

5.3.2   *Why is small body size favoured?*

Environmental grain

The first formal model of species–body size distributions was provided by Hutchinson and MacArthur (1959). They envision an environment composed of an indefinite number of equal-sized mosaic elements of different kinds arranged randomly. Larger-bodied species are considered to require a greater number of contiguous elements, and each qualitatively different combination of elements can only be occupied by a single species (it represents a different niche). Specifying the function relating body size and the number of contiguous elements, and incorporating character displacement, the model predicts that for increasing body size classes the number of species will rise rapidly to a peak and then decline more slowly. Hutchinson and MacArthur test the fit of their model to two mammal assemblages. Its adequacy is questionable, though not wildly imperfect. It remains, however, particularly by present-day standards, a highly simplistic and somewhat questionable caricature. For example, by definition it cannot accommodate spatial turnover in species identities, and the derivation of parameter values is doubtful. Nonetheless, it has provoked significant debate.

From Hutchinson and MacArthur's (1959) paper, May (1978) derived the simple prediction that, above the modal size class, the number of species S in an assemblage should decline roughly as

$$S \propto L^{-2}$$

where L is the characteristic linear dimension of an animal (or, since body mass M is proportional to L^3, $S \propto M^{-2/3}$). On a double logarithmic plot of a species–body size distribution, this means that there should be a linear decline in the number of species in successive body size classes above the mode with a slope of -2 (a 10-fold increase in linear dimension results in a 100-fold decrease in the number of species).

May (1978) indicated this slope on species–body size distributions for a variety of animal assemblages. The fit is not tested and appears highly variable.

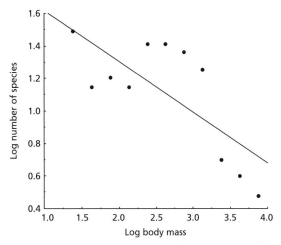

Fig. 5.13 The slope of the frequency distribution of body mass (g) for the breeding bird species of Britain (Fig. 5.10b: $y = -0.31x + 1.93$; $r^2 = 0.54$, $n = 11$, $P = 0.01$). The slope is determined using only size classes in the distribution to the right of the mode (for more details, see text and Loder *et al.* 1997). From data in Appendix III.

Since May's original paper, slope values have been estimated in a number of other studies (see Loder *et al.* 1997 for a review). Published values range between –1.4 and –4.63, and although confidence limits are seldom given, these will tend often to be wide, potentially embracing not only –2, but also, say, –3 (a 10-fold increase in linear dimension results in a 1000-fold decrease in species numbers). The slope values for the frequency distributions of the body masses of the breeding avifaunas of Eastern Wood and Britain illustrated in Figs 5.2b and 5.10b are –0.28 (95% CI –1.04 to 0.49) and –0.31 (95% CI –0.53 to –0.09; Fig. 5.13), respectively. These slopes are not especially close to the predicted value for mass of –0.67, and this value is not encompassed by the confidence intervals of the slope for the British avifauna.

Although there are few cases in which the predicted slope of –2 (or –2/3 for mass) is wildly inaccurate, there are severe problems in fitting such slopes. These include: (i) the relationship between the number of species in body size classes above the mode and body size is unlikely to be linear on log–log axes (e.g. Fig. 5.13); (ii) the value of the regression slope, the size of the associated confidence intervals and the variance explained by the relationship are functions of the number of body size classes employed; and (iii) the value of the regression slope may depend on the treatment of empty size classes and multiple modes (Loder *et al.* 1997). Such issues cast severe doubts over what can be learned from attempting to fit slopes to species–body size distributions.

It has been argued that a simple scaling of numbers of species and body size with some exponent lower than –2 might follow from fractal geometry. Hutchinson and MacArthur's (1959) essential argument was that there are

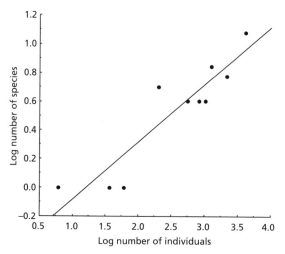

Fig. 5.14 The relationship between the number of species and total number of individuals (summed over all species) in different $\log_{10}$ body mass classes (interval 0.25, from 0.5) for the breeding bird species of Eastern Wood ($y = 0.4x - 0.48$; $r^2 = 0.84$, $n = 10$, $P = 0.0002$). Both axes are $\log_{10}$-transformed. Number of individuals is twice the total number of territories recorded in the wood in the period 1949–79.

more small species than large because the former, being able to subdivide their environment more finely, have more niches available to them (see also May 1978). This idea has been extended by Lawton and others, who have proposed that this relationship might be underpinned by the fractal nature of the environment (Morse *et al.* 1985; Lawton 1986; May 1986; Shorrocks *et al.* 1991; Williamson & Lawton 1991; Gunnarsson 1992; Davenport *et al.* 1996). Smaller species perceive more habitat space, in the same way that finer subdivisions of measurement give a longer measure of the coastline of Britain. More habitat space means more individuals of smaller body size can occupy the habitat and, if the numbers of individuals and species in a body size class are strongly positively correlated, more species. This last assumption seems generally to be borne out by studies finding positive relationships between the number of species and number of individuals in different size classes (Cousins 1994; Hall & Greenstreet 1996; Siemann *et al.* 1996, 1999; Gregory 1998b), and also applies to the avifauna of Eastern Wood (Fig. 5.14). For this last assemblage, the fractal-like branching structure of plants (Morse *et al.* 1985; Fitter & Strickland 1992; West *et al.* 1999) may result in higher habitat availability for the smallest bird species, because only they can forage on the smallest shoots and twigs and nest in the smallest tree-holes (see also Polo & Carrascal 1999).

The fit of animal assemblages to fractal models of habitat structure has been tested only for small-scale assemblages of animals living on plants. The expected increase in number of individuals as the body size of those animals decreases can be estimated from the fractal dimension of the habitat and the expected energy use by an individual animal (see Morse *et al.* 1985 for details). The energy use of individuals also needs to be taken into account, because small-bodied animals use less energy per individual than do large bodied. Assuming that population densities are inversely proportional to rates of individual resource usage (energetic equivalence; see Section 5.5), Morse *et al.* (1985) and Shorrocks *et al.* (1991) both found that the increase in number of individuals with decreasing body length in their assemblages (insects on plants and arthropods in lichen, respectively) fitted the predictions of fractal models. Gee and Warwick (1994) found that the negative slope of the relationship between number of individuals and body size for marine metazoa was steeper in macroalgae with higher fractal dimensions, as expected if habitat size is a significant component of habitat architecture, contributing to the size distribution of animals in a community.

Despite the relatively good agreement of these local assemblages with the predictions of fractal models, there remains an obvious discrepancy between observed and expected patterns: the fractal model predicts that the smallest size class ought to be the mode, yet generally it is not (e.g. Figs 5.2b, 5.4b, 5.5, 5.10b, 5.12). This discrepancy could be accounted for by most body size distributions being taxonomically restricted, whereas the fractal hypothesis should apply to all animals. Thus, while there may be few birds in the smallest body size classes (Fig. 5.4b), the greater amount of habitat available to species of this size is also used by insects and other animal groups: including all of these would shift the mode of the body size distribution to the smallest size class. However, this argument founders on the fact that body size distributions for all terrestrial (Fig. 5.7; May 1978; Tilman & Pacala 1993) and aquatic (Fig. 5.8; Fenchel 1993) animals also show modes displaced from the smallest size class. Fractal models are unlikely to be the sole explanation for the shapes of animal body size distributions.

In this vein, Polo and Carrascal (1999) have argued that the species–body mass distribution for Western Palaearctic passerine birds reflects patterns of habitat use, coupled with structural features of the habitats. Patterns of body mass variation are correlated with habitat use across the species they analyse. Species of open habitats (e.g. grasslands) tend to be larger and to show less variable body masses than species of structurally more complex habitats such as woodland. This effect is also shown in both the specific and phylogenetic components of the data when body mass variation is partitioned using the phylogenetic autoregressive method (Cheverud *et al.* 1985; Gittleman & Kot 1990). This implies that the habitat relationship is not just a consequence of the phylogenetic non-independence of species (it is present in the specific component), but also that habitat use has influenced body mass variation throughout the evolutionary history of passerines (because it is present in the phylogenetic component). Polo and Carrascal argue that these patterns arise because small body mass is required to exploit slender and pliable substrata, such as twigs and fine branches, in complex habitats. Thus, the preponderance of small-bodied birds in the avifauna of Eastern Wood is explained in terms of the wood's habitat complexity. Presumably, the shape of the global body mass distribution for birds follows as a consequence of the relative frequency of different habitat types available to them.

Resource distributions

These hypotheses about environmental grain can be viewed as attempting to match the distribution of animal body sizes to the general quantity of resource likely to be available for species of different sizes. A related body of work attempts to explain the form of consumer species–body size distributions by matching them to the size distribution of the much more specific set of resources that they actually consume. Thus, the body sizes of predators and their metazoan prey tend to be positively correlated (Peters 1983; Vezina 1985; Warren & Lawton 1987; Sabelis 1992; Cohen *et al.* 1993; Gaston *et al.* 1997d), as in some instances do the sizes of herbivorous insects and their host plants (e.g. Kirk 1991; Dixon *et al.* 1995), and parasites and their hosts (Kirk 1991). Differences in species–body size distributions at large and small spatial scales presumably then reflect differences in resource distribution.

The relationship between consumer and resource body sizes is most readily visualized for host specialists (note that within many such taxa, most species are host specialists and a small number are generalists, e.g. Price 1980; Jaenike 1990; Poulin 1998), and in groups that show little variation in life history strategy (like aphids). Where species feed on more than one resource, or where related species utilize rather different resources, the scenario obviously becomes more complex. Thus, while it is easy to see that the sizes of merlins and peregrines relate to the sizes of their prey, it is harder to understand why, on the basis of resource use alone, peregrines, ravens and greater scaup should all be about the same size. Nonetheless, relationships between the body sizes of exploiters and their resources may serve to shift the focus of investigation to the determinants of the species–body size distribution of the latter rather than the former.

Interspecific body size optimization

The theory of optimality provides a somewhat different perspective on the mechanism underlying species–body size distributions. Indeed, optimal body sizes have been explored in a variety of contexts (e.g. Case 1979; Roff 1981; Reiss 1989; Naganuma & Roughgarden 1990; Lundberg & Persson 1993; Dixon & Kindlmann 1994, 1999). The most prominent attempt to extend the theory to species–body size distributions has been that of Brown *et al.* (1993), which has been applied to birds by Maurer (1998b). The starting point is a definition of fitness in terms of energy and the power to do reproductive work. A two-step process is involved: resources must be acquired from the environment and subsequently transformed to reproductive work, once the costs of homeostasis have been met. Brown *et al.* (1993) argued that this energetic definition of fitness shifts the emphasis away from the trade-off between number and quality of offspring that has traditionally preoccupied evolutionary biologists (see Lack 1947 onwards; Stearns 1992). They noted that their model differs from previous ones concerning body size and the allocation of energy, by focusing not only on resource acquisition but also on the physiological constraints affecting energy conversion to reproductive power (but see Chown & Gaston 1997).

The rates at which resources are acquired from the environment (K_0), and are transformed to reproductive work (K_1), both scale allometrically at the level of the individual. Hence reproductive power for an individual of a given size, which is a function of both K_0 and K_1, can be determined from the body mass of the individual (M) as:

$$\frac{dW}{dt} = \frac{C_0 M^{b_0} C_1 M^{b_1}}{C_0 M^{b_0} + C_1 M^{b_1}} \qquad \text{(Eqn 5.1)}$$

where $K_0 = C_0 M^{b_0}$ and $K_1 = C_1 M^{b_1}$.

According to Brown *et al.* (1993), dW/dt '... reflects the energetic limits on the capacity to produce offspring', and is an application of the maximum power principle. An expression for optimal body mass, M^*, can be obtained by assuming that power per unit of body mass is maximized and by re-arranging the equation above as:

$$M^* = \left(\frac{-C_1 b_0}{C_0 b_1} \right)^{1/(b_0 - b_1)} \qquad \text{(Eqn 5.2)}$$

If the values of the allometric constants C_0, C_1, b_0 and b_1 are known, the optimal body size can be predicted. Maurer (1998b) used 0.75 (the same value as commonly claimed for the exponents for the allometry of metabolic rate and growth rate) as the value for b_0, which scales the rate of energy acquisition beyond maintenance needs. He assumed that b_1, which scales the rate of energy transformation, is -0.25, the same as that of the allometric exponent for most biological conversion processes. Brown *et al.* (1993) state that the model is not sensitive to small differences in the exact values of these exponents, which are known to vary both between taxa and between studies (e.g. Peters 1983; Sibly & Calow 1986). The constants C_0 and C_1 will differ more markedly between clades, which explains differences in the shape of the body mass distributions for different taxa, and presumably also their different optima (M^*). For the values of C_0 and C_1 chosen by Maurer (1998b) for birds, a plot of reproductive power against log body mass describes a right-skewed function with its peak at the optimum mass of 33 g. This function is very similar in shape to the species–body size distribution for the world's birds (Fig. 5.15).

Despite its intuitive appeal, a number of problems have been identified with the model proposed by Brown *et al.* (for detailed discussion, see Blackburn &

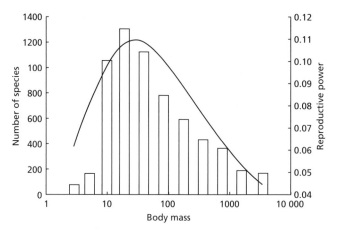

Fig. 5.15 The relationships between reproductive power (solid line; see text for more details) and number of species (bars) and body mass (g) for a large sample of all species of extant birds ($n = 6217$). From Maurer (1998b), with kind permission from Kluwer Academic Publishers.

Gaston 1996f; Chown & Steencamp 1996; Kozłowski 1996; Chown & Gaston 1997; Jones & Purvis 1997; Perrin 1998; Symonds 1999). These can briefly be summarized as follows.

1 Although mortality significantly influences life history evolution and mostly precludes the evolution of a body size that maximizes reproductive power output, it is excluded from the model of Brown *et al.* (Kozłowski 1996).

2 The model assumes that power and not efficiency is maximized, although there are conditions where this is not likely to be the case (Chown & Gaston 1997).

3 The model assumes that resource acquisition and conversion are physiologically limited, although the importance of physiological limitation in ecology remains unclear (Chown & Gaston 1997).

4 Although they appear to be quite similar, the curve of energy conversion efficiency (reproductive power) against body size does not predict the curve of the body size distribution. Rather, it predicts that all species should be the optimum, most efficient size (Blackburn & Gaston 1996f). A separate argument is required to derive the spread of species about the optimum. This is hypothesized to result from competition among those species for a finite set of resources.

In sum, the model in its present form is flawed, and does not actually predict the shape of species–body size distributions. The coincidence between the shape of the function it produces for the relationship between reproductive power and body mass, and the shape of the species–body mass distribution for birds is just that.

Other optimality models exist which address some of these difficulties. Indeed, the model of Brown *et al.* is one of a suite of physiological/life history models examining size at maturity (see Chown & Gaston 1997 for references).

These assume that such a thing as an optimum body size exists for species in a taxonomic assemblage. However, such a viewpoint may not be sensible (Blackburn & Gaston 1996f). A single optimum body size for a taxon seems unlikely given the diversity of life styles, habitats and selective forces acting on a fauna (see also Maurer 1998b). For example, the body size distribution of the world's birds is compiled from species ranging from subtropical to boreal and low Arctic habitats. Given latitudinal variation in avian body mass (Section 5.4), an optimum size for a bird is unlikely to be independent of the latitude at which it lives. In short, it seems unlikely that the variety of environmental influences experienced by birds would all act in concert to produce one optimum body size.

Moreover, the optimum body size depends on the taxonomic level chosen for examination. This has a number of consequences. Assume that the modal size of a bird is optimal for all birds. It follows that all ducks (and indeed all large-bodied birds), for example, are suboptimal, because they are larger than the modal size. It also follows that the smallest ducks are the most optimal, because they are closest to the mode. Identical arguments apply to warblers— most species are suboptimal in size, and only the larger species are optimal. Neither argument seems likely, and both are difficult to reconcile with hypotheses about, for example, the benefits of large body size (e.g. predator avoidance). The alternative is that the optimal size for ducks is not at the smallest size, and that for warblers is not at the largest size. But if this is true, then ducks and warblers have different optima, these are different from the optimum for birds in general, and so there is not a single optimum for all birds. Any argument that makes a special case of species or groups (e.g. swans became very large to escape predation) implicitly admits that there is not a single optimum for all birds. Given these concerns, it is notable that certain distinct subtaxa are frequently omitted from analyses of optimum size arguments (e.g. bats and aquatic species are excluded from mammal size distributions; Brown et al. 1993), because these taxa are suggested to have very different modes of life to other members of the group in question. The problem is repeated at each level in a taxonomic hierarchy. It is not clear that one level should have precedence over any other.

Finally, to test optimum size models, independent evidence that modal species are 'best' in the sense of the model is required. While it is tempting to think that the most common body size in an animal group is the 'best', there is no a priori reason why this should be the case. Even if it were, there are many ways in which this mass could be so: resistance to extinction caused by environmental perturbation (Johst & Brandl 1997; see above) is just one. The presence of a modal mass, whatever its value, provides no evidence that this is an optimum.

The obvious conclusion from arguments about the appropriate taxonomic level for the operation of body size optimization is that, if it occurs at all, each species evolves to the body size that is optimal for its way of life, or at least as close to that optimum as possible given other constraints (see also Dixon & Kindlmann 1994; Dixon *et al.* 1995). Thus, the peregrine is not less optimal than the merlin because it is further from the mode of the avian body mass distribution. This leads to issues of intraspecific body size optimization.

Intraspecific body size optimization

The most significant recent development in understanding the shape of the body size frequency distribution has been through consideration of intraspecific body size optimization, and the publication of a model of this process by Kozłowski and Weiner (1997; see also Cates & Gittleman 1997; Purvis & Harvey 1997). The appeal of this model is that the form of interspecific relationships arises as a consequence of intraspecific model processes, rather than having to be assumed. The advantage of this over interspecific optimization models is that it is within species that natural selection operates: interspecific patterns ought to be consequences, not constraints.

Kozłowski and Weiner's model is based on two simple assumptions. First, species within a taxon differ with respect to the parameters describing the size dependence of production rate (rate of energy assimilation minus rate of respiration) and mortality rate. Kozłowski and Weiner model these rates as power functions, with parameters of the functions chosen at random from normal distributions. These functions are therefore fixed within species, but differ between them.

Second, Kozłowski and Weiner assume that allocation of energy to growth and to reproduction has been optimized independently in each species by natural selection, with the goal of maximizing lifetime offspring production. Under size-dependent mortality, lifetime offspring production is maximized when the amount of energy allocated to growth increases future expected reproductive output by more than that amount. Thus, at the point where a calorie of energy used to grow increases future reproductive output by less than a calorie, the organism should allocate that calorie to current reproduction. In effect, this models evolution moulding the life history of a species by determining the time of the switch from growth to reproduction, with that timing based on the randomly determined production and mortality functions. The size of the species when this switch occurs is its optimum size.

The model described above produces two important results. First, although based on intraspecific allometries, it produces interspecific allometries that generally closely mimic those observed in real taxa. These interspecific relationships have no functional meaning of their own, but instead are the simple consequence of the intraspecific optimization. Second, the interspecific body size distribution resulting from the intraspecific body size optimization is log right-skewed, despite the intraspecific allometries being chosen at random from normally distributed variables. As for interspecific allometries, the interspecific body size distribution has no functional meaning, but is a simple by-product of intraspecific body size optimization.

With regard to the body size distribution, Kozłowski and Weiner's simulations generate a prediction that is, at least in principle, testable. They note that the degree of right-skew in the distribution is largely a consequence of the range of variation in the exponents of the production equations across species. As this is decreased, the body size distribution becomes more log-normal. This suggests that a relationship ought to pertain between interspecific variability in production rate exponents and skew. While difficult to test rigorously, this relationship does at least suggest a reason why body size distributions tend to be right log-skewed for entire taxa, but may less often be skewed for subtaxa (Stanley 1973; Loder 1997), as presumably the latter vary much less in production rate exponents. In that vein, it is interesting that Maurer (1998a) notes that the body mass distributions for most avian orders (20/23) are less skewed than for birds in general. Ten avian orders show significantly lower skew than expected on the basis of the complete bird distribution, whereas only two show significantly higher skew (Table 5.2).

The one significant problem with Kozłowski and Weiner's model is that the results obtained depend on the values (both mean and variance) of the exponents used to simulate the production and mortality functions. The exponents chosen give realistic results, but it is unclear how realistic are the exponents themselves. Nevertheless, Kindlmann *et al.* (1999) showed that the model generates significantly log right-skewed body mass distributions from a wide range of exponent values, but rarely produces significantly log left-skewed distributions.

A more general problem with optimization approaches, be they intra- or interspecific, is that it is difficult to use them to explain non-random differences between body size distributions across scales. They can explain large-scale distributions, but separate factors must be invoked to generate small-scale assemblages that are not simply random samples from those at larger scales. Indeed, the lack of any explicit prediction about differences in patterns in body size distributions at regional and local scales is a feature of all the mechanisms we have examined: none can explain a priori why the avifauna of Eastern Wood is not a random sample of the British woodland avifauna with respect to body size.

Synthesis

In sum, four distinct sets of mechanisms have been proposed to explain why small-bodied species are differentially favoured products of the net effects of speciation and extinction (Table 5.4). One should be wary, however, of regarding these as strictly competing explanations. Indeed, it seems more than likely that environmental grain, resource distributions and some form of body size optimization all contribute to the observed interspecific body size distribution,

Table 5.4 Mechanisms postulated as determining or contributing to why small-bodied species are differentially favoured products of the net effects of speciation and extinction.

Factor	Mechanism
Environmental grain	The relative frequency of niches available to species of different sizes favours small-bodied species
Resource distributions	The relative frequency of resources available to species of different sizes favours small-bodied species
Interspecific body size optimization	Trade-offs optimize the body sizes of different species
Intraspecific body size optimization	Trade-offs optimize the body sizes of individual species

with the first two perhaps constituting a rather different level of explanation from the last.

5.3.3 *Why do small- and large-scale body size distributions differ?*

As far as we are aware, there is only one published mechanism that explains why the body size distributions of species in local assemblages are not random samples of regional body size distributions. This was proposed by Brown and Nicoletto (1991) to explain their results for North American mammal assemblages. The full (verbal) model comprises three separate hypotheses.

First, interspecific competition is assumed to be fiercest among similar-sized species. Competitive exclusion would therefore lead to local assemblages comprising fewer similar-sized species than expected given the number in the regional pool. This would in turn result in more uniform body size distributions for local assemblages than expected, as seen earlier for the case of North American mammals. Moreover, log-uniformity must result from a higher proportion of modal-sized species from the regional distribution being excluded from local assemblages. Because modal-sized species tend to be small bodied, a second consequence of this assumption is that mean body mass should be higher in local than in regional assemblages. This is also a feature of the North American mammalian data.

Second, extinction probabilities are highest for large-bodied species with small geographical ranges, because the large home range size required by these species would result in low population sizes. This would mean that only widespread large-bodied species would be likely to persist. Geographical range size and body size do seem usually to be positively correlated, at least for birds and mammals, although the relationship is seldom strong (Fig. 5.16a; Van Valen 1973b; Brown & Maurer 1987, 1989; Arita *et al.* 1990; Maurer *et al.* 1991; Pagel *et al.* 1991a; Ayres & Clutton-Brock 1992; Cambefort 1994; Taylor & Gotelli 1994; Blackburn & Gaston 1996a; Hecnar 1999; reviewed by Gaston &

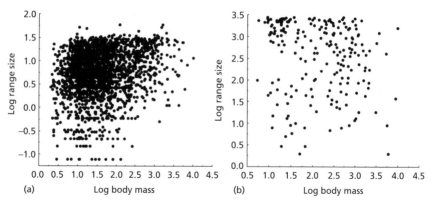

Fig. 5.16 The relationships between log breeding range size and body mass (g) for (a) endemic landbird species in the New World which are not listed as under threat of extinction ($r = 0.19, n = 2423, P < 0.0001$) and (b) bird species in Britain ($r = -0.17, n = 217, P = 0.0065$). For bird species in the New World, range size is the number of squares occupied on the WORLDMAP grid (Williams 1992, 1993), and in Britain is the number of 10×10-km squares occupied on the British National Grid. From data in Appendix III, and sources in Blackburn and Gaston (1996a).

Blackburn 1996a); when analyses are conducted over regions considerably smaller than the entire geographical ranges of the species of concern, then relationships may be positive, negative or non-existent (Fig. 5.16b; McAllister *et al.* 1986; Gaston & Lawton 1988a,b; Owen & Gilbert 1989; Dixon 1990; Sutherland & Baillie 1993; Virkkala 1993; Cambefort 1994; Inkinen 1994). Higher extinction probabilities for large-bodied species with small geographical ranges would also mean that these would exhibit little turnover across habitats and regions, as these species basically have to occur everywhere to maintain a population large enough to avoid extinction. Combined with the first assumption, the result would be few large-bodied species in any given biota.

The third assumption made by Brown and Nicoletto (1991) is that modal-sized species are more specialized. They suggested that the allometric scaling of energy requirements and digestion efficiency requires small-bodied species to utilize higher quality food resources. As these resources are likely to be rarer in the environment than those of poor quality, small-bodied species are forced to specialize to meet their energetic requirements. This in turn restricts the habitats in which these species can successfully forage, leading to higher turn-over of small-bodied species across the environment, but also the potential for more small-bodied species to persist in any given region. Combined with the second assumption, this explains why regional body size distributions tend to be peaked, while the first assumption explains why local distributions tend to be log-uniform.

This explanation for patterns in body size distributions is clearly a hybrid. The second and third strands of the mechanism form a hypothesis for the shape of large-scale distributions, while the first addresses differences between these and local distributions. Therefore, the plausibility of each half can reasonably be judged separately. In fact, there is little evidence to suggest that the explanation for the shape of regional body size distributions is correct. While large-bodied species may indeed have higher extinction rates (Section 5.3.1), there is little reason to believe that small-bodied species are generally more specialized than large-bodied. Moreover, as Brown and Nicoletto point out, an additional explanation is required for the paucity of species of smaller than modal size.

Nevertheless, the failings of Brown and Nicoletto's (1991) hypothesis for the shape of the regional body size distribution are of secondary importance here. Other hypotheses for this shape exist (see above). More relevant is the likelihood that their explanation for the difference between local and regional assemblages is correct. Unfortunately, while the mechanism they propose is plausible, what data there are provide little support. As already discussed, most studies comparing local and regional body size distributions find limited evidence for any non-random differences between them (Section 5.2.2: Smaller spatial scales), although some of these differences may be attributable to the low richness of some local assemblages. The body masses of Eastern Wood birds (Fig. 5.2b) are certainly no more uniformly distributed than are those of the British breeding avifauna as a whole (Fig. 5.10b). Indeed, the statistic quantifying departure from uniformity, the maximum difference between the cumulative relative frequency distribution expected under a uniform distribution and that observed, is greater for the Eastern Wood species (0.28) than for British birds as a whole (0.17). Moreover, the model predicts that the mean body mass of birds in Eastern Wood should be greater than that for the British avifauna, while it is actually less. However, as reiterated throughout the book (and see also Section 5.5.3 below), patterns of change in bird populations in this local assemblage show no evidence of interspecific competition, which is the driving force generating log-uniform distributions at local scales in Brown and Nicoletto's model. Local and regional body size distributions may differ only where competition is a significant structuring force at the local scale. This is, in principle at least, a testable prediction.

5.4 Spatial variation in body mass

One reason why aspects of the body mass distribution of the Eastern Wood avifauna and some other local assemblages may not constitute random samples of their regional pools is that, like species richness and geographical range size, body mass exhibits spatial variation. This was first noted by Carl Bergmann in 1847, which makes it one of the earliest macroecological observations (having recently passed its sesquicentenary). Bergmann (1847; quoted in translation in James 1970) wrote that: '(i)f we could find two species of [homeothermic] animals which would only differ from each other with respect to size, . . . (t)he geographical distribution of the two species would have to be determined by their size. . . . If there are genera in which the species differ only in size, the smaller species would demand a warmer climate, to the exact extent of the size difference. . . . Although it is not as clear as we would like, it is obvious that on the whole the larger species live farther north and the smaller ones farther south.' This observation has been named 'Bergmann's rule' in his honour.

As well as being one of the oldest macroecological observations, Bergmann's rule has been contentious (for example, see Rensch 1938; Scholander 1955; Mayr 1956; Ray 1960; James 1970; McNab 1971; Gittleman 1985; Ralls & Harvey 1985; Geist 1987, 1990; Paterson 1990; Cushman *et al.* 1993; Cotgreave & Stockley 1994; Hawkins 1995; Hawkins & Lawton 1995; Van Voorhies 1996, 1997; Loder 1997; Mousseau 1997; Partridge & Coyne 1997; Chown & Gaston 1999b). Interpretation of evidence for and against the rule has been complicated by confusion surrounding its definition. This has recently been reviewed by Blackburn *et al.* (1999b), who concluded that the rule was best defined as the tendency for a positive association between the body mass of species in a monophyletic higher taxon and the latitude inhabited by those species. We think that this definition is the most useful from the point of view of macroecology, given the current state of understanding, and is the one used here.

Given this definition, Bergmann's rule finds support from a range of taxa, including birds (Blackburn *et al.* 1999b). For example, Cousins (1980, 1989) showed that the mean body mass of landbird species breeding in 10×10-km squares of the British National Grid in north-west Scotland was approximately double that of species breeding in squares in south-east England. Figure 5.17 shows a similar effect, albeit of slightly smaller magnitude, for bird species breeding in seventeen 50×50-km grid squares forming a continuous transect north–south across Britain. At a larger scale, Fig. 5.18 shows the relationship for New World birds (Blackburn & Gaston 1996e), which illustrates a threefold increase in the mean body mass of birds inhabiting different latitudes as one moves from the Equator towards the North Pole. This pattern still holds, albeit that it is considerably weakened, if analysis is performed controlling for the phylogenetic relatedness of taxa (Blackburn & Gaston 1996e).

However, as with most macroecological patterns, exceptions to Bergmann's rule have been found in some faunas. All of these involve insects (Miller 1991; Hawkins 1995; Hawkins & Lawton 1995; Hawkins & DeVries 1996; Loder 1997) or insectivores (Cotgreave & Stockley 1994). This pattern does not deny the existence of latitudinal body mass variation in other taxa, which is very well supported, but does bear on the question of what is causing Bergmann's rule in those taxa that, like birds, do exhibit it.

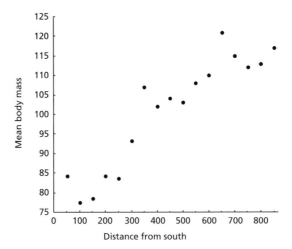

Fig. 5.17 The geometric mean body mass (g) for bird species breeding in 17 quadrats of 50 × 50 km placed in a continuous transect running south–north across Britain, and the distance (km) of the northern edge of those quadrats from the southern end of the transect. The species included are the same as in Fig. 3.11, a total of 179 across the whole transect. The strength of this relationship cannot accurately be assessed by standard statistics, because of spatial autocorrelation in the data (neighbouring quadrats do not contain independent sets of species).

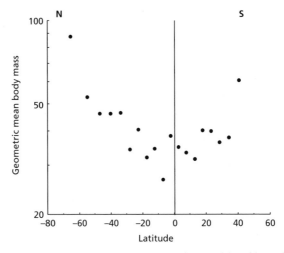

Fig. 5.18 The relationship between geometric mean body mass (g) and latitude (degrees) for bird species in the New World, calculated across latitudinal bands on the WORLDMAP grid (Williams 1992, 1993) using the mid-point method (Section 3.4.3). The Equator is indicated by a vertical line. From Blackburn and Gaston (1996e).

Table 5.5 Mechanisms postulated as determining or contributing to latitudinal gradients in body size (Bergmann's rule).

Factor	Mechanism
Phylogeny I	Random ancestral colonization by a large-bodied species
Phylogeny II	Selective advantage of traits which are accidentally coupled with large body size
Dispersal ability	Small body mass is associated with low dispersal ability
Heat conservation	Increased heat conservation of large-bodied species
Starvation resistance	Increased starvation resistance of large-bodied species

5.4.1   *What determines spatial variation in species body sizes?*

Of the hypotheses that have been suggested to explain Bergmann's rule, five may reasonably be applied to interspecific patterns in birds (Table 5.5).

First, larger body mass at high latitudes in a taxon may result from random ancestral colonization by a large-bodied species, and subsequent diversification from this initial colonist to produce an assemblage of predominantly large-bodied species. This explanation is extremely unlikely. Taxa exhibit Bergmann's rule in areas that were covered by glaciers within the last 20 000 years, far too recently for speciation to have caused the patterns (Cushman *et al.* 1993; Blackburn & Gaston 1996e; Klicka & Zink 1997, 1999; Avise & Walker 1998). Moreover, it is rendered unlikely by phylogenetically controlled analyses of New World birds, which show weak but significant evidence for Bergmann's rule (Blackburn & Gaston 1996e). This result implies repeated and independent evolution of the association between mass and latitude in a large number of taxa, rather than one or two random colonization events.

The second hypothesis is that larger body mass at high latitudes may result from the selective advantage of traits accidentally coupled with body mass. This implies no advantage *per se* to large size at high latitudes, but that since body size is correlated with so many other features of the life histories and biologies of species, it would be quite likely that whichever did cause some species to inhabit high latitudes might by chance be associated with body size. Ultimately, this second, phylogenetic hypothesis has to be correct in some form. It is not the fact that species living at high latitudes tend to weigh more that allows them to live there, any more than it is the lines of latitude painted on the globe that require the body size gradient; rather, it is some feature associated with this additional mass. The question then becomes whether the association is causal or spurious. This hypothesis posits the latter.

The phylogenetically controlled analyses of New World birds (Blackburn & Gaston 1996e) would seem to render this hypothesis unlikely. The significant positive relationship between body mass and latitude within taxa is a consequence of the association evolving independently in a large number of bird taxa, and is thus unlikely to arise through a repeated accidental coupling of body mass to other traits advantageous at high latitudes. Nevertheless, the relationships between body mass and latitude revealed by phylogenetically controlled analyses are generally weak (see also Loder 1997), suggesting that there may be a grain of truth in one or other of the phylogenetic hypotheses (Blackburn & Gaston 1996e), most likely the latter. If body mass is very well correlated with whatever other feature of a species adapts it for life at high latitudes, then even phylogenetically controlled analyses would indicate some association between latitude and body size.

The third mechanism proposed to explain Bergmann's rule is that small body mass is associated with low dispersal ability. Small-bodied species are under-represented at high latitudes because they have failed to disperse there as often as have large-bodied species in the period since these latitudes became habitable. This hypothesis is plausible for birds. In the New World, for example, there is evidence that migrant species tend to be larger bodied and to live further north, on average, than residents (e.g. Blackburn & Gaston 1996e). However, this pattern results because migrants and residents belong to different taxa, and disappears once the effect of phylogeny is removed. Moreover, an association between migratory ability and body size does not necessarily say much about whether the lack of small-bodied species at high latitudes is a consequence of their more limited dispersal abilities. Migration and dispersal are two separate processes (e.g. Paradis *et al.* 1998). A species with a fixed migratory route between well-defined breeding and wintering grounds may actually show little tendency to colonize new areas. Colonization may be as likely to occur through postbreeding or natal dispersal by resident species.

The relationship between dispersal, migratory ability and body size has been investigated in detail for common British birds by Paradis *et al.* (1998). They found that both natal and adult dispersal distances were positively correlated with body mass across species, and that migrants tended to disperse further than residents. These results support the migration ability hypothesis. That said, however, the body mass effect was weak, explaining less than 10% of the variation in both measures of dispersal. Many small-bodied species tended to disperse just as far as most larger-bodied species, while even the smallest bird species can cover long distances on migration (e.g. ruby-throated hummingbird, Pallas's warbler). Even though the normal dispersal distances of most bird species are low—of the order of 5 km (Paradis *et al.* 1998)—that should have enabled even the smallest-bodied, most sedentary bird species to colonize most high latitudes in the 11 000 years since the glaciers last retreated. Moreover, dispersal distances in an environment that a species already inhabits may underestimate rates of spread in unoccupied regions. Paradis *et al.* (1998) calculated arithmetic mean annual adult and natal dispersal distances for the house sparrow in Britain of 1.9 and 1.7 km, respectively, yet Van den

Bosch *et al.* (1992) estimated that this species expanded its range by 10.2 km per year following its introduction into North America. Thus, we think it unlikely that the migration ability hypothesis can realistically explain latitudinal gradients in avian body masses.

The fourth hypothesis for latitudinal body mass gradients we address was that favoured by Bergmann himself. Large body size might allow species to occupy higher latitudes because it increases heat conservation via higher surface area to volume ratios. This mechanism suffers the significant problem that the decrease in cooling rate consequent from larger body mass is negligible in comparison to that gained by increased insulation (e.g. Scholander 1955; Geist 1987, 1990). However, since larger-bodied birds can have thicker and heavier layers of feathers (Scholander *et al.* 1950; Herreid & Kessel 1967; Calder 1984), and hence lower conductance values (Herreid & Kessel 1967; Peters 1983; Calder 1984), large body size may result in better heat conservation for this reason, rather than from the effect on surface area to volume ratios.

A different view of an adaptive relationship between body size and heat was taken by James (1970). She noted that intraspecific body size variation for a number of birds, such as the downy woodpecker, was better related to wet-bulb than dry-bulb temperature. Thus, individuals living in warm humid environments were smaller than individuals living in warm dry environments. This suggested that body size varied in response to the demands of keeping cool, rather than keeping warm, because individuals living in warm dry environments can take advantage of evaporative cooling (birds have no sweat glands, but instead increase evaporation by panting or gular fluttering), whereas individuals living in warm moist environments can only keep cool by decreasing their rate of heat production. Because heat production scales positively with body size, a good way to decrease this production is to be small bodied. James (1970) developed this hypothesis to explain patterns of body size variation within bird species. Whether it also applies across species remains to be tested.

The final hypothesis that we think can plausibly explain Bergmann's rule in birds is that larger body mass increases starvation resistance, because fat reserves increase more rapidly with body size than does metabolic rate (Calder 1984; Lindstedt & Boyce 1985). This may be an advantage at high latitudes where resources are often seasonally scarce. Starvation resistance might also explain the observation that the exceptions to Bergmann's rule have so far been limited to insect and insectivore faunas. Loder (1997) suggested that the lack of a relationship in these groups may be a consequence of the overwintering strategy of insects, most of which spend this season in dormant stages.

Nevertheless, the wider evidence for the starvation resistance hypothesis is not compelling (e.g. Reavey 1992, 1993; Chown & Gaston 1999b). Dunbrack and Ramsay (1993) have questioned the support for the hypothesis in mammals, and their arguments equally apply to birds. They point out that while the

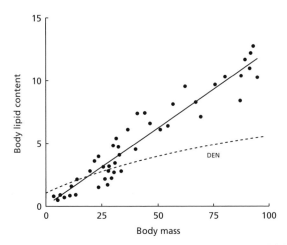

Fig. 5.19 The relationship between body lipid content (g) and body mass (g) for wintering songbird species. The lipid equivalent of daily energy needs (DEN), assuming 1 g of lipid yields 37.7 kJ, is indicated by a dashed line, and shows that the amount of fat surplus to DEN increases with body mass. From Newton (1998).

allometric scaling of fat reserves would imply greater starvation resistance in large-bodied relative to small-bodied species (Fig. 5.19), many other strategies can be adopted to ameliorate the stresses of seasonality. For example, in comparison to their larger-bodied relatives, it may be easier for small species of mammal or bird to exploit stored food reserves, to exploit microclimatic refugia (such as burrows and nests) or to decrease metabolic expenditure through torpor or hibernation. Many small birds exploit torpor on a daily basis (e.g. Anna's hummingbird), and this state can be maintained for several weeks by the common poorwill (Jaeger 1949; Bartholomew *et al.* 1957). Moreover, birds can also ameliorate the stresses of seasonality by short- or long-distance migration. In fact, if food is scarce, the advantage accrued by increasing body mass, and hence absolute food requirements (McNab 1971), is debatable. The jury is thus still out on whether starvation resistance can adequately explain Bergmann's rule.

Of these five hypotheses for interspecific latitudinal variations in body size, none is particularly compelling, at least on the basis of available empirical evidence. However, it is difficult to believe that the pattern does not relate in some way to an adaptive advantage of large body size at high latitudes, or at least a lack of disadvantage.

5.4.2 *Bergmann's rule, species–body size distributions and abundance*

As should be apparent, the explanations which have been proffered for spatial variation in body mass and for the shape of overall body mass distributions are mutually exclusive (compare Table 5.4 and 5.5). We find this remarkable. Where there is spatial variation in body mass, the shapes of body size distributions for large-scale assemblages will certainly be affected by it. This could have at least two separate consequences.

First, the shapes of body size distributions for large-scale assemblages and spatial variation in body mass may be generated by common factors. This is an attractive idea, albeit one that is complicated by the lack of any clear consensus about what those factors might be. As already argued, ultimately the body size distribution should be a consequence of patterns of speciation and extinction, implying that the same should be true of spatial variation in body size. However, this still begs the question of what causes there to be disproportionately more small-bodied species in taxa such as birds, but fewer small-bodied species at high latitudes. Nevertheless, if there is some commonality in these factors, there may be two obvious corollaries. First, any factor that favours large body size mainly at high latitudes may cause small size to be generally favoured, because low latitudes are more extensive and species rich than high latitudes (Section 2.5.2). Second, we might see different patterns in species–body size distributions for taxa that do and do not show Bergmann's rule. As at present we only have large-scale data on the distribution, and spatial variation in, body size for taxa that do show Bergmann's rule, this latter prediction awaits testing.

The second consequence of spatial variation in body mass may be to generate the mismatch between the body mass distributions of local and regional assemblages (Loder 1997). Both North American mammals and British birds show latitudinal gradients in body mass (Cousins 1980, 1989; Zeveloff & Boyce 1988), but smaller-scale assemblages are, in general, not random samples of the regional faunas with respect to mass. The body mass distribution for the fauna of Eastern Wood, for example, differs from random expectation in mean but not in shape. This may be because the site is close to the extreme south-east of England. Because large-bodied bird species tend to breed further north in Britain (Cousins 1980, 1989), they may not be a legitimate part of the source pool from which the Eastern Wood fauna is drawn. An example is the black grouse, which breeds in deciduous woodland (Cramp & Simmons 1980), but which in England no longer occurs south of the Peak District (Gibbons *et al.* 1993). Loder (1997) found that the body sizes of species in local and regional ('life zone') North American butterfly assemblages did not differ in the majority of cases from samples of species from larger source pools. However, the increase in butterfly body size with latitude in North America was only slight, as would be expected if it was this gradient that was generating local–regional differences.

Whether this mechanism can explain the scale-dependent body size variation in North American mammals is less clear. The median body masses of mammal species in biomes differed from random samples of the continental pool, but were always larger than expected. This seems unlikely to be the direction predicted as a consequence of an increase in body mass with latitude. If large-bodied species live only at high latitudes, they should not constitute part of the source pool for more southern biomes, causing most biome median masses to be smaller than expected. However, an alternative scenario is that Bergmann's rule in North American mammals is largely a consequence of the presence of fewer small-bodied species at high latitudes, whereas large-bodied species occupy both high and low latitudes. This would be expected from the relationship between range size and body size, which is often positive, at least in vertebrate taxa (Fig. 5.16; Section 5.3.3; Gaston & Blackburn 1996a). Under this alternative, latitudinal body size variation would lead to an excess of small-bodied species in the continental source pool, which in turn would bias random draws from this pool towards smaller body masses. Unfortunately, we cannot yet judge how realistic is this alternative scenario, because while it is known how mean body mass varies with latitude in North American mammals (Zeveloff & Boyce 1988), it is not known how minimum and maximum masses vary (although see Shepherd 1998). Even if it was, a thorough analysis of the issue would be required.

Finally, spatial variation in body mass can potentially affect observed patterns in the species richness of different regions. We began this chapter noting that the resources available to natural assemblages are finite. Therefore, the number of individual animals that a region can support is also limited. All else being equal, if every individual animal was large bodied, a region ought to be able to support a smaller number of animals than if they were all small bodied. That is because large-bodied individuals use more energy per capita than small-bodied ones. Therefore, animal body size ought to trade off in some way with animal abundance. However, the form of this trade-off will also interact with species richness. A region may predominantly be inhabited by abundant large-bodied species at the expense of low species richness. As both body mass and abundance seem most likely to increase with latitude across species, for whatever reason, their combined effect may be to reduce the species richness of high-latitude regions below that predicted by a direct effect of energy on species number (Blackburn & Gaston 1996c). The precise form of this effect will depend on the relationship between body mass and abundance. It is this relationship that is next considered.

5.5   Abundance–body size relationships

One effect of the allocation of finite resources among the species of a biota may be a trade-off between the body size and abundance of animals. Even if the fact of a trade-off were established, though, this of itself says nothing about how an assemblage constrained by such should be structured. For example, large-bodied species could be more abundant than small-bodied if being large enabled a species to appropriate a higher proportion of available resources.

This might occur, say, if large size was an advantage in contest competition. Alternatively, small body size might be an advantage in scramble competition, or in situations where resources are too scarce to support a viable population of a large-bodied species. Thus, the relationship between body size and abundance in animal assemblages could take a variety of forms. The form it does take has been one of the most contentious issues in macroecology (see reviews in Cotgreave 1993; Blackburn & Lawton 1994; Blackburn & Gaston 1997b, 1999).

5.5.1   *What is the relationship between abundance and body size?*

The first serious attempt at an answer was provided by Damuth (1981), based on data collated from the literature on population density and body mass for 307 species of mammalian terrestrial primary consumers. He found that the density–body mass relationship for these species was linear on logarithmic axes, with a slope of −0.75. This relationship is shown in Fig. 5.20, for a slightly expanded set of species. Thus, density (D) and body mass (M) show a power relationship of the form $D \propto M^{-0.75}$. This is remarkable. Recall that individual energy use (metabolic rate, R) has commonly been argued to scale with body mass as $R \propto M^{0.75}$ (Section 5.1). If these two allometric relationships were combined, then, as Damuth noted, population energy use—the product of population density and individual energy use—should scale as $\propto M^{-0.75}M^{0.75} \propto M^{0}$. This result leads to the unexpected conclusion that species of different body sizes in an assemblage use approximately equal amounts of energy (Damuth 1981). This has since been dubbed the 'energetic equivalence rule' (Nee *et al.* 1991c).

Although widely employed, the term 'energy equivalence' can be misleading. It does not, for example, mean that all species use equivalent amounts of energy. Reference to Fig. 5.20 or any other plot of the abundance–body size relationship will show that there are usually orders of magnitude variation in the abundance attained by species of the same body size. Energy is not equably divided. Instead, energetic equivalence implies that no species has an a priori energetic advantage over any other on the basis of its body size (Damuth 1981). Large-bodied species do not, on average, get a larger slice of the energy 'cake' by virtue of their size. What species gain energetically on the swings of body size they must lose on the roundabouts of abundance.

Damuth's original work has spawned a large number of studies of the relationship between abundance and body size, many of which find similar relationships (Damuth 1981, 1987; Peters & Wassenberg 1983; Peters & Raelson 1984; Marquet *et al.* 1990; Strayer 1994; Ebenman *et al.* 1995). Figure 5.21 shows the relationship for British birds, using data from two different sources. Figure 5.21a comes from a study by Nee *et al.* (1991c), using the population sizes given by Marchant *et al.* (1990). The slope of this relationship matches perfectly that expected if the energetic equivalence rule holds. Figure 5.21b uses the updated

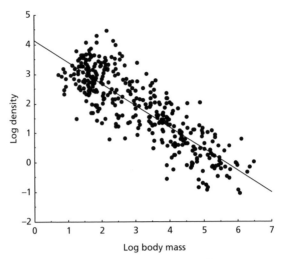

Fig. 5.20 The relationship between density (numbers per km^2) and body mass (g) for a selection of species of mammalian primary consumer ($y = 4.15 - 0.73x$: $r^2 = 0.71, n = 368$, $P < 0.0001$). From data in Damuth (1987).

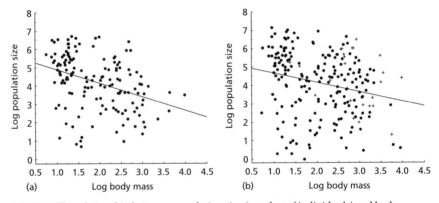

Fig. 5.21 The relationship between population size (number of individuals) and body mass (g) for bird species in Britain for (a) the data used by Nee *et al.* (1991c) ($y = -0.75x + 5.64$; $r^2 = 0.14, n = 147, P < 0.0001$) and (b) the data in Appendix III ($y = -0.51x + 5.20$; $r^2 = 0.06$, $n = 217, P = 0.0004$). Nee *et al.* excluded seabirds and introduced species from their analyses. Excluding these (indicated by +) from (b) results in the relationship: $y = -0.61x + 5.31$ ($r^2 = 0.07, n = 191, P = 0.0002$).

population sizes given in Appendix III. The slope of this relationship is less steep, although the relationship is weak enough that the 95% confidence intervals still embrace −0.75.

However, a significant number of studies find abundance–body size relationships that are not of the simple power form. For example, Brown and Maurer (1987) calculated the abundances of landbirds, using data from the North American Breeding Bird Survey (Robbins *et al.* 1986). Plotted against

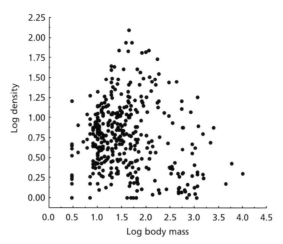

Fig. 5.22 The relationship between density (number of individuals per Breeding Bird Survey census route) and body mass (g) for species of bird in North America. From Brown and Maurer (1987).

body mass, they produce a complex 'polygonal' relationship (Fig. 5.22). Maximum abundance peaks at intermediate body masses, while minimum abundance is constant across all masses. Relationships of this sort have also been shown in a wide variety of taxa (e.g. Gaston 1988; Morse *et al.* 1988; Basset & Kitching 1991; Novotny 1992; Blackburn *et al.* 1993; Cotgreave *et al.* 1993; Blackburn & Lawton 1994; Cambefort 1994; Nilsson *et al.* 1994). While the precise details differ between studies, such relationships are typically characterized by the absence of a marked increase in minimum abundance for small-bodied species, and by a low correlation between abundance and body size. Frequently, peak abundance is at intermediate body sizes, and sometimes the overall relationship is actually positive (Blackburn & Gaston 1997b). Applying the logic of the previous paragraphs to relationships of this sort would suggest that populations of large-bodied species use more energy than do those of small-bodied species.

The linear and polygonal forms of the abundance–body size relationship represent different points on a continuum of possible relationships. Blackburn and Gaston (1997b) collated quantitative data from the literature for around 300 such relationships, in an attempt to reconcile the different forms. Figure 5.23 shows how the regression slopes are distributed for these relationships. Most are negative, but around 25% of all plots show a positive regression slope. The mode of the distribution appears to be around −0.75, although the median and mean slope values are actually −0.58 and −0.51, respectively. In most cases, the coefficient of determination of these relationships was low (Fig. 5.24). Body size explained < 10% of the variation in species abundance in more than one-third of assemblages. These quantitative data show no discontinuities in the distributions of either slope or strength.

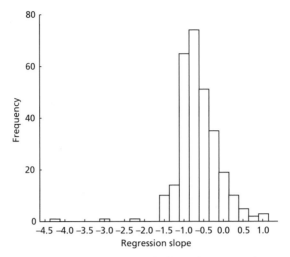

Fig. 5.23 The frequency distribution of model I (ordinary least squares) regression slopes from 291 interspecific plots of log abundance against log body mass for a wide variety of animal assemblages. Horizontal axis labels indicate the lowest slope value in each bar; for example, the first bar includes studies with regression slopes from −4.25 to −4.01. From Blackburn and Gaston (1997b).

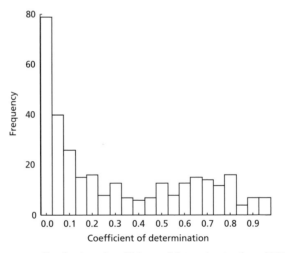

Fig. 5.24 The frequency distribution of coefficients of determination from 326 interspecific plots of log abundance against log body mass for a variety of animal assemblages. Horizontal axis labels indicate the lowest coefficient value in each histogram bar in each case; for example, the first bar includes studies with coefficients from 0 to 0.049. From Blackburn and Gaston (1997b).

Many relationships are difficult to classify unambiguously as either linear or polygonal, while it is unclear how to distinguish between cases that are genuinely polygonal or that show no relationship at all. What, for example, is the correct interpretation of the patterns for British birds (Fig. 5.21)?

5.5.2 *Why do abundance–body size relationships show different forms?*

What determines where interspecific abundance–body size relationships lie along the continuum of observed forms? At least eight different reasons have been proposed (Table 5.6), of which Blackburn and Gaston (1997b) were able to test five. They found that variation in the regression slope between studies is only explained by the type of data used and the scale of study. In general, simple linear negative relationships are obtained for data compiled from a variety of literature sources (Lawton 1989). These are often single- or few-species studies from a range of geographical areas, and densities of different species may have been estimated using very different methodologies. Conversely, polygonal relationships are generally recovered when single areas are sampled to give abundance estimates for all the species in a given taxon that are present, usually using a single or consistent method. The census of the avifauna of Eastern Wood is a good example of this latter study type, and the resulting abundance–body size relationship can reasonably be classed as polygonal (Fig. 5.25). Polygonal relationships also tend to arise from studies performed at local or restricted sites (e.g. Morse *et al.* 1988), whereas linear negative relationships tend to arise as the geographical coverage or spread of data increases. A comparison of abundance–body size relationships for Eastern Wood and Britain (Figs 5.21 & 5.25) is illustrative.

The strength of abundance–body size relationships, as measured by their coefficients of determination, was affected to some degree by four of the five mechanisms that Blackburn and Gaston (1997b) were able to test.

Table 5.6 Mechanisms postulated as determining or contributing to why different empirical abundance–body size relationships have been obtained, and in particular to reconcile why linear and polygonal patterns may both be observed (Blackburn & Gaston 1997b).

Different patterns hypothesized to result from:

---

1 Comparisons of different kinds of data
2 Samples at different spatial scales
3 Differences in the density measures used
4 Differences in the ranges of variation represented by the species included in plots
5 Polygonal relationships being random samples from underlying negative ones
6 Differences in how taxa utilize space
7 Differences in the taxonomic composition of assemblages
8 The effects of migrant species

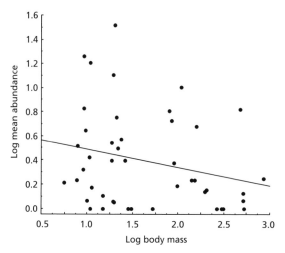

Fig. 5.25 The relationship between mean abundance (average number of territories per year when present) and body mass (g) for the bird species breeding in Eastern Wood in the period 1949–79 ($y = 0.64 - 0.15x$; $r^2 = 0.06$, $n = 45$, $P = 0.10$). From data in Appendices II and III.

Relationships from large-scale studies based on compilation data tended to be stronger than small-scale studies based on uniformly sampled data. In addition, there were effects of the body size range of the organisms in the study (Lawton 1989), and in how taxa utilize space (Juanes 1986). Strong negative relationships are typically obtained from plots spanning several orders of magnitude of body sizes, whereas the ranges spanned by species in weaker relationships are typically less. Polygonal relationships are therefore often suggested to be simply segments of an overall linear negative relationship, which would be recovered if the range of body sizes in the polygonal relationship was increased (Currie 1993). Although it may apply in general, this effect is not apparent from comparison of the birds of Eastern Wood and Britain. The greater body size range exhibited by the entire British avifauna does not increase the strength of the abundance–body size relationship relative to that of Eastern Wood (Figs 5.21 & 5.25). Nevertheless, as pointed out in Chapter 1, Britain is but a small island. Its bird population is less than 0.1% of the estimated world total (200–400 billion individuals; Gaston & Blackburn 1997b), and the commonest species, the wren, has a British breeding population (14.2 million) that is a small fraction of the estimated global population of the red-billed quelea, which is commonly cited as the most abundant species of wild bird (perhaps 750 million breeding and 1500 million post-breeding individuals; Elliott 1989). This may be too small a sample of the global avifauna to detect a general negative abundance–body size relationship.

Weak abundance–body size relationships tend also to be observed for taxa that utilize the environment in three spatial dimensions (Blackburn & Gaston

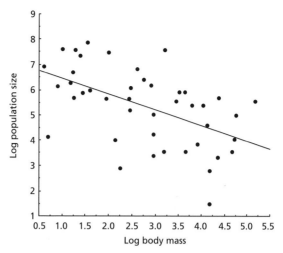

Fig. 5.26 The relationship between population size (number of individuals) and body mass (g) for non-volant wild mammal species in Britain ($y = 7.09 - 0.62x$; $r^2 = 0.28$, $n = 43$, $P = 0.0003$). *Homo sapiens* is excluded. From data in Harris *et al.* (1995).

1997b). As densities are generally measured using areas (at least in the terrestrial environment), the linear negative relationship between density and body size may break down if species are actually utilizing different volumes of space. In that respect, birds generally show weaker abundance–body size relationships than do mammals (Blackburn & Gaston 1997b), which has been argued to result from this effect (Juanes 1986; Cotgreave & Harvey 1992). Certainly, the relationship for British non-volant wild mammals (Fig. 5.26) is stronger than that for British birds (Fig. 5.21). Similar logic suggests that flightless birds should exhibit stronger negative abundance–body size relationships than flighted species. Results presented by Ebenman *et al.* (1995) support the suggestion, although other analyses by Cotgreave and Harvey (1992) do not.

General linear modelling indicated that study scale is the most important factor influencing when abundance–body size relationships of different slope and different explanatory power are likely to arise (Blackburn & Gaston 1997b). There are two reasons why these associations with study scale might exist. First, there is a single 'true' relationship underlying all studies, but either large- or small-scale studies misrepresent the true relationship to some degree. One or other pattern is an artefact of the data. Second, abundance–body size relationships truly show different forms at different scales.

The former of these two possibilities is the one usually assumed to be correct. Relationships of polygonal form are argued to be artefacts of sampling from an underlying abundance–body size relationship of the negative log-linear form. The case for this view is indeed strong. For example, Currie (1993) and Griffiths (1998) showed that polygonal relationships can be generated by random samples within a restricted range of body sizes from a linear negative

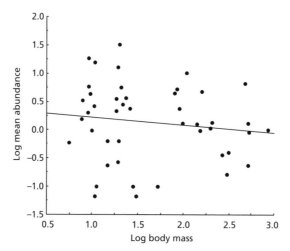

Fig. 5.27 The relationship between mean abundance (average number of territories per year, including years when absent) and body mass (g) for the bird species breeding in Eastern Wood in the period 1949–79 ($y = -0.14x + 0.37$; $r^2 = 0.02$, $n = 45$, $P = 0.39$). From data in Appendices II and III.

relationship. The lower limit to abundance in many local studies is the same across all body sizes, and corresponds to that of a species represented in the sample by a single individual. This suggests that the polygonal form is largely a consequence of the lower abundance limit being set by the limitations of the sampling technique (Blackburn *et al.* 1990; Currie 1993). By contrast, the area encompassed by regional studies makes it unlikely that many density estimates from them will be based on the presence of only one animal, especially if vagrants are excluded (although see Fig. 5.22). Abundances of species in regional studies are more likely to be representative of their true levels.

This random sampling idea does, however, leave unanswered the question of the level of sampling required to reveal the true abundance–body size relationship in a local assemblage. Any relationship that is not simple, linear and negative could be argued to arise from inadequate sampling, in the absence of an objective test. For example, after 30 years of intensive fieldwork, we still find a polygonal relationship between abundance and body size in the avifauna of Eastern Wood (Fig. 5.25). The weakness of this relationship is not just a consequence of the fact that abundance was calculated as the average number of territories in years when present, so constraining the minimum abundance to be 1. If it is calculated as the average number of territories over all years in the sample, we lose the well-defined horizontal minimum abundance, but the relationship is even weaker, its slope even less negative (Fig. 5.27). To what extent would the form of this relationship change given even more intensive local sampling?

In fact, there is little to suggest that the strength of the abundance–body size relationship in local assemblages would be altered markedly by more comprehensive sampling. The tendency for species to be rare at many sites within their range (Section 4.2.1) means that polygonal relationships at local sites might reasonably be expected in the absence of any sampling artefacts. One way in which samples from local sites might distort density estimates is if the densities of all species are calculated over the entire census area. For example, one pair each of sparrowhawk and willow tit may breed in Eastern Wood in a typical year, but whereas the sparrowhawks range through the entire wood, the willow tits may be restricted to a small part (say, damp woodland around the ponds). Both species would apparently have the same density if calculated as numbers in the wood, even though numbers per unit area used are quite different. However, both Carrascal and Tellería (1991) and Gregory and Blackburn (1995) examined how abundance–body size relationships for birds change as the measure of density used is increasingly representative of the area actually used by species. The relationships strengthen as areas become more representative, but in neither case does body mass explain more than 30% of the variation in density. This compares to 74% in Damuth's (1981) mammal data. As has frequently been pointed out (Cotgreave 1993; Currie 1993; Blackburn & Lawton 1994; Silva & Downing 1995), even strong negative abundance–body size relationships usually show densities spanning up to four orders of magnitude at any given body size (e.g. Damuth 1981, 1987). At local scales, this is likely to translate into 'blob-like' relationships, so that body size would not be a good predictor of a species abundance in local assemblages even if all sampling artefacts could be circumvented.

In summary, available evidence favours a broadly linear negative relationship between log abundance and log body size at large spatial scales, such that the large-bodied species occupying a region such as Britain tend, on average, to have lower population sizes and densities than do the small-bodied. Nevertheless, and despite this relationship, body size is a poor predictor of the abundance a species is likely to attain in local assemblages. Why does the abundance–body size relationship not generalize across scales? It is not just a consequence of the narrower range of body sizes of species in local assemblages, because body size range has no consistent effect on the form of the relationship (Blackburn & Gaston 1997b). We think that there are two reasons.

First, regional abundance–body size relationships use abundances averaged across sites, whereas local relationships use individual abundance estimates. As seen when examining the anatomy of the abundance–range size relationship (Section 4.2.1), species do not attain the same abundance at all sites. Abundances show variation. However, this variation is predictable. Minimum abundances are constant across species, whereas maximum abundances are higher for species with higher mean abundance. In absolute terms, common species are rare at some sites, but rare species are rare at all sites. This results in

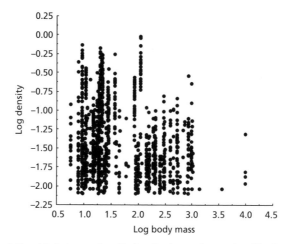

Fig. 5.28 The relationship between density (territories per hectare) and body mass (g) for species of bird on 25 farmland CBC plots in the period 1968–72. Each data point represents one species on one site, so that most species are represented by more than one point. From data in Appendix III, and sources in Gaston *et al.* (1998c).

a triangular abundance–range size relationship when abundances are not averaged, but instead are all plotted on the same graph (Fig. 4.5), because the species with highest mean abundances are also the most widespread. However, since the most abundant species tend also to be those with the smallest body size, we would expect a mirror image of Fig. 4.5 if range size was replaced with body size. In fact, this is basically what we observe (Fig. 5.28), albeit that the blackbird ($\log_{10}$ body mass = 2.03) is noticeably more abundant than expected from its size. It follows that as all small-bodied species are absolutely rare at some sites, most local sites will house some small-bodied species in low numbers. But as large-bodied species should be rare at all sites, triangular abundance–body size relationships will generally result.

This first reason surely contributes towards the different shapes exhibited by local and regional abundance–body size relationships, but it is not a sufficient explanation on its own. This is because many local abundance–body size relationships are positive, or are negative but not significantly so. By contrast, while local abundance–range size relationships are often approximately triangular in shape (e.g. Figs 4.3 & 4.6), they are virtually always significantly positive (Gaston *et al.* 1998c). Yet, the abundance variation argument just outlined applies equally to both relationships. Thus, there must be a further explanation for local and regional pattern variation that explains this difference. A clue is provided by phylogenetically independent analyses of these two patterns.

We saw in the previous chapter that positive abundance–range size relationships tend to be equally strong when controlling for the phylogenetic non-independence of species (Section 4.2.2). By contrast, the abundance–body size relationship does not stand the same rigorous statistical examination. Nee *et al.*

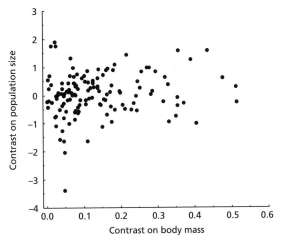

Fig. 5.29 The relationship between population size (number of individuals) and body mass (g) in breeding bird species in Britain, controlling for phylogenetic relatedness using the independent contrast method ($y = 0.64x$; $r^2 = 0.02$, n (number of independent contrasts) $= 132$, $P = 0.09$). The method was implemented by the CAIC computer program (Purvis & Rambaut 1994, 1995), using the CRUNCH option and assuming equal branch lengths. Species were classified following Sibley and Monroe (1990, 1993). From data in Appendix III, which were $\log_{10}$-transformed before analysis.

(1991c) were the first to test for an association between abundance and body size correcting for phylogenetic relatedness. They examined the relationship for British birds pictured here in Fig. 5.21a. Although this relationship is significantly negative when examined across species, the statistical significance is due largely to a difference between passerines and non-passerines. The former are mainly small bodied and abundant, whereas the latter are mainly larger bodied and rarer. Controlling for phylogeny, there is no significant relationship between abundance and body mass in British birds, either for the species and data analysed by Nee *et al.* (1991c), or for the breeding species in Appendix III. Indeed, the relationship for the latter is positive (Fig. 5.29), as it is for 56% of the British bird taxa in this analysis.

The phylogenetically independent analyses of British birds show that, among related taxa, larger body size is often associated with higher abundance. This may contribute to the scale dependence of the abundance–body size relationship. The taxonomic composition of local assemblages is likely to be more uniform than that of regional assemblages. For example, a woodland assemblage like that of Eastern Wood includes several species each in the families Picidae, Sylviidae, Paridae and Corvidae. By contrast, a typical coastal assemblage will contain species in the Procellariidae, Laridae, Sternidae and Alcidae, but few if any species from the families typical of Eastern Wood. Negative abundance–body size relationships may then arise at large scales because different higher taxa occupy different positions in abundance–body size space, as Nee *et al.* noted for passerines and non-passerines. Negative relationships do not pertain at small scales because local assemblages are more homogeneous taxonomically, and the relative abundances of related species are not predictable from their body sizes alone. The abundance–range size relationship, by contrast, is positive however closely related are the species compared.

A linear negative relationship between log abundance and log body size at large spatial scales is the pattern that would be expected if these two traits had to be traded off against each other within a framework of a finite but reasonably equably divided energy supply. However, the existence of this pattern does not constitute evidence that it is indeed mediated by energy requirements. Moreover, the question of the pattern of abundance–body size relationships within taxa, many of which are positive, also requires consideration.

5.5.3   *What generates abundance–body size relationships?*

Energy use by populations and individuals

Although it is not inevitable, it does seem reasonable that the existence of a negative power relationship between abundance and body mass should imply that the relationship is a consequence of energy limitation. Presumably, some total amount of energy is available in the environment, which is divided among species according to some allocation rule to give a frequency distribution of population energy use. This allocation rule could be conceptually similar to Sugihara's (1980) sequential broken stick model (Section 4.3.3), for example. Species then 'choose' to divide their energy into many small- or few large-bodied individuals, independent of how much energy the population has appropriated. If individual energy requirements (e.g. metabolic rate) scaled with body mass across species to the power 0.75, this would result in a power relationship between abundance and body mass of exponent –0.75. There would be no interspecific relationship between population energy use $(D \times R)$ and body mass.

This energetic limitation idea is very seductive, especially given the number of abundance–body mass relationships with exponents around –0.75. There are two significant problems. First, energetic limitation typically assumes that metabolic rate and body mass scale with an exponent of 0.75. For birds (and probably other groups), however, the slope seems consistently to be rather shallower than this. For basal metabolic rate (BMR), Bennett and Harvey (1987) reported an exponent of 0.67, using a major axis analysis and averaging across species within families (because most variation in BMR is accounted for at the level of the family), Koteja (1991) reported exponents of 0.673 for reproducing and 0.624 for non-reproducing individuals, using least squares regression, and Degen and Kam (1995) reported a value of 0.64, also using least squares regression. Field metabolic rate (FMR) is a more appropriate measure of differences in actual energy usage. However, it has been determined a good deal less frequently than BMR, and the relationship between the two is again a topic of some discussion (e.g. Koteja 1991; Degen & Kam 1995; Ricklefs *et al.* 1996). Where the body size scaling of FMR has been calculated for birds, the exponent is typically not dissimilar, though sometimes lower, than that for BMRs. Thus, Bennett and Harvey (1987) report a value of 0.65 using major axis, Nagy (1987) a value of 0.64 using least squares regression, Koteja (1991) exponents of 0.651 for reproducing and 0.530 for non-reproducing individuals, using least squares regression, and Degen and Kam (1995) a value of 0.53, again using least squares regression. On balance, overall an exponent of 0.67 (two-thirds) for scaling of metabolic rate and body size is probably a reasonable first approximation for birds, not 0.75 (see also Brown & Maurer 1986). Of course, regardless of whether BMR or FMR is used, and of which particular value of the exponent, given that all are less than 1, an individual of a large-bodied bird species requires more energy in total to fuel its metabolism than does one of a small-bodied species, although it requires less energy per gram.

If metabolic rate does scale allometrically with an exponent of 0.67, energetic limitation would imply that the slope of abundance–body mass relationships should be −0.67. This figure is closer to the median slope value of −0.58 found by Blackburn and Gaston (1997b) in their compilation of relationships from the literature. Thus, an allometric exponent of 0.67 for metabolic rate would not invalidate energetic limitation, although it would change the slope expected of abundance–body mass relationships. In practice, the variance around such relationships anyway makes it difficult to distinguish between slopes of −0.67 and −0.75.

The second, more fundamental, problem with the idea of energetic limitation is that there is no real evidence that populations are generally energy limited. Certainly, the form of the abundance–body size relationship does not indicate whether or not an assemblage is energy limited. To see why, consider for simplicity a two-species community. One species is twice the size and twice as abundant as the other. If individual energy use scaled as $M^{0.75}$, the commoner species would use in total over three times as much energy as the rarer species. The populations of both species may be limited by the amount of energy available to them, but a positive abundance–body size relationship results. Thus, species may be energy limited whatever shape the abundance–body size relationship takes, depending on how energy is divided between species. This is illustrated further by interpretations of slopes of these relationships that differ from −0.75. Steeper slopes have been taken to indicate that small-bodied species have more energy available to them (they can appropriate larger slices of the total energy 'cake') (e.g. Griffiths 1992, 1998), whereas shallower slopes have been taken to indicate that large-bodied species have more energy available to them (e.g. Brown & Maurer 1986). These relationships take forms that differ from $D \propto M^{-0.75}$, yet energy is still assumed to be mediating abundance–body size relationships in each case.

To prove that the abundance–body size relationship is a result of energy limitation, it is necessary to demonstrate that populations are constrained by the amount of energy available. Such a proof has never, to our knowledge, been presented. Moreover, the ecological literature suggests that the reverse may be true. For example, the abundances of many species of bird are limited by the availability of nest sites, not energy. Williamson (1972) noted that the densities of birds breeding in British woods varied from about 200 to 1800 pairs per square kilometre (pr/km^2). The density of birds breeding in Eastern Wood between 1949 and 1979 varied from 650 to 1312 pr/km^2. Yet, the provision of nestboxes allowed pied flycatchers alone to attain densities exceeding 2000 pr/km^2 in woodland (Fuller 1982). Increases in density after nestbox provision have also been noted for tits and a number of other woodland bird species (e.g. von Haartman 1971; Perrins 1979; East & Perrins 1988; Gustafsson 1988). These often seem to represent genuine population increases, rather than simple movement by individuals into areas with nestboxes, to have minimal effects on the abundances of other bird species, and to result in increases in the total community biomass (see review by Newton 1998). They suggest that energetic limitation does not apply.

In the limit, of course, as argued previously, the total amount of biomass that can be supported on this planet must be constrained by the total amount of available energy. Nevertheless, evidence that assemblages are not limited by energy availability suggests that this ultimate constraint often is not reached. Some other factors may prevent the amount of life from reaching this limit. For example, populations may be limited by some resource other than energy, such as in the nest site example from the previous paragraph. Alternatively, regular disturbances to biotas, for example through the influence of large-scale climate change, may prevent them from reaching the point at which energy is limiting. This would imply that most, if not all, biotas are not in equilibrium (see also Patterson 1999). Instead, they can be viewed as in the process of development, the end point of which would be when all available energy was appropriated. This view would fit with the notion that most assemblages are unsaturated (Section 2.4). It also concurs with the composite explanation for latitudinal species richness gradients based on area, energy and effective evolutionary time (Section 2.5.5). Area and energy determine the ultimate level that richness can attain, but how close a biota gets to that limit depends on the effective time available to get there.

In fact, we think that the energetic hypothesis is correct to the extent that negative abundance–body size relationships are a consequence of variation in the energetic requirements of individuals of species with different average body mass. Given a certain amount of energy appropriated by a species, it follows from the allometric scaling of metabolic rate that that energy can support many small- or few large-bodied individuals, but not many large-bodied individuals. That is not to say, however, that populations are ultimately limited by energy availability. Rather, whatever it is that constrains the population of a species to a certain size results in that species using a certain amount of energy in the environment. Energy use is the consequence of population limitation, rather than the cause.

This view of relationships between log abundance and log body mass explains why many have slopes in the region of -0.67 or -0.75, but why many others do not. Such slopes are expected when all species, on average, have the same amount of energy available to them. However, it may often be the case, by chance or for other reasons, that there is a non-zero relationship between body mass and population energy use across the species in an assemblage: large- or small-bodied species would then utilize more energy, on average (although, of course, energy *per se* is not the limiting factor). Then, the necessary trade-off between abundance and body mass within species would result in abundance–body mass relationships with slopes that differ from the -0.67 or -0.75 expectation.

One situation where there is often a non-zero relationship between body mass and population energy use across species is within taxa. Recall that the linear negative log abundance–log body mass relationship breaks down when the relatedness of species is incorporated into the analysis. Within taxa, rarity is just as likely to be associated with small body size as large.

Nee *et al.* (1991c) found that there seems to be systematic variation within taxa in the slope of the abundance–body size relationship. They showed that, in the British breeding avifauna, positive abundance–body size relationships are common at low taxonomic levels (e.g. across species within a genus), but negative relationships predominate at higher levels (e.g. across superfamilies within orders). This pattern has now been shown for both the British breeding and wintering avifaunas (Nee *et al.* 1991c; Blackburn *et al.* 1994; Gregory 1995), and also seems to hold in the North American breeding avifauna (Table 5.7). Moreover, positive abundance–body size relationships at lower taxonomic levels tend to be associated with the age of the taxon. Tribes that diverged from their nearest closest relative longer ago are more likely to show positive abundance–body size relationships across species (Fig. 5.30; Cotgreave & Harvey 1991; Nee *et al.* 1991c; Blackburn *et al.* 1994; Cotgreave 1994). This last association between tribe age and patterns in abundance is especially interesting, because abundances can change orders of magnitude more rapidly than lineages evolve.

Nee *et al.* (1991c) suggested that these within-taxon patterns could be the result of interspecific competition. They reasoned that this should be most intense between members of an ecological guild, which are often closely related. If large body size is an advantage in interspecific competition, leading

Table 5.7 The number of positive and negative relationships between number of individuals and body mass within taxa at different taxonomic levels, for the breeding avifauna of North America (data were from Terres (1980), Robbins *et al.* (1986), or were kindly supplied by B. Maurer). Analyses either include all species, or exclude seabirds and introduced species (following Nee *et al.* 1991c). For example, 53 genera showed a positive relationship, and 39 a negative relationship, between abundance and mass across their constituent species when all species were included. The penultimate row of the table includes superfamilies within parvorders, parvorders within infraorders, infraorders within suborders, suborders within orders, orders within superorders, superorders within parvclasses, and parvclasses within infraclasses. Within lower taxa (e.g. below the superfamily level), there are significantly more positive than negative relationships between abundance and mass (all species: two-tailed binomial $P = 0.001$; seabirds and introduced species excluded: $P = 0.025$). For 'subtaxa within higher taxa', negative relationships predominate, although there are not significantly more negatives than positives. However, when all species are included, the pattern for 'subtaxa within higher taxa' is significantly different from all other levels of analysis (Monte Carlo contingency test with 5000 iterations, $P = 0.03$), although this is not the case when seabirds and introduced species are excluded ($P = 0.24$).

		All species		Species excluded	
Across	Within	Positive	Negative	Positive	Negative
Species	Genera	53	39	47	37
Genera	Tribes	24	11	21	11
Tribes	Subfamilies	5	1	3	2
Subfamilies	Families	6	4	5	3
Families	Superfamilies	7	0	5	1
Subtaxa	Higher taxa	8	13	9	11
Total		103	68	90	65

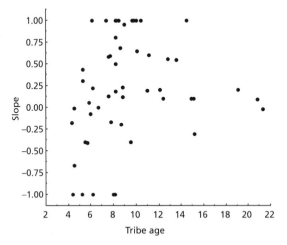

Fig. 5.30 The relationship between the slope of the abundance–body mass relationship across the species in a tribe and the age of that tribe, for breeding birds in North America. There is a significant positive Spearman rank correlation between these variables ($r_S = 0.32$, n (number of tribes) = 52, $P = 0.02$), although slopes are close to zero for the oldest tribes. From data sources in Table 5.7.

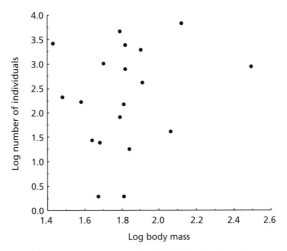

Fig. 5.31 The relationship between number of individuals (total number recorded on all US and Canadian Fish and Wildlife Service Breeding Bird Survey routes for the year 1977) and body mass (g) for species of woodpecker in North America ($n = 19$). From Blackburn and Gaston (1999).

to higher levels of abundance in larger species when competition is important, then positive relationships between abundance and body size would be expected in comparisons between species in the same guild (or at least between species in the same taxon, which are easier to define). Further, if evolutionarily isolated tribes also tend to form complete guilds, then an association between evolutionary isolation and the slope of the abundance–body size relationship would also be expected (Nee *et al.* 1992a; Cotgreave 1994). The example most often quoted is that of the woodpecker tribe (Picidae). This is both phylogenetically isolated and arguably forms a complete guild—all its species have similar nesting and feeding habits not shared by any other species (Cotgreave 1994). The abundance–body size relationship in this tribe is generally positive (Fig. 5.31; Nee *et al.* 1991c; Blackburn *et al.* 1994). In Britain, for example, the two smallest woodpeckers (wryneck and lesser spotted woodpecker) are also the rarest of the four native species.

The competition argument finds support from an ingenious test by Cotgreave (1994), based on an original study by Bock *et al.* (1992) of the abundances of insectivorous birds in a riparian community in Arizona, USA, before and after the erection of nestboxes in an experimental area. The presence of nestboxes caused some species to increase in abundance, and others to decrease. Bock *et al.* argued that those species that increased did so because they were freed from competition for nest sites, while those that decreased did so because of increased competition for food with the species whose abundances increased. In other words, competition was normally more important as a force limiting abundance for those species that increased in abundance after the experimental treatment. Cotgreave (1994) examined the relationship between abundance and body size within tribes in the data published by Bock *et al.* As in other studies, the date of origin of the tribe was positively correlated with the slope of the relationship within a tribe. However, so too was the percentage change in abundance following the experimental treatment, and the proportion of cavity-nesting species; tribes with positive abundance–body size relationships tended to increase in abundance and be hole-nesters, and those with negative relationships all decreased, and tended not to be hole-nesters. This is as would be expected if high competition for nest sites led to positive relationships.

Interspecific competition does have the potential to explain abundance–body size patterns at low taxonomic levels. However, competition is most probable between members of ecological guilds, and there is already evidence that guilds tend to occupy different parts of abundance–body size space (e.g. Peters & Wassenberg 1983; Peters & Raelson 1984; Juanes 1986; Damuth 1987; Silva & Downing 1995). The relative positions of different guilds are more likely to be determined by the ecological requirements of guild membership, rather than competition between members of different guilds. If positive abundance–body size relationships are the result of competition, but competition is most important at low taxonomic levels (e.g. within genera), that would explain both the paucity of positive relationships at higher levels (e.g. across orders), but also why some positive interspecific relationships do exist (Blackburn & Gaston 1997b, 1999; Navarrete & Menge 1997).

Other possibilities—census area and latitude

Although we think that linear negative relationships between log abundance and log body mass can be explained as a consequence of the energy requirements of individual animals in populations with finite resources, other mechanisms have been suggested to explain these patterns (Table 5.8; Blackburn & Gaston 1999). We think that two of these are worth mentioning here, based on interspecific variation in census area, and variation in abundance and body mass with latitude. The others have fundamental flaws (Blackburn & Gaston 1999).

The census area mechanism can only apply to relationships where abundance is measured as density. Thus, it cannot explain the patterns we showed for the birds of Britain (Fig. 5.21). Nevertheless, the artefact on which it is based is important, albeit largely unappreciated by the wider ecological community. Density is normally measured as the number of individuals (or some other relevant unit) in a given area. Thus it has two components, the number of animals and the census area. Variation in either can affect the observed density of a species. The census area mechanism suggests that broadly negative interspecific relationships between density and body size arise because small-bodied species are censused across smaller areas.

Table 5.8 Mechanisms which have been postulated as determining or contributing to observed interspecific relationships between abundance and body size (Blackburn & Gaston 1999).

Factor	Mechanism
Energetic constraint	Constraints on abundance which result from differences in energy demands of individuals of species of different body mass
Census area	Small-bodied species being censused over smaller areas than large-bodied species
Latitudinal gradients	Latitudinal gradients in abundance and body size
Concatenation	Random sampling from the frequency distributions of abundance and body size
Interspecific competition	More intense competition between members of an ecological guild
Differential extinction	Greater likelihood of extinction within taxa of large-bodied species

Consider the interspecific relationship between body mass and abundance for mammalian primary consumers shown in Fig. 5.20 (from data given in Damuth 1987). This plot includes densities of mice and elephants. These are unlikely to have been obtained from censuses over areas of similar size. In fact, within these data there is a strong positive relationship between the body mass of a species and the area over which its density was censused (Blackburn & Gaston 1996g, 1997c). A similar relationship has been shown for mammalian carnivores (Schonewald-Cox *et al.* 1991; Smallwood & Schonewald 1996). For the mammalian primary consumers in Fig. 5.20, the area over which a species is censused is a better (in terms of coefficient of determination) predictor of its abundance than is its body mass (Blackburn & Gaston 1996g, 1997c). The interspecific relationship between density and body size could be the result of the variance in census area. Density estimates tend to be lower from larger census areas, while small-bodied species are rarely censused over such areas. This explanation additionally requires only that densities tend to be measured in areas where a species occurs.

Many factors determine why an area of a particular size is chosen for study of a given species. Smaller areas will be favoured for practical reasons of relative ease of sampling, delineation of study area, control of disturbance and replication. Conversely, areas must in general be large enough that sufficient numbers of individuals occur within them (species are seldom studied in areas in which they are difficult to find), and that populations are not dominated by transient individuals. Trade-offs between these factors are likely to result in different-sized census areas for different-sized species (as well as species with different kinds of population dynamics, trophic habits, habitat usage, etc.). The allometry of density could arise from circular logic, whereby body size influences the spatial extent of study, and the resulting density estimates are related back to body size (Smallwood *et al.* 1996).

The census area mechanism alone is probably insufficient to account for the full range of variation observed in the abundances of organisms. Few would argue that the densities observed for a set of species are entirely the consequence of the area over which they are sampled. We know that meadow pipits generally occur at higher densities than golden eagles. Nevertheless, there is clear evidence that density and census area are not independent. Over moderate to large areas, the densities of individual species should tend to decline as census area increases, because more subareas will be included in which individuals do not occur. Hence, there is no such thing as a single density for a species. Rather, each species has its own (intraspecific) relationship between density and census area (Blackburn & Gaston 1996g, 1997c; for examples and discussion of relationships between density and area for birds, see Odum 1950; Gromadzki 1970; Nilsson 1977, 1986; Ambuel & Temple 1983; Village 1984, 1990; Askins *et al.* 1987; Kostrzewa 1988; Smallwood 1995; Potts 1998; Gaston *et al.* 1999b; Summers-Smith 1999). Examples for four species of British bird are given in Fig. 5.32. The slopes of these relationships are likely to, and indeed do, differ between species, but whenever they differ from zero, density and census area will not be independent.

Given that density and census area are not independent, the question arises whether census areas for species of different size are comparable, in the sense that they result in the measurement of equivalent densities (an issue with far wider ramifications than simply for abundance–body size relationships). There are reasons to doubt that they are comparable in this sense. First, it would be necessary that the trade-offs between the various factors determining the size of a study area result in ecologically equivalent areas being used for the measurement of density both for small and large species. As these factors are predominantly determined by methodological, rather than biological, considerations, this seems exceedingly unlikely. This suggests the possibility that there are systematic differences in the relative use made of census areas by animals of different body sizes. Replotting the density–body size relationship for mammalian primary consumers (Fig. 5.20), statistically controlling for the average area over which density was censused for each species, yielded a relationship that was still negative, but with both the slope and correlation coefficient greatly reduced (Blackburn & Gaston 1996g). The clear implications are that different kinds of densities are being measured for large and small animals, and that the interpretation of patterns of abundance from such interspecific comparisons will be confounded by uncertainty as to what is actually being compared. It is by no means clear how to obtain equivalent densities for interspecific comparisons (Blackburn & Gaston 1996g, 1997c).

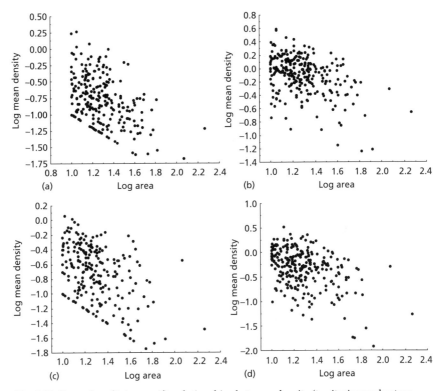

Fig. 5.32 Examples of intraspecific relationships between density (territories per hectare when present) and census area (hectare) for (a) coal tit, (b) robin, (c) blackcap and (d) chaffinch. From Gaston *et al*. (1999).

In summary, the census area mechanism alone is unlikely to generate the negative abundance–body size relationship, at least across very large ranges of body sizes. Nevertheless, it does strongly suggest that the negative relationships typically recovered from investigations where abundance data are compiled from a wide range of studies using a wide range of census methodologies (e.g. Fig. 5.20) may in part be artefacts of the data used (Blackburn & Gaston 1997b). This is a bias that needs to be borne in mind when interpreting any analyses involving interspecific density comparisons. Although it does not negate the arguments presented in the previous section about the likely cause of abundance–body mass relationships, if, as we have just argued seems likely, the densities of small-bodied species are usually systematically overestimated relative to large, their slopes may be systematically underestimated.

The final hypothesis for the general form of abundance–body mass relationships that we shall consider is that they are a consequence of latitudinal variation in both variables. Bird species living at higher latitudes tend, on average, to be larger bodied (Section 5.4). The negative relationship between abundance and body size could perhaps arise if species abundances also showed

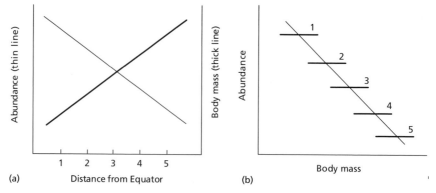

Fig. 5.33 The interspecific relationships between (a) abundance and latitude (negative) and body mass and latitude (positive) and (b) abundance and body mass for a hypothetical taxon. There is no relationship between abundance and body mass across species resident at any of the latitudes marked 1–5 in (a) (thick lines in b), but the systematic variation in mean abundance and body mass across latitudes generates a negative abundance–body mass relationship when species from a range of latitudes are compared (thin line in b).

latitudinal gradients, such that species living at higher latitudes tended, on average, to be less abundant. If abundance–body size relationships were plotted using data compiled from a range of latitudes, a negative abundance–body size relationship could result, even if there were no relationship at any single latitude (Fig. 5.33).

In the previous chapter (Section 4.2.3), we provided evidence that species living at higher latitudes tend, on average, to be more abundant. If so, it is extremely unlikely that latitudinal gradients could cause the negative abundance–body size relationship. The gradients run such that smaller body sizes should be associated with lower population sizes and densities. This would generate a positive abundance–body size relationship.

On the other hand, the existence of latitudinal gradients in abundance and body size implies that the global abundance–body size relationship for a taxon is not representative of the relationship at any given point in space. The global relationship suggests that mean densities should be higher where mean body size is lower, but the latitudinal gradients suggest that low average density and small average body size coincide. The conflict can be resolved if there is a correlation between the latitude that species inhabit and their position in the abundance–body size relationship. For example, if those species from low latitudes tend to fall below the regression line, and species from high latitudes fall above it, then mean density and body mass should both increase with latitude. The negative abundance–body size relationship often observed could then be the composite of a series of latitudinal slices of similar slope but different elevation.

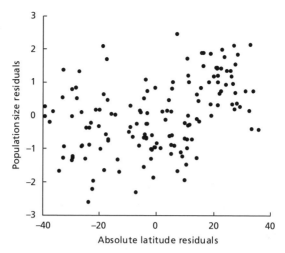

Fig. 5.34 The relationship between population size (number of individuals) and absolute latitudinal mid-point of geographical range (degrees) for the wildfowl species of the world, controlling for the effect of body mass (g) ($r^2 = 0.17$, $n = 147$, $P < 0.0001$). The axes are the residuals of plots of $\log_{10}$ population size against $\log_{10}$ body mass, and of absolute latitudinal mid-point against $\log_{10}$ body mass. Species which live further from the Equator have higher total population sizes, for a given body mass. From data sources in Gaston and Blackburn (1996c).

There is some evidence for this in wildfowl (Gaston & Blackburn 1996c; see also Peters & Raelson 1984). Wildfowl species show no significant relationship between population size and body mass. However, for a given body mass, wildfowl species living further from the Equator have larger global population sizes (Fig. 5.34). In fact, the interspecific population size–body mass relationship becomes weakly, but significantly, negative if the latitude at which species breed is controlled for (Gaston & Blackburn 1996c). Further, wildfowl show positive relationships between population size and latitudinal mid-point of breeding range, and body mass and latitudinal mid-point of breeding range, within taxa (Gaston & Blackburn 1996c); in other words, among closely related taxa, those living closer to the Equator have lower population sizes and body sizes. All the above patterns would be expected if extratropical taxa were consistently larger bodied and more abundant than their tropical relatives.

5.5.4 *Synthesis*

We opened our consideration of the relationship between abundance and body size by noting that regional energy constraints suggest that the body sizes and abundances of organisms should somehow trade off, but that this trade-off could take many possible forms, depending on how the body sizes of organisms affect their ability to appropriate resources from the environment.

The best evidence to date suggests that this form is a negative power relationship for regional scale assemblages, such as the birds of Britain, but that in local sites like Eastern Wood, weakly negative (but sometimes positive) triangular relationships are more typical. Relationships within taxa may take any form, but positive relationships tend to occur in more evolutionarily distinct tribes.

Why negative power relationships between abundance and body size should give local relationships that are typically triangular, but that can take a wide variety of forms, is also controversial (cf. Currie 1993; Blackburn & Gaston 1997b; Griffiths 1998). There seems little doubt that these patterns are not simple sampling artefacts—we recover them from sites as well studied as Eastern Wood! We propose that variation in abundance, and the effect of competition on within-taxon patterns, together can explain this scale dependence, when imposed on an overall negative relationship. Thus, the pattern in local assemblages depends in part, but not in its entirety, on the pattern for the regions in which they are embedded. These ideas remain to be tested.

While interspecific competition can explain patterns within taxa, the question of why the negative log-linear large-scale interspecific abundance–body size relationship on which these within-taxon patterns are imposed should exist remains unanswered. We think the most plausible idea is that the general relationship can be viewed as a consequence of the necessary trade-off between abundance and body mass for species using a given amount of energy.

The fact of finite regional resources implies an additional trade-off between the resources used by individual species and number of species inhabiting a region. The relationships between body size, abundance and latitude that we have just discussed have two clear implications for the form of this trade-off. First, the best current evidence suggests that both abundance and body mass on average increase with latitude. Thus, although it is well known that high latitudes have fewer species than the tropics (e.g. Section 2.5), it seems that they have fewer species than expected on the basis of the amount of biomass residing there. Biomass at high latitudes is divided into larger-bodied, more abundant species. Second, the negative exponent of the power relationship between abundance and body mass suggests that energy may be reasonably equably used by species of different body masses within a region (although the latitudinal gradients imply higher per species energy use in higher latitude regions). These interactions, and their links with other patterns discussed during the course of this book, are developed further in the final chapter.

5.6   Summary

The body sizes of species within an assemblage are bounded, in as much as there are physical and physiological limits to the lower and upper limits which can be achieved by members of a given taxon. In most places those bounds are probably not closely approached. Thus, there seems to be no intrinsic reason to believe that larger species could not belong to the avifauna of Eastern Wood, although whether species smaller than the goldcrest could realistically persist in this area is perhaps more questionable. In practice, the largest and smallest species found in the assemblage are strongly predicated by the sizes of those which belong to the regional pool. However, although reasonably close, the assemblage in Eastern Wood is not simply a random subset of this pool, tending to comprise species which average somewhat smaller than expected on such a basis. Quite what factors modify the likelihood with which species belong to the Eastern Wood assemblage in order to yield this outcome remains obscure. The latitudinal gradient in body sizes of species, with average size decreasing towards lower latitudes, is, nonetheless, a strong contender.

The body size structure of the Eastern Wood avifauna by definition influences the pattern of allocation of energy among the species which reside there, in as much as this allocation is the product of the body sizes of individuals and their numbers. Again, the particular pattern of allocation can only sensibly be interpreted by considering the relationship between abundance and body size which is exhibited by the regional avifauna from which Eastern Wood draws individuals. If the likelihood of a species belonging to the assemblage is in part determined by its abundance, on the one hand (Section 4.3), and its body size, on the other (Section 5.2), then the structure of the local assemblage will be influenced by how abundant are species of different body size.

These views of the body size composition of the Eastern Wood avifauna serve to begin to draw connections between the macroecological patterns which have been discussed in this chapter and some of those which have been mentioned in previous ones. In the next chapter we develop these links at greater length, in exploring the causal, albeit perhaps tangled, threads which run through the book.

6 Synthesis

The truth is rarely pure, and never simple. [Oscar Wilde 1895]

6.1   Introduction

In the end, all ecologists have basically the same goal. They wish to understand the processes that lead to the patterns observed in the distribution and abundance of the species with which they share the planet. Yet, ecologists are equally united in the realization that a system as large and as complicated as that defined by all life on Earth cannot, given present capabilities, be understood in its entirety. For that reason, a variety of simplifying approaches are necessary to enable understanding of the specific aspect of interest. It is believed to be better to achieve some understanding of the simplified system than to have no understanding at all, and the broader hope is that this understanding will give insight into the working of the more complex whole.

In this spirit of simplification, we have distinguished between small- and large-scale approaches to the study of ecology. This is a crude caricature of the great diversity of ecological research, but differentiates between methods by which the simplification of ecological systems can be achieved. Using the small-scale approach, the aim is to understand why species occur where and in the numbers they do, through study of the detailed interactions that constrain the lives of individual plants and animals within the communities they inhabit. Their responses are studied to conspecifics and to interspecific competitors, the effects are examined of predators, parasites and pathogens, and tests are made of their environmental needs and tolerances. The system is simplified by ignoring the wider context within which the fragment of interest resides, and the fact that all the competitors, predators, parasites and pathogens each sit at the centre of a web of equally complex interactions with the biotic and abiotic environments, as in turn do their competitors, predators, parasites and pathogens. The fact is ignored that the entire community sits within a broader environmental context that influences what interactions will occur.

By contrast, the large-scale approach simplifies the study of ecological systems by ignoring the fine details of species interactions, and instead focuses on the patterns and processes that define the broader environmental context.

It aims to identify the processes structuring assemblages by examining patterns in the responses of large numbers of species to these processes.

The principal aim of this book has been to promote an understanding of why large-scale ecology, or macroecology, is an important part of the general programme of ecological research. We have attempted to show that a broader context is required. To this end, and with special reference to the avifauna of Britain, large-scale patterns have been examined in species richness, spatial distributions, body sizes and abundances, and we have considered some of the processes thought to be driving them. We have shown how these patterns can relate to and influence the structure of the small-scale ecological assemblages with which inevitably ecologists are almost invariably more immediately familiar. This has been done in the belief that the macroecological viewpoint will contribute to the general understanding of ecology. Ecological systems are continuous across all scales, from the individual organism to the entire planet. Therefore, a complete understanding of these systems is going to require observations made from a range of viewpoints, or at a variety of scales. Hopefully, it will eventually be possible to integrate these for that more complete understanding.

A key feature of these arguments has been this continuity of ecological systems, and that the divisions inserted for the sake of study are arbitrary. The truth of this is perhaps best evidenced in the fact that while a distinction may be drawn between large- and small-scale ecology, a rather broad set of scales of study may be gathered under the macroecological umbrella. Yet, the arbitrary nature of such divisions is equally true within the field of macroecology (as indeed it is within 'microecology'). This book has been divided into separate chapters, with subheadings dealing with specific features of the variables with which each is concerned. Ultimately, however, these divisions are manufactured for our convenience, and for that of the reader. The patterns discussed, and the processes contributing to them, are all interwoven in the structure of the global community.

The aim of this last chapter is to draw together links between the patterns and processes, and to show how these might fit into an integrated whole. Although we have conducted some integration thus far, previous chapters have taken something of a 'pattern-by-pattern' approach. This is akin to the majority of investigations in macroecology to date. One possible reason for the generally limited attention which has been paid in the literature to the links between different macroecological patterns is that, as should now perhaps be apparent, much of the discussion of their respective causes has commonly been rooted in rather different theoretical frameworks (Gaston & Blackburn 1999). Thus, for interspecific abundance–body size relationships attention has predominantly been directed to energetic explanations (e.g. Damuth 1981), for interspecific abundance–range size relationships to explanations rooted in niches and metapopulation dynamics (e.g. Brown 1984, 1995; Hanski et al.

1993; Maurer 1999) and for interspecific range size–body size relationships to explanations based on minimum viable population sizes (e.g. Brown & Maurer 1987). As we have shown, in almost every case, the full breadth of explanations that have been considered is in fact much wider than this, but most suggested mechanisms have simply been ignored by most workers.

It would be an impossible, as well as tedious, task to attempt to integrate all of the macroecological patterns we have discussed thus far in the context of each of the different theoretical frameworks which have been employed. Instead, here we will examine how a number of macroecological patterns might be linked into a coherent whole with regard to simple constraints on the subdivision of available energy. This has been a common thread through much of the discussion to this point (Sections 2.5.3, 4.3.3, 5.3.2, 5.5). To some extent our attempts will inevitably appear simplistic. It is hard enough to understand the complexities and implications of the macroecological patterns considered separately, let alone attempt to weave these into a coherent and seamless whole. Nevertheless, there are many obvious interactions and dependences between the variables that have been examined.

Throughout, we draw on earlier considerations about birds, and some other animal groups, to construct a general set of arguments for animals more generally. In trying to paint on the broadest of canvases we will be less concerned than we have been to this point with the specifics of the avifauna of Britain. More details of the following arguments will be given elsewhere (T.M. Blackburn and K.J. Gaston, in preparation).

6.2   Knitting patterns

6.2.1   *Energy and biomass*

Consider a hypothetical land area sufficiently large and isolated that the predominant determinants of the number of species found there are the processes of speciation and extinction; the effects of immigration and emigration are negligible. This is analogous to Rosenzweig's (1995, p. 264) definition of a 'province', and we will use this term to refer to the hypothetical area. This scale is explicitly larger than that encompassed by Britain, which is just one part of such a province. We assume, as is inevitable, that the processes of speciation and extinction are the ultimate determinants of the assemblage of species found in the province, but give little further explicit consideration to them in what follows. Here, we are interested in exploring the likely macroecological features of species in the assemblage that speciation and extinction have produced.

Imagine a set of species within the province, exploiting all of the resources available. Each species entering our imaginary assemblage appropriates a proportion of the resources available in the system. These resources potentially encompass a wide variety of factors necessary for the maintenance and propagation of life, but for simplicity we refer here to available energy. The species may be primary producers obtaining energy directly from solar radiation, or may be consumers, like the birds on which we have focused, deriving it 'second hand' from producers. Whichever, we assume that the energy available to a species imposes a limit on it at some point and, for the moment, that the amount of energy appropriated by each species is fixed. This does not mean that we believe energy *per se* to be the key factor limiting species populations (see also Section 5.5.3). The model makes no assumptions about what limits any given species, which may be any of a variety of factors (e.g. resource availability, competition, predation, disease). All the model requires is that once the population of a species is limited, then that species utilizes a certain amount of the energy available in the environment. This is trivially true.

The species in the province use the energy available to them to fuel their productivity. In effect, they convert it into biomass. For each species, total biomass is dictated by energy availability and the energy requirements of its individuals. The latter depend on metabolic rates, which in turn depend on, among other things, the body mass of individuals. Across a wide range of organisms, metabolic rate is a power function of body mass, with the exponent being positive but less than 1 (Section 5.5.3). This means that large-bodied organisms use less energy per gram of body mass than do small-bodied organisms (although the form of relationship may be different when comparison is between small- and large-bodied individuals within the same species). Total biomass for each species is the product of the number of individuals and their body masses.

6.2.2 *Population size and body mass*

From these interactions, it follows that each species uses energy to support a certain amount of biomass, that this biomass may be divided among many small-bodied or fewer larger-bodied individuals, and that the total biomass of each species supported will be greater if individuals are large bodied. The constant energy assumption means that while body mass and population size can take any of a large number of values, they are constrained to be negatively related for each species: population size can only increase if the energy requirements of the individuals, and hence their body mass, decrease. If individual energy requirements scale allometrically to the power x, it follows that the expected population size for a species would scale allometrically to the power $-x$. Therefore, population size and body mass for a given species are constrained to fall somewhere along a straight line with slope $-x$ in log–log space.

The elevation of this population size–body mass constraint slope for each species is proportional to the total amount of energy it can utilize. In the special case where the total amount of energy available is divided equally between all species in a system, and they have identical allometric scaling of metabolic rate, they will all lie on the same log population size–log body mass curve.

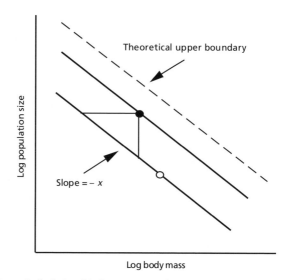

Fig. 6.1 The theoretical relationship between population size and body mass on logarithmic axes. Thick lines connect points of equal energy use. Species 1 (filled circle) appropriates a greater amount of available energy than species 2 (open circle). A given amount of energy can be converted either into many small-bodied or fewer large-bodied individuals, resulting in the negative slopes of magnitude –*x* for these lines, where *x* is the allometric exponent of individual energy use. The thin lines bound the part of the lower thicker line where species 2 would have to lie for the interspecific population size–body mass relationship between these two species to be positive. The dashed line is the theoretical upper boundary on which a single species using all available energy would lie. In practice, species are likely to move vertically as the energy available to them changes.

Therefore, for a given amount of energy used by species, a linear negative interspecific relationship of slope –*x* is expected between log population size and log body mass. Otherwise, the interspecific population size–body mass relationship between the species in this system can take any form, and may be positive. By definition, positive relationships occur whenever both the body mass and population size of the species using less energy are lower than those of the species using more energy. An example for an assemblage of two species is given in Fig. 6.1. A positive relationship will occur whenever a species using the same amount of energy as species 1 falls on that part of its population size–body mass trade-off function bounded by the horizontal and vertical lines connecting that function to the position occupied by species 2 on its trade-off function. The length of this bounded region increases as the difference between the elevations of two trade-off functions increases. If both trade-off functions have similar elevation, then the species on the lower function will only fall in the bounded region if it is of slightly smaller body mass. Thus, positive population size–body mass relationships between two species are more likely when available energy is less equally divided, or when the two species are similar in body size.

The constraint applied to the population size–body mass relationship by the amount of energy available to any given species sets a theoretical negative upper boundary to this relationship (Fig. 6.1). This is the slope along which a single species utilizing all available energy in the province would lie. In practice, the upper boundary will be defined by the population sizes and body masses of those species in a multispecies assemblage that appropriate most available energy. Nevertheless, the trade-off between population size and body mass for any given amount of energy means that this limit will also be negative.

For a given taxon, the range of body mass values possible will ultimately be limited by design constraints. There are physical limits on any given way of life (e.g. Calder 1984; Schmidt-Nielsen 1984; Reiss 1989). This forces population size–body mass relationships to lie within certain body mass limits (Brown & Maurer 1987; Maurer 1999), restricting the ways in which a given amount of biomass can be subdivided into individuals.

Whether, as well as constraints on the upper boundary and the lower and upper body mass, there is a necessary constraint on the lower boundary of population size–body mass relationships is less clear (Lawton 1990; Silva & Downing 1994; Blackburn & Gaston 1997b). In theory, some species in an assemblage may appropriate very little energy, so that the elevation of the population size–body mass trade-off line on which they must lie is very low. However, because no extant species may have a population size less than a single individual, there may be few attainable population size–body mass combinations for such species. They would inevitably have to have both low population sizes and small body masses. Since species with small populations are vulnerable to extinction, there may be a minimum population size below which long-term persistence is unusual. Whether this varies across species is unclear. If it does, it seems most likely that small-bodied species would require larger population sizes to persist. However, some studies suggest that medium-sized species are most extinction resistant for a given population size (Section 5.3.1, e.g. Johst & Brandl 1997). Thus, it remains to be resolved whether there is any necessary constraint on the lower boundary of population size–body mass relationships, beyond that imposed by the requirement that at least one individual must exist of any extant species.

Within the constraints described above, the members of a multispecies assemblage can lie almost anywhere with respect to each other in a population size–body mass plot. Whether the overall interspecific population size–body mass relationship is negative depends on the relative positions of all species along the slopes on which they lie by virtue of their energy availability. However, unless energy is particularly inequably divided among the species in an assemblage, negative interspecific relationships are quite likely. With equable energy division and a reasonable spread of body masses relative to population size variation for a given body mass, the interspecific slope should be close to $-x$.

If metabolic rate scales with body mass with an exponent of 0.75 (we use this conventionally accepted figure although, as noted in Section 5.5.3, the evidence perhaps better supports 0.67, at least for birds), the allometry of population size ought to be a power function with exponent −0.75. However, there are at least two caveats to this conclusion. First, the exponent 0.75 for the allometry of metabolic rate is calculated across species. As we have just framed our arguments in terms of the population size–body mass trade-off for a species using a fixed amount of energy, might the inverse of an intraspecific metabolic rate exponent not better approximate the exponent of the population size–body mass trade-off? In fact, the interspecific exponent is more likely to be appropriate in this case. In effect, intraspecific relationships can be viewed as showing how energy use varies for individuals of different body mass in the static case where the species is occupying a point in population size–body mass space. However, the constraint lines imagined in this space connect points for which the average body mass of the species differs. A species moving along such a constraint line would necessarily change its average body mass, and hence its metabolic rate. We modelled these lines in terms of the changes that they would imply for a species moving along them, but this was purely for illustrative purposes. In real ecological systems, the constraint lines will connect different species appropriating the same amount of available energy at a given point in time, rather than the same species at different points in time. Thus, they are probably better modelled by interspecific allometries.

Second, the theoretical trade-off between population size and body mass applies only to situations where available energy is constant. Therefore, strictly its exponent should be the inverse of the allometric scaling of metabolic rate across species using equal amounts of energy. As interspecific allometries are not plotted with respect to species total energy usage, we do not know what value this exponent will take. Nevertheless, there is reason to believe that it will not differ greatly from that observed across all species. The elevation of constraint lines will be similar for species using similar amounts of energy. Therefore, where energy is reasonably equably divided among species, and there is a wide spread of body masses for all levels of energy use, the exponents of relationships across species using different amounts of energy will not differ greatly from those across species for which energy use is constant (Fig. 6.2). The general tendency for the allometric exponent of metabolic rate to be around 0.75 in a variety of studies suggests that the use of this figure may not be unreasonable.

An interspecific allometric exponent of 0.75 for metabolic rate implies that that for population size ought to be −0.75. Exponents approximating this value have indeed been found for some assemblages, including British birds (Fig. 5.21a), but not for others (e.g. Gaston & Blackburn 1996c). There are at least four reasons why observed exponents might differ from the predicted value. First, an allometric exponent of 0.75 might not be the most appropriate;

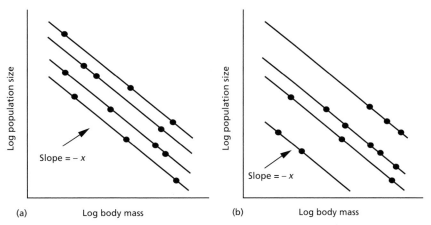

Fig. 6.2 A hypothetical example illustrating when interspecific relationships between population size and body mass (a) will or (b) will not have slopes approximating the theoretical value of –*x* that should pertain across species using equal amounts of energy. Thick lines connect points of equal energy use. This energy can be converted either into many small-bodied or fewer large-bodied individuals, resulting in the negative slopes of magnitude –*x* for these lines (see Fig. 6.1 and text). Energy is more equably divided among species, and there is a wider spread of body masses for all levels of energy use, in (a) relative to (b). The interspecific slope in (a) will closely approximate –*x*, while that in (b) will take a different value.

indeed, for birds it probably is not (Section 5.5.3). Second, energy may be particularly inequably divided in an assemblage. Third, where species actually lie along constraint lines may not be independent of energy appropriated. Fourth, the species may display a limited range of body masses or abundances. In the second two cases, the differences arise because the slope prediction strictly applies only across species using equal amounts of energy, and the distribution of species across the interspecific relationship violates the conditions described in the previous paragraph whereby the slope of the interspecific relationship will approximate the theoretical prediction. In the fourth case, the restricted set of species results in a poor estimate of the true form of the relationship.

Note, however, that an interspecific allometric exponent of –0.75 for population size does not mean that populations are energy limited (Section 5.5.3). Rather, all it shows is that the allometric exponent of per gram energy use (metabolic rate) is 0.75. Whatever limits the resources available to species, be it, say, competition, prey abundance or nest sites, once they have appropriated the energy that this limiting factor allows, then they can only allocate it to more small- or fewer large-bodied individuals, and the form of this trade-off must be mediated by metabolic rate. Thus, the exponent of the interspecific population size–body mass relationship says nothing about the factors that limit population sizes.

6.2.3   *Range size*

The populations of species in this hypothetical multispecies assemblage must be distributed across space. Consider first a two-species case, in which both species appropriate equal amounts of energy. We assume that there is a single relationship across all species for the allometric scaling of metabolic rate. Then, if the body masses of the species are equal, so too must be their population sizes, and hence their range sizes. If species 2 is larger than species 1, then the shape of the population size–range size relationship depends on the allometric scaling of average individual area requirements. Range size is the product of average individual area requirements and population size, which both vary with body mass. If the absolute values of the allometric exponents of average individual area requirements and population size are equal, so that individual area requirements increase with body mass at the same rate that population size decreases, then the range sizes of both species will be constant, and the population size–range size relationship will have a slope of 0. If the absolute value of the allometric exponent of average individual area requirements is less than that for population size, so that individual area requirements increase with body mass more slowly than population size decreases, then the population size–range size relationship will be positive: the larger-bodied species will have both smaller population size and range size. However, if the absolute value of the allometric exponent of average individual area requirements is greater than that for population size, then the population size–range size relationship will be negative: the larger-bodied species will have smaller population size but larger range size.

When species differ in energy appropriated, the situation is more complicated. Population size is inversely related to mass across species using a constant amount of energy, but the product of population size and mass is greater for species using more energy. Therefore, the allometric scaling of distributional extent depends on population size and the allometric scaling of average individual area requirements (Fig. 6.3). Now, whether the population size–range size relationship for our two-species assemblage is positive or negative depends on exactly how much energy a species appropriates, the allometric exponent of average individual area requirements and the body masses of the species. It is clear from Fig. 6.3, however, that positive relationships will occur for many parameter values, and especially when species do not differ greatly in body mass. If individual area requirements scale with the same absolute allometric exponent as population size, then population size–range size relationships are always positive (the more abundant species is always more widespread), except in the special case that the species use exactly the same amount of energy, when the slope of the relationship is zero (Fig. 6.3).

If the scaling constraints applied to the average individual area requirements of species are relaxed, then absolute limits can be defined on where

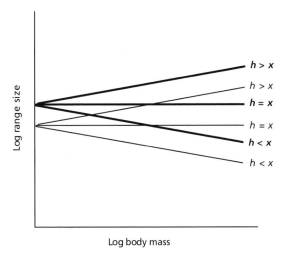

Fig. 6.3 The allometric scaling of geographical range size for two hypothetical species in an imaginary province. Species 2, represented by thick lines, appropriates more energy than species 1, represented by thin lines. The term h is the absolute value of the power exponent of the relationship between average individual area requirements and body mass, while x is the absolute value of the power exponent of the relationship between population size and body mass (which equals the allometric scaling exponent of metabolic rate). Because geographical range size is the product of population size and average individual area requirements, the slope of its relationship with mass for a species depends on the values of h and x, while the elevation of its relationship with mass depends on how much energy the species appropriates (and hence its population size for a given mass). If $h = x$ for both species, the population size–range size relationship is always positive. If $h > x$ for both species, positive population size–range size relationships will definitely pertain for all values of body mass for species 1 with range sizes less than the minimum for species 2. Conversely, if $h < x$ for both species, positive population size–range size relationships will definitely pertain for all values of body mass for species 2 with range sizes greater than the maximum for species 1. Otherwise, the slope of the population size–range size relationship depends on the exact value of h, the amount of energy the species appropriate and their respective body masses.

species can lie in population size–range size space (Fig. 6.4). First, there is an upper limit to range size set by the size of the province. Second, there is a lower boundary to distributional extent set by the minimum area into which populations of different sizes can be squeezed. Unless individuals can be stacked on top of each other, this area will inevitably increase with population size. Third, there is a lower limit to the population size–range size relationship set by the maximum amount of area that can be occupied by populations of a given size. This probably increases with population size in the real world, although in theory a single individual using all the available energy in the province would have a home range size encompassing the entire area. Finally, there is an upper limit to the population size that the province can maintain, which equals the total number of individuals that would occupy the province if the assemblage

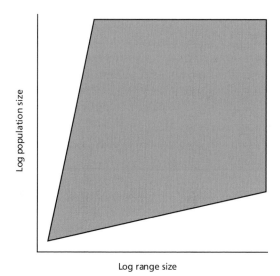

Fig. 6.4 Representation of the shape of population size–range size space within which species are absolutely constrained to lie. See text for more details.

consisted of only a single species of body mass that minimized individual energy use.

These constraints imply that a set of species occupying random points in population size–range size space would likely show a positive relationship, as is indeed the case (Section 4.2). Moreover, constraints on energy allocation within this space suggest that positive relationships are more likely between closely related species. These will differ to some degree in the amount of energy appropriated, but will be physically and physiologically quite similar: the relative with the larger population size should then have the greater spatial extent (Fig. 6.3).

As Fig. 6.3 shows, the relationship between range size and body mass can take a variety of forms, both for a given amount of appropriated energy, or across species appropriating different amounts. Ultimate constraints are placed on the relationship by the size of the inhabited province and again by the minimum and maximum body mass attainable by species in a taxon. The extent of the range required to house a minimum viable population of a species probably also increases with body mass, suggesting that the lower bound of the range size–body mass constraint space will be positive (Brown & Maurer 1987). This observation does carry some caveats, however. First, the sign of this boundary slope will depend on precisely how minimum viable population size and average individual area requirements vary with body mass. Second, some species in an assemblage may not have viable abundances, so falling below the constraint line. This is particularly likely at present given the extinction crisis (May *et al.* 1995). Third, the concept of viability requires reference to a time frame. Extinction is inevitable, so even huge populations have a finite probability of becoming extinct over any time period. The concept of a minimum viable population is therefore somewhat nebulous, even when extinctions are at the 'normal' background level. Nevertheless, we can envisage a population size below which time to extinction will be short relative to the average lifespan of a species in the taxon in question, and that the range size required to house this population increases with body mass. Then, the boundary constraints described imply that a set of species occupying random points in range size–body mass space might show a positive relationship, but that this form is only marginally more likely than any other. This does indeed seem to be the case (Section 5.3.3).

6.2.4   *Density*

Although much of the preceding discussion on the determinants of range size has been framed in terms of average individual area requirements, this requirement is actually the inverse of population density. Thus, we can consider how density may vary in our hypothetical assemblage. In fact, it is relatively unconstrained. While it necessarily varies allometrically as the inverse of average individual area requirements, the exponent is not restricted to any particular value.

Population density is, however, constrained in one important way. If, for a given amount of appropriated energy, average individual area requirements scale allometrically with exponent x, where $-x$ is the allometric exponent for population size, then range size is mass invariant. As the average individual area requirement is the inverse of population density, density must scale as mass to the power $-x$ in this case. Thus, population size and density both scale with mass in the same way. But, if range size varies with body mass, then population size and density must scale with different allometric exponents, for a given amount of appropriated energy. In a multispecies assemblage, with energy reasonably equably divided among the species then, as already noted, the interspecific population size–body mass exponent will be close to $-x$. However, the interspecific population density–body mass exponent will only take the same value if range size does not vary systematically with body mass. If range size increases with mass, it must do so because individual area requirements increase (and hence density declines) with body mass at a faster rate than population size declines. A similar argument applies if range size decreases with mass.

A tendency for positive interspecific relationships to pertain between range size and body mass, at least at the largest spatial scales (Section 5.3.3), implies the form of the allometric scaling of density. As was shown above, range size is mass invariant if average individual area requirements scale allometrically with exponent x, where $-x$ is the allometric exponent for population size.

Positive range size–body mass relationships thus imply an allometric exponent of greater than x for average individual area requirements, and thus of less than $-x$ for its inverse, density. Because x is typically taken to be approximately 0.75, the allometric exponent for population density should be less than -0.75. The precise form of the relationship between population density and body mass has been the subject of intense debate. Slopes close to -0.75 on log–log plots have been claimed for a variety of taxa, but the issue remains unresolved (Section 5.5). Some of the reasons why slopes of -0.75 might not be observed for log population size–log body mass slopes also apply here. Nevertheless, the generally weak and variable nature of the relationship between range size and body mass suggests that population density–body mass slopes of around -0.75 may not be unreasonable.

6.2.5   *Species richness*

We have assumed that in a hypothetical multispecies assemblage, the energy available in the province as a whole is divided among the species, and that relationships between macroecological variables result from constraints on how individual species use the energy they appropriate. A logical consequence is that there is a limit to the number of species that the province will support. The maximum richness attainable equals the amount of available energy divided by the amount of energy required to support the minimum biomass that constitutes a viable population of a species. The precise value of this latter quantity depends on how minimum viable population size and individual energy requirements trade off against body mass. Thus, the province would have lower species richness if the elevation of the allometric relationships of either minimum viable population size or individual energy requirements were higher, or if the total amount of energy available in the province was lower. This latter situation would arise either if the amount of energy available per unit area was lower, if the province covered less area, or both.

Although we assume that maximum species richness is ultimately limited by the total amount of energy available in a province, it is doubtful that this upper limit is ever actually reached in natural systems. Provincial species richness may be depressed below the theoretical maximum dictated by energy availability because of a variety of effects acting in ecological and evolutionary time, including seasonality in energy availability, periodicity in the global climate and occasional global catastrophes. Nevertheless, there is evidence that species richness is correlated with both energy availability per unit area and geographical area at large spatial scales (Sections 2.5.2, 2.5.3), the product of the last two being the provincial energy availability. If some factors do indeed act to depress species richness below the theoretical maximum (e.g. Section 2.5.4), it seems that they do not act sufficiently differentially with respect to provincial energy availability to disrupt its relationship to species richness.

6.2.6   *From macro to micro*

So far in this chapter we have attempted to show how a number of the patterns of principal interest to macroecologists might fit together into a single framework that agrees with current empirical evidence about the general form of such relationships. Ultimate constraints on where species can lie in parameter space are provided by energy availability, and by the energy required to maintain viable populations (or the area required from which to harvest that energy). However, most species do not lie close to these ultimate boundaries, strongly implying that other factors are determining the position of species in parameter space. For example, while we have shown that population size and range size are likely to be positively related, real relationships are not a simple consequence of the random distribution of individuals across the environment (e.g. Wright 1991; Gaston *et al.* 1997a, 1998e; Venier & Fahrig 1998). Similarly, J. Harte and T.M. Blackburn (in preparation) show that the bivariate relationship between distributional extent (E) and size of the grid cells over which distribution is mapped (A, where $E = WA$, and W = number of grid cells occupied) can potentially take many forms depending on the distribution of individuals across the environment, but that if species are distributed according to the principle of self-similarity, then the form of that relationship is tightly constrained. It is the search for these additional, proximate constraining factors that has dominated much of the literature pertaining to some (but not all) macroecological patterns, which have constituted the bulk of Chapters 3–5.

Although the constraint model that has been outlined is framed in terms of constant energy use by species, in the real world this is not constant. Rather, it will fluctuate as do the effects of whatever factors happen to limit the population of a species. In addition, some macroecological traits of species are more plastic than others. The population size and range size of a species will in general change much more quickly than will its body mass (e.g. Gaston & Blackburn 1997a; Blackburn *et al.* 1998a; Gaston 1998). In real ecological time, species will not move along constraint lines of the type illustrated in Figs 6.1 and 6.3. Instead, they will tend to move between them as the energy appropriated by their populations changes. Moreover, where one of the axes refers to body mass, movement in constraint space will tend to be perpendicular to it. The model is not supposed to mimic nature in this sense, but rather to illustrate the constraints placed on populations using a given amount of energy.

In the case of birds, this amount may actually be rather low relative to that which is potentially available in the environment at large. Indeed, avian assemblages may, directly and indirectly, commonly exploit a rather small proportion of the primary production in a locality (e.g. Wiens 1973; Holmes & Sturges 1975; Brockie & Mooed 1986). The bird assemblage of the forested watersheds of the Hubbard Brook Experimental Forest was estimated to consume 0.17% of the net annual productivity of these ecosystems. Such low

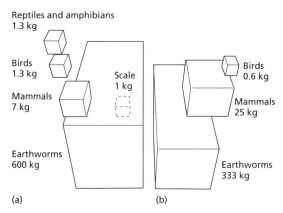

Reptiles and amphibians
1.3 kg

Birds
1.3 kg

Mammals
7 kg

Scale
1 kg

Birds
0.6 kg

Mammals
25 kg

Earthworms
600 kg

Earthworms
333 kg

(a)   (b)

Fig. 6.5 Comparative animal biomass in (a) deciduous oak–hornbeam forest in Belgium and (b) evergreen broadleaf–podocarp forest of the Orongorongo Valley in New Zealand. From Brockie and Mooed (1986).

levels of exploitation are reflected in the generally low biomass of birds in many vegetation types compared with other animal groups (Fig. 6.5). The estimated total biomass of all breeding bird species in Britain (13 212 tonnes) is only one-twelfth that of all wild mammal species (158 021 tonnes; Greenwood *et al.* 1996).

Many more connections among macroecological patterns could have been made, and indeed already have been made in the literature. For example, Harte and colleagues (Harte & Kinzig 1997; Harte *et al.* 1999; J. Harte & T.M. Blackburn, in preparation) have drawn explicit links between species–area relationships, turnover, the species–abundance distribution and the form of the abundance–range size relationship. The first and third of these patterns are linked by Hubbell's (1997) neutral model (Section 4.3.4). Interactions between species–area relationships and turnover, between turnover and range size, and between range size and latitude, suggest that these may all be connected. We could go on.

Both in terms of the connections that we have made, and those that we have not, the structure that we have created is undoubtedly crude. Most macroecological studies to date have considered relationships between no more than two or three variables. There are sensible pragmatic reasons for this. It is hard enough to understand the mechanisms driving pairwise interactions, without introducing the additional complications of linking in relationships to other patterns. The process is not helped by the normally high degree of unexplained variation around even well-defined macroecological patterns, which for the most part we have ignored in developing the model framework above. In addition, the tendency to consider small numbers of variables, usually in the same combinations, means that many potential links between patterns have not even been considered. Thus, the more links one attempts to make, the deeper one descends into the realm of speculation. For that reason, we have not tried here to draw all the connections that otherwise could be made. Even within these limits, however, it is clear that any attempt to fit what is known of these patterns into a single seamless structure on the basis of current knowledge will inevitably produce crude results.

Nevertheless, we hope that our attempt might stimulate other researchers to consider the problem. It is clearly an important one, because the major macroecological patterns must be related. As noted at the start of this chapter, divisions between the different patterns are arbitrary separations made for scientific convenience. In reality, ecological systems are continuous. Ultimately, all the different macroecological patterns we have described derive, for defined areas, from information on the same set of individual organisms. For the specific example that we have focused on, that means the approximately 125 million individual breeding (or 116 million wintering) birds that inhabit Britain. Put in these terms, it is obvious that the patterns that have been considered must be linked.

Moreover, once the dependence of wider patterns on the individuals that go to make up the regional fauna or flora is appreciated, it becomes just as apparent that large- and small-scale patterns and processes must also be connected. All the individual birds of which the British avifauna is comprised inhabit local communities. Therefore, interactions at local scales will influence patterns observed at larger scales. But, equally, large-scale patterns and processes will influence local assemblage structure. One cannot understand the ecology of local assemblages in terms of local processes alone. In that spirit, we now return to the avifauna of Eastern Wood, to see how the macroecological context has helped us to understand the structure of this exemplary local assemblage. We do so physically as well as metaphorically.

6.3   Eastern Wood revisited

We returned to Eastern Wood exactly one year to the day after our first visit. Any hopes we might have entertained about our second visit being on a more typically mild April day had been dashed by one of a run of cold nights, but at least this time there was no snow lying, and the clear skies that had allowed the temperature to fall so low overnight persisted to produce a beautiful, if chilly, morning for birdwatching. The sunshine that made us happy to be in the open air seemed to have a similarly stimulating effect on the avifauna, and as we entered the wood there was appreciably more bird song and visible activity than there had been the year before at exactly the same time of day. This was reflected in the steeper initial slope of the species accumulation curve for 1999 compared to 1998 (Fig. 6.6), and after two hours in the wood we had seen five more species than in the comparable period the previous year. Nevertheless, that higher rate did not translate into a greatly inflated species total by the time

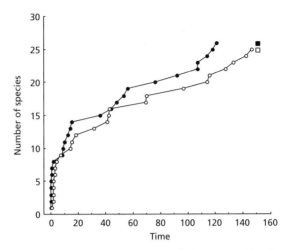

Fig. 6.6 The cumulative number of bird species recorded in Eastern Wood with time (minutes) on the mornings of 16 April 1998 (open circles) and 16 April 1999 (filled circles). The term 0 minutes is 07.30 hours. The square points indicate the total number of species seen in each visit.

we left the wood, as we saw no new species in the last half hour. Although the average number of bird species found by intensive censuses to be breeding in Eastern Wood (Figs 1.3 & 1.21) suggests that we did not detect every species that may have been present, we did seem to be close to the limit of the number of species that could easily be seen there. That number was almost identical to that reached on the same paths in the same time the previous year.

The similarity between the avifaunas of Eastern Wood recorded on our two visits is not limited just to species number. Of the 29 species observed in total over the two visits, 22 were seen on both. We missed dunnock, willow warbler and bullfinch on the second visit, but instead recorded mandarin, stock dove, treecreeper and goldcrest. Any of these species would have been easy to overlook. The mandarin and stock dove bred in only six and five, respectively, of the 30 years between 1949 and 1979 in which Eastern Wood was censused, and in only one of those years was more than one pair of either present. Goldcrests bred in 11 census years, but never numbered more than three pairs. The remaining four species all bred in most census years. However, the willow warbler population in the wood had declined substantially over the census period, and by the end was only a sporadic member of the breeding bird assemblage (Fig. 1.17). The average breeding population sizes when present of dunnock, treecreeper and bullfinch were 3.17, 2.1 and 2.5 pairs, respectively. We failed to find only two of the 22 most common breeding species in Beven's censuses, and the date on which we visited the wood was arguably too early to stand a reliable chance of finding one of those (garden warbler).

Table 6.1 Geometric means and their 95% confidence intervals for body mass (g), breeding range size in Britain (number of 10 × 10-km squares occupied) and breeding population size (number of individuals) for the bird species assemblage recorded in Eastern Wood in 2.5 hours on the morning of 16 April in two different years.

Trait	Geometric mean	95% confidence intervals
1998		
Body mass	47.20	25.18–88.51
Range size	2177.7	1981–2399
Population size	1 185 768	576 766–2 432 204
1999		
Body mass	53.95	27.73–104.95
Range size	1958.8	1600–2399
Population size	851 138	389 045–1 857 805

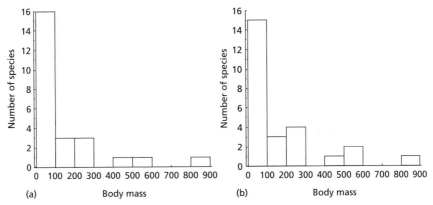

Fig. 6.7 The frequency distributions of body masses (g) of bird species recorded in Eastern Wood on the morning of 16 April in (a) 1998 and (b) 1999.

Given the similarity in species identities between the avifaunas of Eastern Wood we recorded on our two visits, it is unsurprising that there is also similarity in specific traits (Table 6.1). In both cases, most of the species were small-bodied (Fig. 6.7), were relatively widely distributed across Britain (Fig. 6.8), but had low British population sizes relative to the most abundant species (Fig. 6.9). Thus, despite some differences in fine detail, the patterns we recorded in the structure of the bird assemblage of Eastern Wood on the morning of 16 April 1998 were essentially repeated in the bird assemblage at the same site one year later. Some of the species were different, as undoubtedly were many of the individuals, but what differences there were had only minor effects on the trait values for the community as a whole.

A core theme of this volume has been to demonstrate how this apparent structure in a local assemblage is determined, at least in part, by the web of

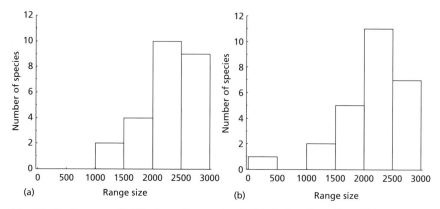

(a) Range size

(b) Range size

Fig. 6.8 The frequency distributions of range sizes in Britain (number of 10×10-km squares occupied on the British National Grid; Gibbons *et al.* 1993) of bird species recorded in Eastern Wood on the morning of 16 April in (a) 1998 and (b) 1999.

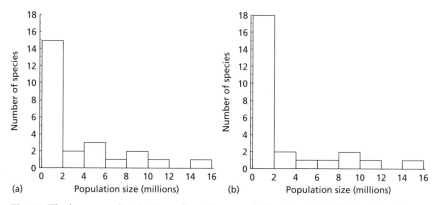

(a) Population size (millions)

(b) Population size (millions)

Fig. 6.9 The frequency distributions of breeding population sizes in Britain (number of individuals) of bird species recorded in Eastern Wood on the morning of 16 April in (a) 1998 and (b) 1999.

interactions sketched thus far in this chapter. These interactions constitute the regional context in which the local avifauna is set, and the description of this web reinforces the notion that to regard the local assemblage independently of this context is to fail to understand many of the reasons for the way in which it is itself structured.

In arguing that local assemblages must be placed in their regional context, one must at all times be aware that regional assemblages have no reality independent of the localities within which individual animals reside. A regional assemblage is a collection of local assemblages. This implies that there is a circular chain of reasoning about the determinants of local and regional patterns in assemblages, which is difficult to break. Thus, regarding the distinction between local- and large-scale approaches to understanding ecology as

concerning different points on a continuum of scales is perhaps overly simplistic. Nevertheless, the influence of local on regional and regional on local
patterns clearly requires approaches based on both.

Perhaps the most fundamental interaction between local and regional
patterns is that species richnesses are positively correlated between the two.
Therefore, the size of the regional pool sets a severe constraint on what local
richness can achieve. The size of the regional pool is determined by the balance
of speciation and extinction, both processes which operate at scales above
that of the local site. Large-scale variation in speciation and extinction rates
depends on a number of factors, of which the most important are the area
encompassed by a region, and its geographical position.

The size of the sample of the regional pool which occurs at a local site
depends fundamentally on the area of that site. Eastern Wood contains
approximately the number of breeding bird species that would be expected on
the basis of other similar woodlands in the same region, and which sample the
same regional pool. However, the way in which Eastern Wood samples this
pool is not random. We showed in Chapter 2 that random samples from the
entire British avifauna, equal in size to the number of territories ever recorded
in Eastern Wood, generate null assemblages with far more species than have
ever been recorded there (around 116). Restricting the species pool to woodland bird species reduces this difference, but the richness of null assemblages
is still too high (range 52–63, compared to 45 observed).

The reason or reasons why local species richness is not simply a random
draw of the wider regional assemblage has not yet been resolved. In general,
however, there are two broad explanations for such a mismatch. First, the
regional pool may be incorrectly specified. The effects of changing the specification of the regional pool have been shown for Eastern Wood by examining the similarity of its structure to a random draw from the entire avifauna of
Britain and the avifauna of deciduous woodlands in Britain (Chapters 2–5),
and in the case of abundance to the pool of species known to have bred in Eastern Wood since 1949. For the great majority of the statistics calculated, finer
definition of the regional pool results in closer matches between random samples and the real assemblage. For example, we have just discussed the effect
on the species richness of random assemblages of reducing the species pool
from all British breeding birds to those species that breed in deciduous woodland. The species richness difference between random and real assemblages
disappears entirely if the pool is restricted to species that have been observed
breeding in Eastern Wood from 1949 to 1997—in other words, not just those
species breeding in the period 1949–79 of interest to us (although this pool
definition is so restricted that the lack of difference is hardly surprising).

In fact, the correct specification of the regional pool is exceedingly difficult.
Clearly, the entire British avifauna is an inappropriate species pool for Eastern
Wood, because a number of seabirds and open-country species are highly

unlikely ever to breed in its mainly woodland habitat. That is why we defined a second pool of British species associated with woodland. However, the probability that many of these woodland species will ever be recorded from Eastern Wood is also very low. For example, the black grouse breeds in broken woodland near moorland (Ehrlich *et al.* 1994). In the last century, this species bred throughout southern Britain, as far south as Hampshire and Cornwall (Gibbons *et al.* 1993). Presumably, the mixture of woodland and heathland of which the greater area of Bookham Common is comprised would have been suitable for black grouse, which may have bred in the area (if not actually in Eastern Wood itself). However, this species has suffered a catastrophic population decline in southern Britain over the past century (Gibbons *et al.* 1993). In England, it is now found no further south than Staffordshire (over 200 km from Eastern Wood). Thus, while this species cannot be precluded from occurring in Eastern Wood on the basis of its habitat preferences, it is highly unlikely that it will ever do so, at least in our lifetime. Another example is the golden pheasant, a Chinese species introduced to southern Britain. Although this prefers thicket-stage conifer woodland, in southern England it can be found in hazel coppice under oak (Gibbons *et al.* 1993). Thus, it is certainly a potential colonist of Eastern Wood, and indeed breeds little over 50 km from this site. Nevertheless, golden pheasants are very sedentary birds, and their British population size is low (Appendix III). It is more likely to occur in Eastern Wood than the black grouse, but its appearance is nevertheless still unlikely. By contrast, the jackdaw has never been observed breeding in Eastern Wood (or at least had not been so recorded by the census teams up to 1997). However, we recorded this species just 30 minutes after entering the wood for the first time. Why it has never been recorded breeding there is unclear. Its current presence is unlikely to be a simple consequence of regional population increases, as the density of jackdaws in British woodlands has declined over the past 30 years (Gregory & Marchant 1995, although its density on farmland has increased). Perhaps, jackdaws do not breed there because of a lack of suitable nesting holes in mature trees. This situation will surely change as the woodland ages (recall from Chapter 1 that mature trees were removed from the wood until the 1950s as part of its management; Beven 1976). If so, jackdaws may nest in Eastern Wood before too long.

These examples clearly demonstrate that the probability that bird species will breed in Eastern Wood forms a continuum, from those that will never do so (e.g. most of the birds of the world), through those that reasonably could (e.g. black grouse, golden pheasant), to those that almost certainly will (e.g. jackdaw), and finally to those that do (robin, blue tit, hawfinch). Clearly, dividing this continuum into species that are in the source pool for Eastern Wood and those that are not involves an arbitrary decision about where to draw the line. Yet, as Gotelli and Graves (1996) point out, and we have already shown, species pool designation can have a strong effect on the outcome of null model tests. Gotelli and Graves suggest that a source pool 'should include all species that have a *reasonable* probability of occurring' at a site (their italics). However, this simply raises the question of what is reasonable. Moreover, any division will inevitably cause species with similar probabilities of occurrence at a site to be both in and out of the pool.

Ultimately, only species that have actually occurred at a site at some point can be stated unequivocally to be in its species pool, and even then one has to assume that the status of the species and the composition of the site have not drastically changed since they were seen to occur. Such a restricted definition may be of limited utility when trying to understand how assemblage structure differs from a null expectation, because such a high proportion of that structure is assumed in the null.

The second reason for the mismatch between random draws from the regional pool and the structure of the avifauna of Eastern Wood is that sampling is not random, but instead is modified. There are many factors which might obviously contribute to such modification. For example, the habitat composition of the wood and the habitat requirements of the birds will affect assemblage composition. Some of these effects will be subsumed by the definition of the species pool, but not all. Thus, even if the regional bird species pool for Eastern Wood is defined as those species that breed in woodland, fine details of the wood's habitat structure could still prevent certain species from breeding there. The example of the failure of the jackdaw to breed in Eastern Wood is a case in point.

The modifying factors causing differences between real and random assemblages may not only be small-scale features of the site, but may also be larger-scale features of the landscape in which a site is positioned. Principal among these is the isolation of a site from other similar habitats (Section 2.3), which affects the likelihood that species with different dispersal abilities will locate and colonize it. Such an effect is central to explanations of species richness patterns in the equilibrium theory of island biogeography (MacArthur & Wilson 1967; Section 2.2.1), and to Hubbell's (1997) explanation for differences between local and regional (metacommunity) species–abundance distributions (Section 4.3.4). Thus, while mismatches occur between Eastern Wood and null assemblages, such that the form of the former is not simply a consequence of large-scale patterns in the structural variables, some of those mismatches are likely themselves to be consequences of large-scale factors.

Aside from species richness, mismatches between the Eastern Wood avifauna and equivalent assemblages drawn randomly from the British woodland bird species pool occur for abundances (Sections 4.1, 4.3.1) and body sizes (Section 5.2.2). The average body mass of species in random assemblages tends to be slightly larger than for species recorded breeding in Eastern Wood. The situation with abundance is more complicated. Species breeding in Eastern Wood have higher abundances than British birds not breeding there, but lower abundances than would be expected if the assemblage was a random sample of woodland birds, where probability of sampling was weighted by species abundance. Random draws designed to mimic the species abundance distribution (by drawing an equivalent number of territories from the entire British woodland avifauna; Section 4.3.1) generate assemblages with lower mean and maximum species abundances than observed in Eastern Wood. To some extent, this last mismatch is a consequence of the species richness difference between real and random assemblages. Removing this difference, by restricting the species pool just to species that have bred in Eastern Wood at some point in the period 1949–97, results in random species–abundance distributions that share a number of features with that actually observed for Eastern Wood (Table 4.4).

The other mismatches can be removed if we consider range sizes. If the position of a site in the landscape is an important determinant of which species are likely to colonize the site, or alternatively, which species from the regional pool the site samples, then how species are distributed across that landscape is likely to be an important determinant of which are sampled. In particular, we would expect the most widely distributed species to be most likely to encounter any given site. This expectation is upheld by the Eastern Wood avifauna, which consists of British bird species with higher than average range sizes (Section 3.2.2). Moreover, the average range size of the Eastern Wood avifauna is exactly what would be expected were the assemblage a random sample of British woodland birds with sampling probability proportional to range size (Section 3.2). Thus, as expected, more widespread woodland species are more likely to be sampled by Eastern Wood, and this likelihood is proportional to their range sizes (Blackburn & Gaston, in press).

The range sizes of species are usually positively correlated with their population sizes, and this relationship is relatively strong for British birds (Fig. 4.4). Thus, we might expect that the population sizes and range sizes of Eastern Wood species should show similar patterns with respect to the wider British fauna. However, the results of the random draws discussed above show that the range size of a species is a better predictor of how likely it is to be sampled from the regional pool by a site than is its abundance. High abundance in Britain does not guarantee that a species will encounter Eastern Wood if all those individuals reside in the north. By contrast, the range of a widespread species is likely to include Eastern Wood however abundant it is.

In fact, weighting the probability of sampling species from the woodland pool by their range sizes generates null assemblages with average properties that are indistinguishable from the Eastern Wood avifauna, removing differences between real and random assemblages in mean body masses and population sizes (Table 6.2). For body masses, this effect probably derives from a weak negative correlation between range size and body mass in the British avifauna (Sutherland & Baillie 1993). Given that more widespread species tend to

Table 6.2 Results of models simulating the abundances of bird species in Eastern Wood as random draws of individuals from the pool of breeding deciduous woodland bird species in Britain. The probability that a species was selected from the pool in a simulation was weighted by its range size; $n = 45$ for all real values. Simulated values derive from 1000 iterations. None of the real values differ significantly from the results of the simulations. See text for further details.

Statistic	Real value	Simulated mean ± SD	Simulated range
Arithmetic mean range size	1965	1969 ± 60	1750–2120
Geometric mean range size	1766	1774 ± 118	1243–2052
Arithmetic mean population size	2 100 322	1 917 673 ± 183 466	1 229 370–2 226 840
Geometric mean population size	472 402	438 938 ± 86 137	161 863–692 484
Arithmetic mean body mass	123.05	146.49 ± 21.88	89.39–250.78
Geometric mean body mass	43.49	48.23 ± 5.61	35.16–68.10

occur in Eastern Wood, the body masses of these species will be slightly lower than expected from the species–body mass distribution for British birds, as observed (Section 5.2.2). This may well answer the question posed in Chapter 5 about why body mass distributions of local assemblages differ from those expected on the basis of random draws from regional assemblages. Thus, given the number of bird species breeding there, and their broad habitat requirements (in the definition of the species pool), many features of the structure of the Eastern Wood bird assemblage can be predicted from knowledge of the extent of the species distribution in the regional landscape (Blackburn & Gaston, in press).

As has long been recognized, one of the key issues in ecology is the extent to which local assemblages can be modelled accurately as random draws from regional species pools, or conversely the extent to which factors modify the similarity of real assemblages to randomly simulated ones. In consequence, there has been much, and at times vociferous, debate as to the construction of appropriate random, or null, models (this large and varied literature has been reviewed by Gotelli & Graves 1996). For present purposes, what is crucial is that almost invariably some, often much, and occasionally most of the basic structure of local assemblages, such as Eastern Wood, can be explained by such models. Local assemblage structure and the regional context are inseparable.

6.4   Human interference

Throughout this book the possible impacts of human activities on the patterns discussed have largely been ignored. Nonetheless, it is plain that these activities are having a profound effect on the Earth's biota. This is evidenced, for example, by the high proportion of the areas of biomes which have experienced

Table 6.3 The land cover composition of Britain as of 1990. From Barr *et al.* (1993).

Land cover class	Area	%
Continuous urban	2 603	1.1
Suburban	13 169	5.5
Tilled land	51 313	21.4
Managed grassland	65 672	27.3
Rough grass/marsh	4 307	1.8
Bracken	3 603	1.5
Heath/moor grass	20 203	8.4
Open shrub heath/moor	27 868	11.6
Dense shrub heath/moor	7 220	3.0
Bog	4 309	1.8
Deciduous/mixed woodland	12 329	5.1
Coniferous woodland	7 722	3.2
Inland bare	2 566	1.1
Saltmarsh	389	0.2
Coastal bare	1 421	0.6
Inland water	1 714	0.7
Sea/estuary	7 683	3.2
Unclassified	6 133	2.6
Total	240 222	100.0

human disturbance (Table 1.2; Hannah *et al.* 1995), the dramatic changes which have occurred in land use over the last three centuries (e.g. a 19% reduction in the extent of forests and woodlands, an 8% reduction in grasslands and pasture, and a 466% increase in croplands brought under cultivation; Richards 1990) and the numbers of species which are presently listed as being threatened with extinction in the near future (1111 species of birds alone; Collar *et al.* 1994).

Human impacts have been particularly severe over much of the temperate north, with the landscape of Britain to most intents and purposes having been shaped by these activities; this island (or strictly group of islands) is not only small and damp (Section 1.6), but also has one of the most disturbed floras and faunas. Table 6.3 shows the current habitat composition of the British landscape. Prior to human colonization, Britain was mainly covered by woodland, but this habitat now covers less than 9% of the country. By contrast, urban, suburban and agricultural land (tilled land and managed grassland) now cover more than 50% of the island. Only one-third of the land cover of Britain now consists of what can be termed seminatural vegetation (Barr *et al.* 1993). Despite greater general awareness of the aesthetic, scientific and conservation value of natural habitats, their transformation is an ongoing process. Thus, there was a 25% decline in area of native pinewood between 1950 and 1986 (Bain 1987, cited in Hill *et al.* 1990). Barr *et al.* (1993) estimate that 49 000 km of hedgerows were removed in Britain between 1984 and 1990, and that in addition to the 200 000 km estimated to have been lost between 1946 and 1974 (Pollard *et al.* 1974).

With regard to macroecology, the important question about human activities is whether their effects have distorted patterns and processes to the extent that those we observe today bear little or no relation to those that would have pertained before the rise of human influence. If they have, then for many and perhaps most regions of the world it would be foolish to consider macroecological patterns outside of this human context, and the particular nature of human activities and their consequences will have to be accommodated. The obvious arguments in favour of the necessity of such an approach are that no other single species has had such a profound and widespread influence on the abundance and distribution of flora and fauna, and that the scale of activities exhibited by humans is unprecedented. Indeed, we have often heard it argued that at the very least macroecological investigations should not be centred on assemblages from northern Europe, or much of North America, because of the pervasive influence of human activities.

However, it does not follow that human activities necessarily have the fundamental influence which has been claimed on the underlying processes that shape macroecological patterns. There are at least two reasons. First, these activities have not served to destroy, or even markedly alter, the broad environmental patterns which must contribute to explanations of many macroecological patterns. These patterns include the latitudinal gradients in means and variances of temperature, and in the relative areas of the major biomes. More generally, the basic laws of mathematics and physics that mediate many of the interactions between animals and their environments remain unchanged.

Second, human activities have not served to alter the fundamental life history traits exhibited by species. Patterns of investment by species in reproduction, trading off current and future investment against a background of adult and juvenile mortality, remain largely unchanged. Great tits still lay frequent large broods and tend to be short lived, while albatrosses lay single eggs, do not nest in every year, and tend to be long-lived. That is not to say, however, that human activities have had no influence on species life histories. Recent climatic warming, which is widely believed to be driven by human-caused atmospheric changes, appears to have allowed a number of British bird species to lay eggs earlier in the year (Crick *et al.* 1997). Severe reductions in British populations of the sparrowhawk caused by the negative effects of pesticides on reproductive success resulted in increases in the body masses of one prey species, the great tit. Larger great tits apparently survive better through harsh winter weather than do smaller individuals, but are more susceptible to sparrowhawk predation. Thus, the absence of sparrowhawks from some areas of Britain removed this selective pressure from the great tit population, allowing average body masses in the population to increase (Gosler *et al.* 1995).

Nevertheless, these are no more than minor alterations to the species overall life history strategies, which remain essentially unaffected.

Hence, most of the fundamental large-scale features of the environment, and the broad features of the biologies of species that determine their interaction with that environment, have remained unchanged in the face of human activities. Of course, this is not to say that such activities have had no impact on macroecological patterns. The large numbers of bird species which are presently at high risk of extinction (about 10% of all extant bird species) are largely a direct or indirect consequence of these activities, predominantly through habitat loss and degradation (Collar *et al.* 1994). These species on average have smaller populations, smaller range sizes, and are larger-bodied than those which are not at risk of extinction (Sections 3.3.6, 5.3.1, 5.5.3), and are doubtless biased with regard to a number of other ecological variables which are associated with these. If the categorizations of likelihood of extinction are correct, then this raises the spectre that humans are modifying the patterns of extinction, which ultimately must play a role (in conjunction with speciation) in shaping, for example, species abundance, species range size and species body size distributions, latitudinal gradients in these variables, and so forth. Human activities may also affect speciation rates. This is something about which rather little is known, although it is easy to see that if the likelihood of speciation is a function of geographical range size (Section 3.3.6), then the reductions in range size which many species have experienced will impact on speciation rates. However, although it is now more widely appreciated that evolution can be a rapid process (e.g. Thompson 1998), speciation will still be slow relative to extinction for those taxa that currently are most extinction prone (e.g. birds and mammals). Human impacts on macroecological patterns are most likely to derive from elevated extinction rates in the taxa most amenable to macroecological study.

The key point in this discussion is that while human activities are likely to have influenced the factors that drive macroecological patterns, they are unlikely to have influenced what those factors actually are. This suggests that macroecologists will still be able to gain insight into the processes that shape large-scale patterns in the abundance and range size of species, even if the patterns themselves have been somewhat modified by human activities. For example, consider the avian species–body mass distribution (Section 5.2). Bird species at risk of extinction, and those known to have been driven extinct in historical times, tend to be large bodied (Gaston & Blackburn 1995b). This implies that the present-day avian species–body mass distribution will differ quantitatively from that pertaining before the global spread of modern humans, although evidence from the fossil record suggests that such distributions have always shown a similar tendency to be right log-skewed (Section 5.2.2). However, since the shape of species–body mass distributions must ultimately be driven by the processes of speciation and extinction, human activities have not altered the processes at work. They clearly have altered the absolute rates of these processes, as well as the relationship of these rates to body mass, but not the processes themselves. Human activities also will have had little effect on the processes proposed to cause small (but not the smallest) body masses to be favoured by the interaction between speciation and extinction (Section 5.2.3). They may have affected environmental grain and resource distributions through habitat destruction or fragmentation, but this would only serve to modify the action of these mechanisms (if they were shown to be the ones acting), not remove their effect. It is difficult to see how human activities would have affected the processes of interspecific and intraspecific body size optimization at all.

Thus, human activities seem likely to be able to modify the patterns and processes generating macroecological patterns, but not alter the processes themselves. However, even if one takes the extreme view that human activities are now the principal drivers of macroecological patterns, to the extent that the patterns differ qualitatively from what would have pertained prior to human influence, that does not negate the macroecological approach. All it argues is that the process driving macroecological patterns is human activity. This is as valid a process as any other, and no less amenable to testing using the approaches taken throughout this volume. Whatever the cause of large-scale ecological patterns, they will still influence the structure of local assemblages, a structure we will not understand without taking these patterns into account. Hence, while we do not believe that human activities cause the macroecological patterns with which this book has been concerned, it would not affect our advocacy of this approach to ecological questions if they had.

This consideration of the effects of human influence leads to an important conclusion. If one argues that patterns can be modified without affecting the processes underlying them, then it is clear that the pattern alone provides very weak evidence for the action of the process. That the process invoked to explain a pattern can do so proves nothing. Indeed, extensive influence of human activities could potentially cause mismatches between pattern and process even in cases where there is a real link between the two. Moreover, arguing from pattern to process is, as McArdle (1996) notes, 'a dangerous thing to do, since any one pattern is rarely explicable by only one process'. Therefore, additional evidence for an association between pattern and process is required. Partly with the aim of stimulating such evidence, we have argued elsewhere that a major goal of macroecology at the present time must be to expand a programme of rigorous tests of macroecological hypotheses (Gaston & Blackburn 1999).

On balance, then, we suspect that human activities have the effect of modifying the outworkings of the rules which govern the occurrence of macroecological patterns, rather than changing those rules altogether. This does not, of course, detract from the severity of human impacts and the desperate need to reduce their negative effects on the global biota. This need is only intensified by the knowledge that local communities are affected by regional events, so that sites removed from human activities, and even those explicitly protected from them, can still be impacted by such activities.

6.5   Final words

Science is a cultural activity. Although as scientists we do our best to remain objective and impartial in the face of data, we cannot help but be influenced to some degree by our cultural context. We live in a time when technology has advanced to the point that information can be transmitted virtually instantaneously around the globe, when it is as easy to find out what is happening on the other side of the planet as on the other side of a village, and when satellite images allow us to view our planet from a perspective, and on a scale, never previously possible. Is it a coincidence, then, that despite a venerable history of study, the field of macroecology has only really taken off in the last few years, in tandem with the globalization of human perspective? We do not think that it is.

The broadening of human perspective could hardly have been more timely. We live in an age when many of the most serious issues facing humankind concern phenomena at large spatial scales. For example, environmental change, human population growth and the spread of disease are all global problems. Climate change will not just affect those countries that do nothing to stop the emissions of the greenhouse gases thought to cause it. Global problems will require global solutions.

We hope that we have convinced the reader that the need for a large-scale perspective applies equally to ecology. This has never been in doubt from the practical viewpoint, where responses to bird declines and threats of extinction, for example, have required knowledge of events on wintering grounds and migration routes, as well as in breeding areas. General acknowledgement in the wider ecological community of the role of large-scale patterns and processes in local assemblage patterns and processes has been slower to develop. Yet, the influence of large-scale patterns and processes is clear for all to see. The next time you visit your local wood, field or lake, consider the influence of the region in which that site is located on the individual birds you see and the species to which they belong. Remember that the number of species you are likely to observe depends on regional richness levels, that whether or not a species is likely to be found at the site depends not just on the suitability of the site, but also on the species distribution and abundance in the surrounding region, and that the body sizes of the species inhabiting the site are related to the latitude at which it lies. Thus, the study of macroecology is not abstracted from the real world. Rather, it helps us to understand what causes the world around us to look the way it does.

References

Abramsky, Z. & Rosenzweig, M.L. (1984) Tilman's predicted productivity–diversity relationship shown by desert rodents. *Nature* **309**, 150–151.

Adams, J.M. & Woodward, F.I. (1989) Patterns in tree species richness as a test of the glacial extinction hypothesis. *Nature* **339**, 699–701.

Allsopp, P.G. (1997) Probability of describing an Australian scarab beetle: influence of body size and distribution. *Journal of Biogeography* **24**, 717–724.

Al-Mufti, M.M., Sydes, C.L., Furness, S.B., Grime, J.P. & Band, S.R. (1977) A quantitative analysis of shoot phenology and dominance in herbaceous vegetation. *Journal of Ecology* **65**, 759–791.

Ambrose, S.J. (1994) The Australian bird count: a census of the relative abundance of common land birds in Australia. In: *Bird Numbers 1992. Distribution, Monitoring and Ecological Aspects. Proceedings of the 12th International Conference of the IBCC and EOAC, Noordwijkerhout, The Netherlands* (eds W.J.M. Hagemeijer & T.J. Verstrael), pp. 595–606. Statistics Netherlands, Voorburg/Heerlen and SOVON, Beek-Ubbergen, The Netherlands.

Ambuel, B. & Temple, S.A. (1983) Area-dependent changes in the bird communities and vegetation of southern Wisconsin forests. *Ecology* **64**, 1057–1068.

Anderson, R.M., Gordon, D.M., Crawley, M.J. & Hassell, M.P. (1982) Variability in the abundance of animal and plant species. *Nature* **296**, 245–248.

Anderson, S. (1977) Geographic ranges of North American terrestrial mammals. *American Museum Novitates* **2629**, 1–15.

Anderson, S. (1984a) Geographic ranges of North American terrestrial birds. *American Museum Novitates* **2785**, 1–17.

Anderson, S. (1984b) Areography of North American fishes, amphibians and reptiles. *American Museum Novitates* **2802**, 1–16.

Anderson, S. (1985) The theory of range size (RS) distributions. *American Museum Novitates* **2833**, 1–20.

Andrén, H. (1994a) Effects of habitat fragmentation on birds and mammals in landscapes with different proportions of suitable habitat: a review. *Oikos* **71**, 355–366.

Andrén, H. (1994b) Can one use nested subset pattern to reject the random sample hypothesis? *Oikos* **70**, 489–491.

Angermeier, P.L. (1995) Ecological attributes of extinction-prone species: loss of freshwater fishes of Virginia. *Conservation Biology* **9**, 143–158.

Angermeier, P.L. & Winston, M.R. (1998) Local vs. regional influences on local diversity in stream fish communities of Virginia. *Ecology* **79**, 911–927.

Anonymous (1997) *The Times Atlas of the World*, concise edition. Times Books, London.

Arita, H.T. & Figueroa, F. (1999) Geographic patterns of body-mass diversity in Mexican mammals. *Oikos* **85**, 310–319.

Arita, H.T., Robinson, J.G. & Redford, K.H. (1990) Rarity in Neotropical forest mammals and its ecological correlates. *Conservation Biology* **4**, 181–192.

Arneberg, P., Skorping, A., Grenfell, B. & Read, A.F. (1998) Host densities as determinants of abundance in parasite communities. *Proceedings of the Royal Society, London, B* **265**, 1283–1289.

Arnold, A.J., Kelly, D.C. & Parker, W.C. (1995) Causality and Cope's rule; evidence from the planktonic Foraminifera. *Journal of Paleontology* **69**, 203–210.

Arrhenius, O. (1921) Species and area. *Journal of Ecology* **9**, 95–99.

Arrhenius, O. (1923) On the relation between species and area: a reply. *Ecology* **4**, 90–91.

Ashby, C.B. (1990) Geoffrey Beven, M.D., B.Sc., F.Z.S., M.B.O.U. 1914–90. *London Naturalist* **69**, 147–155.

Askins, R.A., Philbrick, M.J. & Sugeno, D.S. (1987) Relationships between the regional abundance of forest and the composition of forest bird communities. *Biological Conservation* **39**, 129–152.

Atmar, W. & Patterson, B.D. (1993) The measure of order and disorder in the distribution of species in fragmented habitat. *Oecologia* **96**, 373–382.

Atmar, W. & Patterson, B.D. (1995) *The Nestedness Temperature Calculator: A Visual Basic Program, Including 294 Presence–absence Matrices.* AICS Research, University Park, New Mexico, and The Field Museum, Chicago, Illinois.

Avery, M.I. & Haines-Young, R.H. (1990) Population estimates for the dunlin *Calidris alpina* derived from remotely sensed satellite imagery of the flow country of northern Scotland. *Nature* **344**, 860–862.

Avise, J.C. & Walker, D. (1998) Pleistocene phylogeographic effects on avian populations and the speciation process. *Proceedings of the Royal Society, London, B* **265**, 457–463.

Ayres, J.M. & Clutton-Brock, T.H. (1992) River boundaries and species range size in Amazonian primates. *American Naturalist* **140**, 531–537.

Baillie, S.R. & Peach, W.J. (1992) Population limitation in Palaearctic-African migrant passerines. *Ibis* **134** (Suppl. 1), 120–132.

Bain, C. (1987) *Native Pinewoods in Scotland: A Review.* RSPB, The Lodge, Sandy, Bedfordshire.

Bakker, R.T. (1977) Tetrapod mass extinctions: a model of the regulation of speciation rates and immigration by cycles of topographic diversity. In: *Patterns of Evolution* (ed. A. Hallam), pp. 439–468. Elsevier, Amsterdam.

Barlow, C., Wacher, T. & Disley, T. (1997) *A Field Guide to the Birds of the Gambia and Senegal.* Pica Press, Mountfield, East Sussex.

Barr, C.J., Bunce, R.G.H., Clarke, R.T. *et al.* (1993) *Countryside Survey 1990: Main Report.* Department of the Environment, London.

Bart, J. & Klosiewski, S.P. (1989) Use of presence–absence to measure changes in avian density. *Journal of Wildlife Management* **53**, 847–852.

Bartholomew, G.A., Howell, T.R. & Cade, T.J. (1957) Torpidity in the white-throated swift, Anna hummingbird and poor-will. *Condor* **59**, 145–155.

Baskin, Y. (1997) Center seeks synthesis to make ecology more useful. *Science* **275**, 310–311.

Basset, Y. & Kitching, R.L. (1991) Species number, species abundance and body length of arboreal arthropods associated with an Australian rainforest tree. *Ecological Entomology* **16**, 391–402.

Begon, M., Harper, J.L. & Townsend, C.R. (1996) *Ecology: Individuals, Populations and Communities.* Blackwell Scientific Publications, Oxford.

Bellamy, P.E., Hinsley, S.A. & Newton, I. (1996a) Factors influencing bird species numbers in small woods in south-east England. *Journal of Applied Ecology* **33**, 249–262.

Bellamy, P.E., Hinsley, S.A. & Newton, I. (1996b) Local extinctions and recolonisations of passerine bird populations in small woods. *Oecologia* **108**, 64–71.

Bengtsson, J., Baillie, S.R. & Lawton, J.H. (1997) Community variability increases with time. *Oikos* **78**, 248–256.

Bennett, P.M. & Harvey, P.H. (1987) Active and resting metabolism in birds: allometry, phylogeny and ecology. *Journal of Zoology, London* **213**, 327–363.

Bentley, P. (1995) Understanding map projections, bird distribution, migration and vagrancy. *Birding World* **8**, 231–239.

Bergmann, C. (1847) Ueber die Verhältnisse der Wärmeökonomie der Thiere zu ihrer Grösse. *Gottinger Studien* **3**, 595–708.

Beven, G. (1945) An area census in Zululand. *Ostrich* **16**, 1–18.

Beven, G. (1976) Changes in breeding bird populations of an oak-wood on Bookham Common, Surrey, over twenty-seven years. *London Naturalist* **55**, 23–42.

Bibby, C.J., Phillips, B.N. & Seddon, A.J.E. (1985) Birds of restocked conifer plantations in Wales. *Journal of Applied Ecology* **22**, 619–633.

Björkland, M. (1997) Are 'comparative methods' always necessary? *Oikos* **80**, 607–612.

Blackburn, T.M., Brown, V.K., Doube, B.M., Greenwood, J.J.D., Lawton, J.H. & Stork, N.E. (1993) The relationship between body size and abundance in natural animal assemblages. *Journal of Animal Ecology* **62**, 519–528.

Blackburn, T.M. & Gaston, K.J. (1994a) The distribution of body sizes of the world's bird species. *Oikos* **70**, 127–130.

Blackburn, T.M. & Gaston, K.J. (1994b) Animal body size distributions change as more species are described. *Proceedings of the Royal Society, London, B* **257**, 293–297.

Blackburn, T.M. & Gaston, K.J. (1994c) Animal body size distributions: patterns, mechanisms and implications. *Trends in Ecology and Evolution* **9**, 471–474.

Blackburn, T.M. & Gaston, K.J. (1995) What determines the probability of discovering a species?: a study of South American oscine passerine birds. *Journal of Biogeography* **22**, 7–14.

Blackburn, T.M. & Gaston, K.J. (1996a) Spatial patterns in the geographic range sizes of bird species in the New World. *Philosophical Transactions of the Royal Society, London, B* **351**, 897–912.

Blackburn, T.M. & Gaston, K.J. (1996b) Spatial patterns in the species richness of birds in the New World. *Ecography* **19**, 369–376.

Blackburn, T.M. & Gaston, K.J. (1996c) A sideways look at patterns in species richness, or why there are so few species outside the tropics. *Biodiversity Letters* **3**, 44–53.

Blackburn, T.M. & Gaston, K.J. (1996d) The distribution of bird species in the New World: patterns in species turnover. *Oikos* **77**, 146–152.

Blackburn, T.M. & Gaston, K.J. (1996e) Spatial patterns in the body sizes of bird species in the New World. *Oikos* **77**, 436–446.

Blackburn, T.M. & Gaston, K.J. (1996f) On being the right size—different definitions of 'right'. *Oikos* **75**, 551–557.

Blackburn, T.M. & Gaston, K.J. (1996g) Abundance–body size relationships: the area you census tells you more. *Oikos* **75**, 303–309.

Blackburn, T.M. & Gaston, K.J. (1997a) The relationship between geographic area and the latitudinal gradient in species richness in New World birds. *Evolutionary Ecology* **11**, 195–204.

Blackburn, T.M. & Gaston, K.J. (1997b) A critical assessment of the form of the inter-specific relationship between abundance and body size in animals. *Journal of Animal Ecology* **66**, 233–249.

Blackburn, T.M. & Gaston, K.J. (1997c) Who is rare? Artefacts and complexities in rarity determination. In: *The Biology of Rarity* (eds W.E. Kunin & K.J. Gaston), pp. 48–60. Chapman & Hall, London.

Blackburn, T.M. & Gaston, K.J. (1998) Some methodological issues in macroecology. *American Naturalist* **151**, 68–83.

Blackburn, T.M. & Gaston, K.J. (1999) The relation between animal abundance and body size: a review of the mechanisms. *Advances in Ecological Research* **28**, 181–210.

Blackburn, T.M. & Gaston, K.J. (in press) Local avian assemblages as random draws from regional pools. *Ecography*.

Blackburn, T.M., Gaston, K.J., Greenwood, J.J.D. & Gregory, R.D. (1998b) The anatomy of the interspecific abundance–range size relationship for the British avifauna: II. Temporal dynamics. *Ecology Letters* **1**, 47–55.

Blackburn, T.M., Gaston, K.J. & Gregory, R.D. (1997a) Abundance–range size relationships in British birds: is unexplained variation a product of life history? *Ecography* **20**, 466–474.

Blackburn, T.M., Gaston, K.J. & Lawton, J.H. (1998a) Patterns in the geographic ranges of the world's woodpeckers. *Ibis* **140**, 626–638.

Blackburn, T.M., Gaston, K.J. & Loder, N. (1999b) Geographic gradients in body size: a clarification of Bergmann's rule. *Diversity and Distribution* **5**, 165–174.

Blackburn, T.M., Gaston, K.J., Quinn, R.M., Arnold, H. & Gregory, R.D. (1997b) Of mice and wrens: the relation between abundance and geographic range size in British mammals and birds. *Philosophical Transactions of the Royal Society, London, B* **352**, 419–427.

Blackburn, T.M., Gaston, K.J., Quinn, R.M. & Gregory, R.D. (1999a) Do local abundances of British birds change with proximity to range edge? *Journal of Biogeography* **26**, 493–505.

Blackburn, T.M., Gates, S., Lawton, J.H. & Greenwood, J.J.D. (1994) Relations between body size, abundance and taxonomy of birds wintering in Britain and Ireland. *Philosophical Transactions of the Royal Society, London, B* **343**, 135–144.

Blackburn, T.M., Harvey, P.H. & Pagel, M.D. (1990) Species number, population density and body size in natural communities. *Journal of Animal Ecology* **59**, 335–346.

Blackburn, T.M. & Lawton, J.H. (1994) Population abundance and body size in animal assemblages. *Philosophical Transactions of the Royal Society, London, B* **343**, 33–39.

Blackburn, T.M., Lawton, J.H. & Gregory, R.D. (1996) Relationships between abundances and life histories of British birds. *Journal of Animal Ecology* **65**, 52–62.

Blackburn, T.M., Lawton, J.H. & Perry, J.N. (1992) A method of estimating the slope of upper bounds of plots of body size and abundance in natural animal assemblages. *Oikos* **65**, 107–112.

Blake, J.G. & Karr, J.R. (1987) Breeding birds of isolated woodlots: area and habitat relationships. *Ecology* **68**, 1724–1734.

Blakers, N., Davies, S.J.J.F. & Reilly, P.N. (1984) *The Atlas of Australian Birds*. Melbourne University Press, Carlton.

Bland, R.L. & Tully, J. (1993) *Atlas of Breeding Birds in Avon*. Privately published.

Blondel, J. & Mourer-Chauviré, C. (1998) Evolution and history of the Western Palaearctic avifauna. *Trends in Ecology and Evolution* **13**, 488–492.

Bock, C.E. (1987) Distribution–abundance relationships of some Arizona landbirds: a matter of scale? *Ecology* **68**, 124–129.

Bock, C.E., Cruz, A., Grant, M.C., Aid, C.S. & Strong, T.R. (1992) Field experimental evidence for diffuse competition among southwestern riparian birds. *American Naturalist* **140**, 815–828.

Boecklen, W.J. (1986) Effects of habitat heterogeneity on the species–area relationships of forest birds. *Journal of Biogeography* **13**, 59–68.

Boecklen, W.J. (1997) Nestedness, biogeographic theory, and the design of nature reserves. *Oecologia* **112**, 123–142.

Boecklen, W.J. & Nocedal, J. (1991) Are species trajectories bounded or not? *Journal of Biogeography* **18**, 647–652.

Boecklen, W.J. & Simberloff, D. (1986) Area-based extinction models in conservation. In: *Dynamics of Extinction* (ed. D.K. Elliott), pp. 247–276. Wiley, New York.

Boeken, B. & Shachak, M. (1998) The dynamics of abundance and incidence of annual plant species during colonization in a desert. *Ecography* **21**, 63–73.

Böhning-Gaese, K. & Oberrath, R. (1999) Phylogenetic effects on morphological, life-history, behavioural and ecological traits of birds. *Evolutionary Ecology Research* **1**, 347–364.

Bond, R.R. (1957) Ecological distribution of breeding birds in the upland forests of southern Wisconsin. *Ecological Monographs* **27**, 351–384.

Bonham, P.F. & Robertson, J.C.M. (1975) The spread of the Cetti's warbler in north-west Europe. *British Birds* **68**, 393–408.

Boström, U. & Nilsson, S.G. (1983) Latitudinal gradients and local variations in species richness and structure of bird communities on raised peat-bogs in Sweden. *Ornis Scandinavica* **14**, 213–226.

Boulière, F. (1975) Mammals, small and large: the ecological implications of size. In: *Small Mammals: Their Productivity and Population Dynamics* (eds F.B. Golley, K. Petrusewicz & L. Ryszkowski), pp. 1–8. Cambridge University Press, Cambridge.

Boycott, A.E. (1919) On the size of things, or the importance of being rather small. *Contributions to Medical and Biological Research* **1**, 226–234.

Brewer, A. & Williamson, M. (1994) A new relationship for rarefaction. *Biodiversity and Conservation* **3**, 373–379.

Brockie, R.E. & Mooed, A. (1986) Animal biomass in a New Zealand forest compared with other parts of the world. *Oecologia* **70**, 24–34.

Brooke, R.K., Lockwood, J.L. & Moulton, M.P. (1995) Patterns of success in passeriform bird introductions on Saint Helena. *Oecologia* **103**, 337–342.

Brown, J.H. (1971) Mammals on mountaintops: non-equilibrium insular biogeography. *American Naturalist* **105**, 467–478.

Brown, J.H. (1978) The theory of insular biogeography and the distribution of boreal birds and mammals. *Great Basin Naturalist Memoirs* **2**, 209–227.

Brown, J.H. (1984) On the relationship between abundance and distribution of species. *American Naturalist* **124**, 255–279.

Brown, J.H. (1995) *Macroecology*. University of Chicago Press, Chicago.

Brown, J.H. & Kodric-Brown, A. (1977) Turnover rates in insular biogeography: effect of immigration on extinction. *Ecology* **58**, 445–449.

Brown, J.H., Marquet, P.A. & Taper, M.L. (1993) Evolution of body size: consequences of an energetic definition of fitness. *American Naturalist* **142**, 573–584.

Brown, J.H. & Maurer, B.A. (1986) Body size, ecological dominance and Cope's Rule. *Nature* **324**, 248–250.

Brown, J.H. & Maurer, B.A. (1987) Evolution of species assemblages: effects of energetic constraints and species dynamics on the diversification of the American avifauna. *American Naturalist* **130**, 1–17.

Brown, J.H. & Maurer, B.A. (1989) Macroecology: the division of food and space among species on continents. *Science* **243**, 1145–1150.

Brown, J.H., Mehlman, D.W. & Stevens, G.C. (1995) Spatial variation in abundance. *Ecology* **76**, 2028–2043.

Brown, J.H. & Nicoletto, P.F. (1991) Spatial scaling of species composition: body masses of North American land mammals. *American Naturalist* **138**, 1478–1512.

Brown, J.H., Stevens, G.C. & Kaufman, D.W. (1996) The geographic range: size, shape, boundaries and internal structure. *Annual Review of Ecology and Systematics* **27**, 597–623.

Brown, J.W. (1987) The peninsular effect in Baja California: an entomological assessment. *Journal of Biogeography* **14**, 359–365.

Brown, J.W. & Opler, P.A. (1990) Patterns of butterfly species richness in peninsular Florida. *Journal of Biogeography* **17**, 615–622.

Brown, L.H., Urban, E.K. & Newman, K. (1982) *The Birds of Africa*, Vol. 1. Academic Press, London.

Brown, M. & Dinsmore, J.J. (1988) Habitat islands and the equilibrium theory of island biogeography: testing some predictions. *Oecologia* **75**, 426–429.

Brualdi, R.A. & Sanderson, J.G. (1999) Nested species subsets, gaps, and discrepancy. *Oecologia* **119**, 256–264.

Brucker, J.W., Gosler, A.G. & Heryet, A.R. (1992) *Birds of Oxfordshire*. Pisces Publications, Newbury, Berkshire.

Brussard, P.F. (1984) Geographic patterns and environmental gradients: the central-marginal model in *Drosophila* revisited. *Annual Review of Ecology and Systematics* **15**, 25–64.

Buckley, R.C. (1982) The habitat–unit model of island biogeography. *Journal of Biogeography* **9**, 339–344.

Burbidge, A.A. & McKenzie, N.L. (1989) Patterns in the modern decline of Western Australia's vertebrate fauna: causes and conservation implications. *Biological Conservation* **50**, 143–198.

Burgman, M.A. (1989) The habitats of scarce and ubiquitous plants: a test of the model of environmental control. *American Naturalist* **133**, 228–239.

Burgman, M.A., Ferson, S. & Akçakaya, H.R. (1993) *Risk Assessment in Conservation Biology*. Chapman & Hall, London.

Burnett, M.R., August, P.V., Brown, J.H. Jr & Killingbeck, K.T. (1998) The influence of geomorphological heterogeneity on biodiversity. I. A patch-scale perspective. *Conservation Biology* **12**, 363–370.

Burton, J.F. (1995) *Birds and Climate Change*. Christopher Helm, London.

Buzas, M.A. & Culver, S.J. (1991) Species diversity and dispersal of benthic Foraminifera. *Bioscience* **41**, 483–489.

Cade, T.J. & Woods, C.P. (1997) Changes in distribution and abundance of the loggerhead shrike. *Conservation Biology* **11**, 21–31.

Calder, W.A. (1984) *Size, Function and Life History*. Harvard University Press, Cambridge, Massachusetts.

Caley, M.J. (1997) Local endemism and the relationship between local and regional diversity. *Oikos* **79**, 612–615.

Caley, M.J. & Schluter, D. (1997) The relationship between local and regional diversity. *Ecology* **78**, 70–80.

Cambefort, Y. (1994) Body size, abundance and geographical distribution of Afrotropical dung beetles (Coleoptera: Scarabaeidae). *Acta Oecologica* **15**, 165–179.

Carrascal, L.M., Bautista, L.M. & Lázaro, E. (1993) Geographical variation in the density of the white stork *Ciconia ciconia* in Spain: influence of habitat structure and climate. *Biological Conservation* **65**, 83–87.

Carrascal, L.M. & Tellería, J.L. (1991) Bird size and density: a regional approach. *American Naturalist* **138**, 777–784.

Carter, R.N. & Prince, S.D. (1981) Epidemic models used to explain biogeographical distribution limits. *Nature* **293**, 644–645.

Carter, R.N. & Prince, S.D. (1985) The geographical distribution of the prickly lettuce (*Lactuca serriola*). 1. A general survey of its habitats and performance in Britain. *Journal of Ecology* **73**, 27–38.

Carter, R.N. & Prince, S.D. (1988) Distribution limits from a demographic viewpoint. In: *Plant Population Ecology* (eds A.J. Davy, M.J. Hutchings & A.R. Watkinson), pp. 165–184. Blackwell Scientific Publications, Oxford.

Case, T.J. (1979) Optimal body size and an animal's diet. *Acta Biotheoretica* **28**, 54–69.

Castro, G. (1989) Energy costs and avian distributions—limitations or chance?—a comment. *Ecology* **70**, 1181–1182.

Cates, S.E. & Gittleman, J.L. (1997) Reading between the lines—is allometric scaling useful? *Trends in Ecology and Evolution* **12**, 338–339.

Caughley, G. (1987) The distribution of Eutherian body weights. *Oecologia* **74**, 319–320.

Caughley, G. & Krebs, C.J. (1983) Are big mammals simply little mammals writ large? *Oecologia* **59**, 7–17.

Ceballos, G. & Navarro, L.D. (1991) Diversity and conservation of Mexican mammals. In: *Latin American Mammalogy: History, Biodiversity and Conservation* (eds M.A. Mares & D.J. Schmidly), pp. 167–198. University of Oklahoma Press, Norman, Oklahoma.

Cherrett, J.M. (1989) Key concepts: the results of a survey of our members' opinions. In: *Ecological Concepts: The Contribution of Ecology to an Understanding of the Natural World* (ed. J.M. Cherrett), pp. 1–16. Blackwell Scientific Publications, Oxford.

Cheverud, J.M., Dow, M.M. & Leutenegger, W. (1985) The quantitative assessment of phylogenetic constraints in comparative analyses: sexual dimorphism in body weight among primates. *Evolution* **39**, 1335–1351.

Chown, S.L. & Gaston, K.J. (1997) The species–body size distribution: energy, fitness and optimality. *Functional Ecology* **11**, 365–375.

Chown, S.L. & Gaston, K.J. (1999a) Patterns in procellariiform diversity as a test of species–energy theory in marine systems. *Evolutionary Ecology Research* **1**, 365–373.

Chown, S.L. & Gaston, K.J. (1999b) Exploring links between physiology and ecology at macro-scales: the role of respiratory metabolism in insects. *Biological Reviews* **74**, 87–120.

Chown, S.L. & Steenkamp, H.E. (1996) Body size and abundance in a dung beetle assemblage: optimal mass and the role of transients. *African Entomology* **4**, 203–212.

Cody, M.L. (1975) Towards a theory of continental species diversities: bird distributions over Mediterranean habitat gradients. In: *Ecology and Evolution of Communities* (eds

M.L. Cody & J.M. Diamond), pp. 214–257. Harvard University Press, Cambridge, Massachusetts.

Cohen, J.E., Pimm, S.L., Yodzis, P. & Saldana, J. (1993) Body sizes of animal predators and animal prey in food webs. *Journal of Animal Ecology* **62**, 67–78.

Coleman, B.D., Mares, M.A., Willig, M.R. & Hsieh, Y.-H. (1982) Randomness, area, and species richness. *Ecology* **63**, 1121–1133.

Collar, N.J., Crosby, M.J. & Stattersfield, A.J. (1994) *Birds to Watch 2. The World List of Threatened Birds*. Birdlife International, Cambridge.

Collins, J.J. & Chow, C.C. (1998) It's a small world. *Nature* **393**, 309–410.

Collins, S.L. & Glenn, S.M. (1990) A hierarchical analysis of species' abundance patterns in grassland vegetation. *American Naturalist* **135**, 633–648.

Collins, S.L. & Glenn, S.M. (1991) Importance of spatial and temporal dynamics in species regional abundance and distribution. *Ecology* **72**, 654–664.

Collins, S.L. & Glenn, S.M. (1997) Effects of organismal and distance scaling on analysis of species distribution and abundance. *Ecological Applications* **7**, 543–551.

Colwell, R.K. & Futuyma, D.J. (1971) On the measurement of niche breadth and overlap. *Ecology* **52**, 567–576.

Colwell, R.K. & Hurtt, G.C. (1994) Nonbiological gradients in species richness and a spurious Rapoport effect. *American Naturalist* **144**, 570–595.

Colwell, R.K. & Winkler, D.W. (1984) A null model for null models in biogeography. In: *Ecological Communities: Conceptual Issues and the Evidence* (eds D.R. Strong, D. Simberloff, L.G. Abele & A.B. Thistle), pp. 344–359. Princeton University Press, Princeton, New Jersey.

Connor, E.F. & McCoy, E.D. (1979) The statistics and biology of the species–area relationship. *American Naturalist* **113**, 791–833.

Conroy, C.J., Demboski, J.R. & Cook, J.A. (1999) Mammalian biogeography of the Alexander Archipelago of Alaska: a north temperate nested fauna. *Journal of Biogeography* **26**, 343–352.

Cook, R.E. (1969) Variation in species density of North American birds. *Systematic Zoology* **18**, 63–84.

Cook, R.R. (1995) The relationship between nested subsets, habitat subdivision, and species diversity. *Oecologia* **101**, 204–210.

Cook, R.R. & Quinn, J.F. (1995) The influence of colonization in nested species subsets. *Oecologia* **102**, 413–424.

Cook, R.R. & Quinn, J.F. (1998) An evaluation of randomization models for nested species subsets analysis. *Oecologia* **113**, 584–592.

Coope, G.R. (1995) Insect faunas in ice age environments: why so little extinction?. In: *Extinction Rates* (eds J.H. Lawton & R.M. May), pp. 55–74. Oxford University Press, Oxford.

Cornell, H.V. (1985a) Species assemblages of cynipid gall wasps are not saturated. *American Naturalist* **126**, 565–569.

Cornell, H.V. (1985b) Local and regional richness of cynipine gall wasps on California oaks. *Ecology* **66**, 1247–1260.

Cornell, H.V. & Karlson, R.H. (1996) Species richness of reef-building corals determined by local and regional processes. *Journal of Animal Ecology* **65**, 233–241.

Cornell, H.V. & Lawton, J.H. (1992) Species interactions, local and regional processes, and limits to the richness of ecological communities: a theoretical perspective. *Journal of Animal Ecology* **61**, 1–12.

Cotgreave, P. (1993) The relationship between body size and population abundance in animals. *Trends in Ecology and Evolution* **8**, 244–248.

Cotgreave, P. (1994) The relation between body size and abundance in a bird community: the effects of phylogeny and competition. *Proceedings of the Royal Society, London, B* **256**, 147–149.

Cotgreave, P. & Harvey, P.H. (1991) Bird community structure. *Nature* **353**, 123.

Cotgreave, P. & Harvey, P.H. (1992) Relationships between body size, abundance and phylogeny in bird communities. *Functional Ecology* **6**, 248–256.

Cotgreave, P. & Harvey, P.H. (1994a) Associations among biogeography, phylogeny and bird species diversity. *Biodiversity Letters* **2**, 46–55.

Cotgreave, P. & Harvey, P.H. (1994b) Evenness of abundance in bird communities. *Journal of Animal Ecology* **63**, 365–374.

Cotgreave, P., Middleton, D.A.J. & Hill, M.J. (1993) The relationship between body size and population size in bromeliad tank faunas. *Biological Journal of the Linnean Society* **42**, 367–380.

Cotgreave, P. & Stockley, P. (1994) Body size, insectivory and abundance in assemblages of small mammals. *Oikos* **71**, 89–96.

Cousins, S. (1980) On some relationships between energy and diversity models of ecosystems. *Acta XVII Congressus Internationalis Ornithologici* 1051–1055.

Cousins, S.H. (1989) Species richness and the energy theory. *Nature* **340**, 350–351.

Cousins, S.H. (1994) Taxonomy and functional biotic measurement, or, will the Ark work? In: *Systematics and Conservation Evaluation* (eds P.L. Forey, C.J. Humphries & R.I. Vane-Wright), pp. 397–419. Clarendon Press, Oxford.

Cowley, M.J.R., Thomas, C.D., Thomas, J.A. & Warren, M.S. (1999) Flight areas of British butterflies: assessing species status and decline. *Proceedings of the Royal Society, London, B* **266**, 1587–1592.

Cowlishaw, G. & Hacker, J.E. (1997) Distribution, diversity, and latitude in African primates. *American Naturalist* **150**, 505–512.

Cox, G.W. (1961) The relation of energy requirements of tropical finches to distribution and migration. *Ecology* **42**, 253–266.

Cramp, S. (ed.) (1985) *Birds of the Western Palaearctic*, Vol. IV. *Terns to Woodpeckers*. Oxford University Press, Oxford.

Cramp, S. (ed.) (1988) *Birds of the Western Palaearctic*, Vol. V. *Tyrant Flycatchers to Thrushes*. Oxford University Press, Oxford.

Cramp, S. (ed.) (1992) *Birds of the Western Palaearctic*, Vol. VI. *Warblers*. Oxford University Press, Oxford.

Cramp, S. & Perrins, C.M. (eds) (1993) *Birds of the Western Palaearctic*, Vol. VII. *Flycatchers to Shrikes*. Oxford University Press, Oxford.

Cramp, S. & Perrins, C.M. (eds) (1994a) *Birds of the Western Palaearctic*, Vol. VIII. *Crows to Finches*. Oxford University Press, Oxford.

Cramp, S. & Perrins, C.M. (eds) (1994b) *Birds of the Western Palaearctic*, Vol. IX. *Buntings and New World Warblers*. Oxford University Press, Oxford.

Cramp, S. & Simmons, K.E.L. (eds) (1977) *Birds of the Western Palaearctic*, Vol. I. *Ostrich to Ducks*. Oxford University Press, Oxford.

Cramp, S. & Simmons, K.E.L. (eds) (1980) *Birds of the Western Palaearctic*, Vol. II. *Hawks to Bustards*. Oxford University Press, Oxford.

Cramp, S. & Simmons, K.E.L. (eds) (1983) *Birds of the Western Palaearctic*, Vol. III. *Waders to Gulls*. Oxford University Press, Oxford.

Cresswell, J.E., Vidal-Martinez, V.M. & Crichton, N.J. (1995) The investigation of saturation in the species richness of communities: some comments on methodology. *Oikos* **72**, 301–304.

Crick, H.Q.P., Dudley, C., Glue, D.E. & Thomson, D.L. (1997) UK birds are laying eggs earlier. *Nature* **388**, 526.

Crowe, T.M. & Crowe, A.A. (1982) Patterns of distribution, diversity and endemism in Afrotropical birds. *Journal of Zoology, London* **198**, 417–442.

Curnutt, J.L., Pimm, S.L. & Maurer, B.A. (1996) Population variability of sparrows in space and time. *Oikos* **76**, 131–144.

Curran, P.J., Foody, G.M. & van Gardingen, P.R. (1997) Scaling-up. In: *Scaling-Up: From Cell to Landscape* (eds P.R. van Gardingen, G.M. Foody & P.J. Curran), pp. 1–5. Cambridge University Press, Cambridge.

Currie, D.J. (1991) Energy and large-scale patterns of animal- and plant-species richness. *American Naturalist* **137**, 27–49.

Currie, D.J. (1993) What shape is the relationship between body size and population density? *Oikos* **66**, 353–358.

Currie, D.J. & Fritz, J.T. (1993) Global patterns of animal abundance and species energy use. *Oikos* **67**, 56–68.

Cushman, J.H., Lawton, J.H. & Manly, B.F.J. (1993) Latitudinal patterns in European ant assemblages: variation in species richness and body size. *Oecologia* **95**, 30–37.

Cutler, A. (1991) Nested faunas and extinction in fragmented habitats. *Conservation Biology* **5**, 496–505.

Cutler, A. (1994) Nested biotas and biological conservation: metrics, mechanisms, and the meaning of nestedness. *Landscape and Urban Planning* **5**, 496–505.

Damuth, J. (1981) Population density and body size in mammals. *Nature* **290**, 699–700.

Damuth, J. (1987) Interspecific allometry of population density in mammals and other animals: the independence of body mass and population energy use. *Biological Journal of the Linnean Society* **31**, 193–246.

Damuth, J.D. (1992) Taxon-free characterization of animal communities. In: *Terrestrial Ecosystems Through Time: Evolutionary Paleoecology of Terrestrial Plants and Animals* (eds A.K. Behrensmeyer, J.D. Damuth, W.A. DiMichele, R. Potts, H.-D. Sues & S.L. Wing), pp. 183–203. University of Chicago Press, Chicago.

Daniels, R.J.R., Hegde, M., Joshi, N.V. & Gadgil, M. (1991) Assigning conservation value: a case study from India. *Conservation Biology* **5**, 464–475.

Daniels, R.J.R., Joshi, N.V. & Gadgil, M. (1992) On the relationship between bird and woody plant species diversity in the Uttara Kannada district of south India. *Proceedings of the National Academy of Sciences, USA* **89**, 5311–5315.

Davenport, J., Pugh, P.J.A. & McKechnie, J. (1996) Mixed fractals and anisotropy in sub-antarctic marine macroalgae from south Georgia: implications for epifaunal biomass and abundance. *Marine Ecology–Progress Series* **136**, 245–255.

Davis, R. & Dunford, C. (1987) An example of contemporary colonization of montane islands by small, nonflying mammals in the American southwest. *American Naturalist* **129**, 398–406.

Degen, A.A. & Kam, M. (1995) Scaling of field metabolic rate to basal metabolic rate ratio in homeotherms. *Écoscience* **2**, 48–54.

Dekker, R.W.R.J. (1989) Predation and the western limits of megapode distribution (Megapodiidae; Aves). *Journal of Biogeography* **16**, 317–321.

Dennis, B. & Patil, G.P. (1988) Applications in ecology. In: *Lognormal Distributions:*

Theory and Applications (eds E.L. Crow & K. Shimizu), pp. 303–330. Marcel Dekker, New York.

Dennis, M.K. (1996) *Tetrad Atlas of the Breeding Birds of Essex.* The Essex Birdwatching Society.

Dial, K.P. & Marzluff, J.M. (1988) Are the smallest organisms the most diverse? *Ecology* **69**, 1620–1624.

Diamond, J.M. (1984) 'Normal' extinctions of isolated populations. In: *Extinctions* (ed. M.H. Nitecki), pp. 191–246. University of Chicago Press, Chicago.

Diamond, J. (1986) Overview: laboratory experiments, field experiments, and natural experiments. In: *Community Ecology* (eds J. Diamond & T.J. Case), pp. 3–22. Harper & Row, New York.

Diamond, J. (1998) *Guns, Germs and Steel. A Short History of Everybody for the Last 13 000 Years.* Vintage, London.

Díaz, M., Carbonell, R., Santos, T. & Tellería, J.L. (1998) Breeding bird communities in pine plantations of the Spanish plateaux: biogeography, landscape and vegetation effects. *Journal of Applied Ecology* **35**, 562–574.

Diaz-Uriarte, R. & Garland, T. (1996) Testing hypotheses of correlated evolution using phylogenetically independent contrasts: sensitivity to deviations from Brownian motion. *Systematic Biology* **45**, 27–47.

Dixon, A.F.G. (1990) Ecological interactions of aphids and their host plants. In: *Aphid– Plant Genotype Interactions* (eds R.K. Campbell & R.D. Eikenbary), pp. 7–19. Elsevier, Amsterdam.

Dixon, A.F.G. & Kindlmann, P. (1994) Optimum body size in aphids. *Ecological Entomology* **19**, 121–126.

Dixon, A.F.G. & Kindlmann, P. (1999) Cost of flight apparatus and optimum body size of aphid migrants. *Ecology* **80**, 1678–1690.

Dixon, A.F.G., Kindlmann, P. & Jarosik, V. (1995) Body size distribution in aphids: relative surface area of specific plant structures. *Ecological Entomology* **20**, 111–117.

Dixon, A.F.G., Kindlmann, P., Leps, J. & Holman, J. (1987) Why are there so few species of aphids, especially in the tropics. *American Naturalist* **129**, 580–592.

Dobzhansky, T. (1950) Evolution in the tropics. *American Scientist* **38**, 209–221.

Donald, P.F. & Fuller, R.J. (1998) Ornithological atlas data: a review of uses and limitations. *Bird Study* **45**, 129–145.

Donoghue, A.M., Quicke, D.L.J. & Brace, R.C. (1986) Turnstones apparently preying on sea anemones. *British Birds* **79**, 91.

Dony, J.G. (1970) *Species–area Relationships.* Unpublished Report to the Natural Environment Research Council.

Doyle, A.C. (1887) *A Study in Scarlet.* Beeton's Christmas Annual.

Due, A.D. & Polis, G.A. (1986) Trends in scorpion diversity along the Baja California peninsula. *American Naturalist* **128**, 460–468.

Dunbrack, R.L. & Ramsay, M.A. (1993) The allometry of mammalian adaptations to seasonal environments: a critique of the fasting endurance hypothesis. *Oikos* **66**, 336–342.

Duncan, R.P. (1997) The role of competition and introduction effort in the success of passeriform birds introduced to New Zealand. *American Naturalist* **149**, 903–915.

Duncan, R.P., Blackburn, T.M. & Veltman, C.J. (1999) Relationships between range size and life history traits in introduced New Zealand and British breeding birds. *Journal of Animal Ecology* **68**, 963–975.

Dunning, J.B. (1992) *CRC Handbook of Avian Body Masses.* CRC Press, Boca Raton, Florida.

Dymond, J.N., Fraser, P.A. & Gantlett, S.J.M. (1989) *Rare Birds in Britain and Ireland*. T. & A.D. Poyser, Calton.

Eadie, J.M., Broekhoven, L. & Colgan, P. (1987) Size ratios and artifacts: Hutchinson's rule revisited. *American Naturalist* **129**, 1–17.

East, M.L. & Perrins, C.M. (1988) The effect of nest-boxes on breeding populations of birds in broadleaved temperate woodlands. *Ibis* **130**, 393–401.

Ebenhard, T. (1988) Introduced birds and mammals and their ecological effects. *Swedish Wildlife Research* **13**, 1–107.

Ebenman, B., Hedenström, A., Wennergren, U., Ekstam, B., Landin, J. & Tyrberg, T. (1995) The relationship between population density and body size: the role of extinction and mobility. *Oikos* **73**, 225–230.

Eggleton, P. (1994) Termites live in a pear-shaped world: a response to Platnick. *Journal of Natural History* **28**, 1209–1212.

Ehrlich, P.R., Dobkin, D.S., Wheye, D. & Pimm, S.L. (1994) *The Birdwatchers Handbook. A Guide to the Natural History of the Birds of Britain and Europe*. Oxford University Press, Oxford.

Elgar, M.A. & Harvey, P.H. (1987) Basal metabolic rates in mammals: allometry, phylogeny and ecology. *Functional Ecology* **1**, 25–36.

Elkins, N. (1995) *Weather and Bird Behaviour*, 2nd edn. Poyser, London.

Elliott, C.C.H. (1989) The pest status of the quelea. In: *Quelea quelea: Africa's Bird Pest* (eds R.L. Bruggers & C.C.H. Elliott), pp. 17–34. Oxford University Press, Oxford.

Elliott, J.M. (1994) *Quantitative Ecology and the Brown Trout*. Oxford University Press, Oxford.

Enghoff, H. & Báez, M. (1993) Evolution of distribution and habitat patterns in endemic millipedes of the genus *Dolichoiulus* (Diplopoda: Julidae) on the Canary Islands, with notes on distribution patterns of other Canarian species swarms. *Biological Journal of the Linnean Society* **49**, 277–301.

Erwin, D.H. (1996) Understanding biotic recoveries: extinction, survival and preservation during the End-Permian mass extinction. In: *Evolutionary Paleobiology* (eds D. Jablonski, D.H. Erwin & J.H. Lipps), pp. 398–418. University of Chicago Press, Chicago.

Erwin, T.L. (1982) Tropical forests: their richness in Coleoptera and other arthropod species. *Coleopterists Bulletin* **36**, 74–75.

Erwin, T.L. (1991) How many species are there?: revisited. *Conservation Biology* **5**, 330–333.

Evans, L.G.R. (1996) *The Ultimate Site Guide to Scarcer British Birds*. LGRE Productions, Amersham.

Evans, L.G.R. (1997a) Comparative lists of European countries. *Rare Birds* **3**, 327.

Evans, L.G.R. (1997b) *The Definitive Checklist of the Birds of the Western Palaearctic*. LGRE Productions, Amersham.

Felsenstein, J. (1985) Phylogenies and the comparative method. *American Naturalist* **125**, 1–15.

Fenchel, T. (1993) There are more small than large species? *Oikos* **68**, 375–378.

Ferry, C. & Frochot, B. (1990) Bird communities of the forests of Burgundy and the Jura (eastern France). In: *Biogeography and Ecology of Forest Bird Communities* (ed. A. Keast), pp. 183–195. SPB Academic Publishing, The Hague.

Finlayson, J.C. (1999) Species abundance across spatial scales. *Science* **283**, 1979.

Fischer, A.G. (1960) Latitudinal variations in organic diversity. *Evolution* **14**, 64–81.

Fischer, A.G. (1981) Climatic oscillations in the biosphere. In: *Biotic Crises in Ecological and Evolutionary Time* (ed. M.H. Nitecki), pp. 103–131. Academic Press, New York.

Fisher, R.A., Corbet, A.S. & Williams, C.B. (1943) The relation between the number of species and the number of individuals in a random sample of an animal population. *Journal of Animal Ecology* **12**, 42–58.

Fitter, A.H. & Strickland, T.R. (1992) Fractal characterization of root system architecture. *Functional Ecology* **6**, 632–635.

Fjeldså, J. (1994) Geographical patterns for relict and young species of birds in Africa and South America and implications for conservation priorities. *Biodiversity and Conservation* **3**, 207–226.

Flegg, J.J.M. & Bennett, T.J. (1974) The birds of oak woodlands. In: *The British Oak* (eds M.G. Morris & F.H. Perring), pp. 324–340. E. W. Classey, Faringdon, Berkshire.

Flessa, K.W. & Jablonski, D. (1996) The geography of evolutionary turnover: a global analysis of extant bivalves. In: *Evolutionary Paleobiology* (eds D. Jablonski, D.H. Erwin & J.H. Lipps), pp. 376–397. University of Chicago Press, Chicago.

Flessa, K.W. & Thomas, R.H. (1985) Modelling the biogeographic regulation of evolutionary rates. In: *Phanerozoic Diversity Patterns. Profiles in Macroevolution* (ed. J.W. Valentine), pp. 355–376. Princeton University Press, Princeton, New Jersey.

Ford, H.A. (1987) Bird communities in habitat islands in England. *Bird Study* **34**, 205–218.

Fowler, C.W. & MacMahon, J.A. (1982) Selective extinction and speciation: their influence on the structure and functioning of communities and ecosystems. *American Naturalist* **119**, 480–498.

Fox, B.J. (1983) Mammal species diversity in Australian heathlands: the importance of pyric succession and habitat diversity. In: *Mediterranean-Type Ecosystems: The Role of Nutrients* (eds F.J. Kruger, D.T. Mitchell & J.U.M. Jarvis), pp. 473–489. Springer-Verlag, Berlin.

France, R. (1992) The North American latitudinal gradient in species richness and geographical range of freshwater crayfish and amphipods. *American Naturalist* **139**, 342–354.

Fraser, P. (1997) How many rarities are we missing? Weekend bias and length of stay revisited. *British Birds* **90**, 94–101.

Fraser, R.H. & Currie, D.J. (1996) The species richness–energy hypothesis in a system where historical factors are thought to prevail: coral reefs. *American Naturalist* **148**, 138–159.

Freemark, K.E. & Merriam, H.G. (1986) Importance of area and habitat heterogeneity to bird assemblages in temperate forest fragments. *Biological Conservation* **36**, 115–141.

Freitag, R. (1969) A revision of the species of the genus *Evarthrus* LeConte (Coleoptera: Carabidae). *Quaestiones Entomologicae* **5**, 89–120.

Fretwell, S.D. & Lucas, H.L. (1970) On territorial behaviour and other factors influencing habitat distribution in birds. *Acta Biotheoretica* **19**, 16–36.

Fuller, R.J. (1982) *Bird Habitats in Britain*. T. & A.D. Poyser, Calton.

Fuller, R.J., Gregory, R.D., Gibbons, D.W. *et al.* (1995) Population declines and range contractions among lowland farmland birds in Britain. *Conservation Biology* **9**, 1425–1442.

Futuyma, D.J. (1998) Wherefore and whither the Naturalist? *American Naturalist* **151**, 1–6.

Gardezi, T. & da Silva, J. (1999) Diversity in relation to body size in mammals: a comparative study. *American Naturalist* **153**, 110–123.

Gaston, K.J. (1988) Patterns in the local and regional dynamics of moth populations. *Oikos* **53**, 49–57.

Gaston, K.J. (1990) Patterns in the geographical ranges of species. *Biological Reviews* **65**, 105–129.

Gaston, K.J. (1991a) Body size and probability of description: the beetle fauna of Britain. *Ecological Entomology* **16**, 505–508.

Gaston, K.J. (1991b) The magnitude of global insect species richness. *Conservation Biology* **5**, 283–296.

Gaston, K.J. (1991c) Estimates of the near-imponderable: a reply to Erwin. *Conservation Biology* **5**, 564–566.

Gaston, K.J. (1991d) How large is a species' geographic range? *Oikos* **61**, 434–437.

Gaston, K.J. (1994a) *Rarity.* Chapman & Hall, London.

Gaston, K.J. (1994b) Measuring geographic range sizes. *Ecography* **17**, 198–205.

Gaston, K.J. (1996a) The multiple forms of the interspecific abundance–distribution relationship. *Oikos* **76**, 211–220.

Gaston, K.J. (1996b) Biodiversity–latitudinal gradients. *Progress in Physical Geography* **20**, 466–476.

Gaston, K.J. (1996c) Species–range-size distributions: patterns, mechanisms and implications. *Trends in Ecology and Evolution* **11**, 197–201.

Gaston, K.J. (1998) Species–range size distributions: products of speciation, extinction and transformation. *Philosophical Transactions of the Royal Society, London, B* **353**, 219–230.

Gaston, K.J. (1999) Implications of interspecific and intraspecific abundance–occupancy relationships. *Oikos* **86**, 195–207.

Gaston, K.J. & Blackburn, T.M. (1994) Are newly discovered species small bodied? *Biodiversity Letters* **2**, 16–20.

Gaston, K.J. & Blackburn, T.M. (1995a) The frequency distribution of bird body weights: aquatic and terrestrial species. *Ibis* **137**, 237–240.

Gaston, K.J. & Blackburn, T.M. (1995b) Birds, body size and the threat of extinction. *Philosophical Transactions of the Royal Society, London, B* **347**, 205–212.

Gaston, K.J. & Blackburn, T.M. (1996a) Range size–body size relationships: evidence of scale dependence. *Oikos* **75**, 479–485.

Gaston, K.J. & Blackburn, T.M. (1996b) The tropics as a museum of biological diversity: an analysis of the New World avifauna. *Proceedings of the Royal Society, London, B* **263**, 63–68.

Gaston, K.J. & Blackburn, T.M. (1996c) Global scale macroecology: interactions between population size, geographic range size and body size in the Anseriformes. *Journal of Animal Ecology* **65**, 701–714.

Gaston, K.J. & Blackburn, T.M. (1997a) Age, area and avian diversification. *Biological Journal of the Linnean Society* **62**, 239–253.

Gaston, K.J. & Blackburn, T.M. (1997b) How many birds are there? *Biodiversity and Conservation* **6**, 615–625.

Gaston, K.J. & Blackburn, T.M. (1999) A critique for macroecology. *Oikos* **84**, 353–368.

Gaston, K.J., Blackburn, T.M., Greenwood, J.J.D., Gregory, R.D., Quinn, R.M. & Lawton, J.H. (in press) Abundance–occupancy relationships: what, how, why and implications. *Journal of Applied Ecology.*

Gaston, K.J., Blackburn, T.M. & Gregory, R.D. (1997b) Interspecific abundance–range size relationships: range position and phylogeny. *Ecography* **20**, 390–399.

Gaston, K.J., Blackburn, T.M. & Gregory, R.D. (1997c) Abundance–range size relationships of breeding and wintering birds in Britain: a comparative analysis. *Ecography* **20**, 569–579.

Gaston, K.J., Blackburn, T.M. & Gregory, R.D. (1998d) Interspecific differences in intraspecific abundance–range size relationships of British breeding birds. *Ecography* **21**, 149–158.

Gaston, K.J., Blackburn, T.M. & Gregory, R.D. (1999a) Intraspecific abundance–range size relationships: case studies of six bird species in Britain. *Diversity and Distributions* **5**, 197–212.

Gaston, K.J., Blackburn, T.M. & Gregory, R.D. (1999b) Does variation in census area confound density comparisons? *Journal of Applied Ecology* **36**, 191–204.

Gaston, K.J., Blackburn, T.M., Gregory, R.D. & Greenwood, J.J.D. (1998c) The anatomy of the interspecific abundance–range size relationship for the British avifauna: I. Spatial patterns. *Ecology Letters* **1**, 38–46.

Gaston, K.J., Blackburn, T.M. & Lawton, J.H. (1997a) Interspecific abundance–range size relationships: an appraisal of mechanisms. *Journal of Animal Ecology* **66**, 579–601.

Gaston, K.J., Blackburn, T.M. & Lawton, J.H. (1998e) Aggregation and interspecific abundance–range size relationships. *Journal of Animal Ecology* **67**, 995–999.

Gaston, K.J., Blackburn, T.M. & Loder, N. (1995) Who gets described first?: the case of North American butterflies. *Biodiversity and Conservation* **4**, 119–127.

Gaston, K.J., Blackburn, T.M. & Spicer, J.I. (1998b) Rapoport's rule: time for an epitaph? *Trends in Ecology and Evolution* **13**, 70–74.

Gaston, K.J. & Chown, S.L. (1999a) Geographic range size and speciation. In: *Evolution of Biological Diversity* (eds A.E. Magurran & R.M. May), pp. 236–259. Oxford University Press, Oxford.

Gaston, K.J. & Chown, S.L. (1999b) Why Rapoport's rule does not generalise. *Oikos* **84**, 309–312.

Gaston, K.J., Chown, S.L. & Styles, C.V. (1997d) Changing size and changing enemies: the case of the mopane worm. *Acta Oecologica* **18**, 21–26.

Gaston, K.J. & Curnutt, J.L. (1998) The dynamics of abundance–range size relationships. *Oikos* **81**, 38–44.

Gaston, K.J. & Hudson, E. (1994) Regional patterns of diversity and estimates of global insect species richness. *Biodiversity and Conservation* **3**, 493–500.

Gaston, K.J. & Kunin, W.E. (1997) Rare–common differences: an overview. In: *The Biology of Rarity: Causes and Consequences of Rare–Common Differences* (eds W.E. Kunin & K.J. Gaston), pp. 3–11. Chapman & Hall, London.

Gaston, K.J. & Lawton, J.H. (1988a) Patterns in the distribution and abundance of insect populations. *Nature* **331**, 709–712.

Gaston, K.J. & Lawton, J.H. (1988b) Patterns in body size, population dynamics, and regional distribution of bracken herbivores. *American Naturalist* **132**, 662–680.

Gaston, K.J. & Lawton, J.H. (1989) Insect herbivores on bracken do not support the core-satellite hypothesis. *American Naturalist* **134**, 761–777.

Gaston, K.J. & Lawton, J.H. (1990) Effects of scale and habitat on the relationship between regional distribution and local abundance. *Oikos* **58**, 329–335.

Gaston, K.J. & McArdle, B.H. (1993) All else is not equal: temporal population variability and insect conservation. In: *Perspectives on Insect Conservation* (eds K.J. Gaston, T.R. New & M.J. Samways), pp. 171–184. Intercept, Andover.

Gaston, K.J., Quinn, R.M., Blackburn, T.M. & Eversham, B.C. (1998a) Species–range size distributions in Britain. *Ecography* **21**, 361–370.

Gaston, K.J. & Spicer, J.I. (1998) *Biodiversity. An Introduction.* Blackwell Science, Oxford.

Gaston, K.J. & Williams, P.H. (1996) Spatial patterns in taxonomic diversity. In: *Biodiversity: A Biology of Numbers and Difference* (ed. K.J. Gaston), pp. 202–229. Blackwell Science, Oxford.

Gates, S., Gibbons, D.W., Lack, P.C. & Fuller, R.J. (1994) Declining farmland bird species: modelling geographical patterns of abundance in Britain. In: *Large-Scale Ecology and Conservation Biology* (eds P.J. Edwards, R.M. May & N.R. Webb), pp. 153–177. Blackwell Scientific Publications, Oxford.

Gauld, I.D. (1986) Latitudinal gradients in ichneumonid species-richness in Australia. *Ecological Entomology* **11**, 155–161.

Gauld, I.D. & Gaston, K.J. (1995) The Costa Rican hymenopteran fauna. In: *The Hymenoptera of Costa Rica* (eds P.E. Hanson & I.D. Gauld), pp. 13–19. Oxford University Press, Oxford.

Gee, M.J. & Warwick, R.M. (1994) Body-size distributions in a marine metazoan community and the fractal dimensions of macroalgae. *Journal of Experimental Marine Biology and Ecology* **178**, 247–259.

Geist, V. (1987) Bergmann's rule is invalid. *Canadian Journal of Zoology* **65**, 1035–1038.

Geist, V. (1990) Bergmann's rule is invalid: a reply to J.D. Paterson. *Canadian Journal of Zoology* **68**, 1613–1615.

Gibbons, D.W., Avery, M.L. & Brown, A.F. (1996) Population trends of breeding birds in the United Kingdom since 1800. *British Birds* **89**, 291–305.

Gibbons, D., Gates, S., Green, R.E., Fuller, R.J. & Fuller, R.M. (1995) Buzzards *Buteo buteo* and ravens *Corvus corax* in the uplands of Britain: limits to distribution and abundance. *Ibis* **137**, S75–S84.

Gibbons, D.W., Reid, J.B. & Chapman, R.A. (1993) *The New Atlas of Breeding Birds in Britain and Ireland: 1988–91*. T. & A.D. Poyser, London.

Gilbert, F. & Owen, J. (1990) Size, shape, competition, and community structure in hoverflies (Diptera: Syrphidae). *Journal of Animal Ecology* **59**, 21–39.

Gill, R.E. (1986) What won't turnstones eat? *British Birds* **79**, 402–403.

Gilpin, M.E. & Diamond, J.M. (1976) Calculation of immigration and extinction curves from the species–area–distance relation. *Proceedings of the National Academy of Sciences, USA* **73**, 4130–4134.

Gittleman, J.L. (1985) Carnivore body size—ecological and taxonomic correlates. *Oecologia* **67**, 540–554.

Gittleman, J.L. & Kot, M. (1990) Adaptation: statistics and a null model for estimating phylogenetic effects. *Systematic Zoology* **39**, 227–241.

Gittleman, J.L. & Luh, H.-K. (1992) On comparing comparative methods. *Annual Review of Ecology and Systematics* **23**, 383–404.

Gittleman, J.L. & Luh, H.-K. (1993) Phylogeny, evolutionary models, and comparative methods: a simulation study. In: *Phylogenetics and Ecology* (eds P. Eggleton & D. Vane-Wright), pp. 103–122. Academic Press, London.

Gittleman, J.L. & Purvis, A. (1998) Body size and species richness in carnivores and primates. *Proceedings of the Royal Society, London, B* **265**, 113–119.

Gleason, H.A. (1922) On the relation between species and area. *Ecology* **3**, 158–162.

Gleason, H.A. (1925) Species and area. *Ecology* **6**, 66–74.

Gleason, H.A. (1926) The individualistic concept of the plant association. *Bulletin of the Torrey Botanical Club* **53**, 7–26.

Gleason, H.A. (1929) The significance of Raunkiaer's law of frequency. *Ecology* **10**, 406–408.

Gonzalez, A., Lawton, J.H., Gilbert, F.S., Blackburn, T.M. & Evans-Freke, I. (1998) Metapopulation dynamics, abundance, and distribution in a microecosystem. *Science* **281**, 2045–2047.

Goodall, D.W. (1952) Quantitative aspects of plant distribution. *Biological Reviews* **27**, 194–245.

Gosler, A.G., Greenwood, J.J.D. & Perrins, C. (1995) Predation risk and the cost of being fat. *Nature* **377**, 621–623.

Goss-Custard, J.D. (1993) The effect of migration and scale on the study of bird populations: 1991 Witherby Lecture. *Bird Study* **40**, 81–96.

Gotelli, N.J. (1991) Metapopulation models: the rescue effect, the propagule rain, and the core-satellite hypothesis. *American Naturalist* **138**, 768–776.

Gotelli, N.J. (1995) *A Primer of Ecology.* Sinauer Associates, Sunderland, Massachusetts.

Gotelli, N.J. & Graves, G.R. (1990) Body size and the occurrence of avian species on land-bridge islands. *Journal of Biogeography* **17**, 315–325.

Gotelli, N.J. & Graves, G.R. (1996) *Null Models in Ecology.* Smithsonian Institution Press, Washington DC.

Gotelli, N.J. & Simberloff, D. (1987) The distribution and abundance of tallgrass prairie plants: a test of the core-satellite hypothesis. *American Naturalist* **130**, 18–35.

Graham, R.W. (1992) Late Pleistocene faunal changes as a guide to understanding effects of greenhouse warming on the mammalian fauna of North America. In: *Global Warming and Biological Diversity* (eds R.L. Peters & T.E. Lovejoy), pp. 76–87. Yale University Press, New Haven.

Graham, R.W., Lundelius, E.L., Graham, M.A. *et al.* (1996) Spatial response of mammals to Late Quaternary environmental fluctuations. *Science* **272**, 1601–1606.

Gray, J.S. (1987) Species–abundance patterns. In: *The Organisation of Communities: Past and Present* (eds J.H.R. Gee & P.S. Giller), pp. 53–67. Blackwell Scientific Publications, Oxford.

Green, R.E. (1996) Factors affecting the population density of the corncrake *Crex crex* in Britain and Ireland. *Journal of Applied Ecology* **33**, 237–248.

Green, R.E. (1997) The influence of numbers released on the outcome of attempts to introduce exotic bird species to New Zealand. *Journal of Animal Ecology* **66**, 25–35.

Greenslade, P.J.M. (1968) Island patterns in the Solomon Islands bird fauna. *Evolution* **22**, 751–761.

Greenwood, J.J.D., Baillie, S.R., Gregory, R.D., Peach, W.J. & Fuller, R.J. (1994) Some new approaches to conservation monitoring of British breeding birds. *Ibis* **137**, S16–S28.

Greenwood, J.J.D., Gregory, R.D., Harris, S., Morris, P.A. & Yalden, D.W. (1996) Relations between abundance, body size and species number in British birds and mammals. *Philosophical Transactions of the Royal Society, London, B* **351**, 265–278.

Gregory, R.D. (1994) Species abundance patterns of British birds. *Proceedings of the Royal Society, London, B* **257**, 299–301.

Gregory, R.D. (1995) Phylogeny and relations among abundance, geographical range and body size of British breeding birds. *Philosophical Transactions of the Royal Society, London, B* **349**, 345–351.

Gregory, R.D. (1998a) An intraspecific model of species' expansion, linking abundance and distribution. *Ecography* **21**, 92–96.

Gregory, R.D. (1998b) Biodiversity and body size: patterns among British birds. *Ecography* **21**, 87–91.

Gregory, R.D. (2000) Abundance patterns of European breeding birds. *Ecography* **23**, 201–208.

Gregory, R.D. & Blackburn, T.M. (1995) Abundance and body size in British birds: reconciling regional and ecological densities. *Oikos* **72**, 151–154.

Gregory, R.D. & Blackburn, T.M. (1998) Macroecological patterns in British breeding birds: covariation of species' geographical range sizes at differing spatial scales. *Ecography* **21**, 527–534.

Gregory, R.D. & Gaston, K.J. (2000) Explanations of commonness and rarity in British breeding birds: separating resource use and resource availability. *Oikos* **88**, 515–526.

Gregory, R.D., Greenwood, J.J.D. & Hagemeijer, E.J.M. (1998) The EBCC Atlas of European Breeding Birds: a contribution to science and conservation. *Biologia E Conservazione Della Fauna*, **102**, 38–49.

Gregory, R.D. & Marchant, J.H. (1995) Population trends of jays, magpies, jackdaws and carrion crows in the United Kingdom. *Bird Study* **43**, 28–37.

Griffiths, D. (1986) Size–abundance relations in communities. *American Naturalist* **127**, 140–166.

Griffiths, D. (1992) Size, abundance, and energy use in communities. *Journal of Animal Ecology* **61**, 307–315.

Griffiths, D. (1997) Local and regional species richness in North American lacustrine fish. *Journal of Animal Ecology* **66**, 49–56.

Griffiths, D. (1998) Sampling effort, regression method, and the shape and slope of size–abundance relations. *Journal of Animal Ecology* **67**, 795–804.

Griffiths, D. (1999) On investigating local–regional species richness relationships. *Journal of Animal Ecology* **68**, 1051–1055.

Grime, J.P. (1973) Control of species diversity in herbaceous vegetation. *Journal of Environmental Management* **1**, 151–167.

Grinnell, J. (1922) The role of the 'accidental'. *Auk* **39**, 373–380.

Gromadzki, M. (1970) Breeding communities of birds in mid-field afforested areas. *Ekologia Polska* **18**, 1–44.

Groombridge, B. (1992) *Global Diversity: Status of the Earth's Living Resources*. Chapman & Hall, London.

Guest, J.P., Elphick, D., Hunter, J.S.A. & Noumar, D. (1992) *The Breeding Bird Atlas of Cheshire and Wirral*. Cheshire and Wirral Ornithological Society.

Guillet, A. & Crowe, T.M. (1985) Patterns of distribution, species richness, endemism and guild composition of water-birds in Africa. *African Journal of Ecology* **23**, 89–120.

Guillet, A. & Crowe, T.M. (1986) A preliminary investigation of patterns of distribution and species richness of southern African waterbirds. *South African Journal of Wildlife Research* **16**, 65–81.

Gunnarsson, B. (1992) Fractal dimension of plants and body size distributions in spiders. *Functional Ecology* **6**, 636–641.

Guo, Q. & Berry, W.L. (1998) Species richness and biomass: dissection of the hump-shaped relationship. *Ecology* **79**, 2555–2559.

Gustafsson, L. (1988) Inter- and intraspecific competition for nest-holes in a population of the collared flycatcher *Ficedula albicollis*. *Ibis* **130**, 11–15.

Gyllenberg, M. & Hanski, I. (1992) Single-species metapopulation dynamics; a structured model. *Theoretical Population Biology* **42**, 35–61.

von Haartman, L. (1971) Population dynamics. In: *Avian Biology*, Vol. 1 (eds D.S. Farner & J.R. King), pp. 391–459. Academic Press, New York.

Haffer, J. (1988) Avian species richness in tropical South America. *Studies of Neotropical Fauna and Environment* **25**, 157–183.

Hagan, J.M. & Johnston, D.W. (eds) (1992) *Ecology and Conservation of Neotropical Migrant Landbirds.* Smithsonian Institution Press, Washington.

Hagemeijer, W.J.M. & Blair, M.J. (1997) *The EBCC Atlas of European Breeding Birds.* T. & A.D. Poyser, London.

Haila, Y. (1983) Land birds on northern islands: a sampling metaphor for insular colonization. *Oikos* **41**, 334–351.

Haila, Y. (1988) Calculating and miscalculating density: the role of habitat geometry. *Ornis Scandinavica* **19**, 88–92.

Haila, Y. & Hanski, I.K. (1993) Birds breeding on small British islands and extinction risks. *American Naturalist* **142**, 1025–1029.

Haila, Y., Hanski, I.K. & Raivio, S. (1993) Turnover of breeding birds in small forest fragments: the 'sampling' colonization hypothesis corroborated. *Ecology* **74**, 714–725.

Haila, Y. & Järvinen, O. (1983) Land bird communities on a Finnish island: species impoverishment and abundance patterns. *Oikos* **41**, 255–273.

Haila, Y., Nicholls, A.O., Hanski, I.K. & Raivio, S. (1996) Stochasticity in bird habitat selection: year-to-year changes in territory locations in a boreal forest bird assemblage. *Oikos* **76**, 536–552.

Hall, S.J. & Greenstreet, S.P. (1996) Global diversity and body size. *Nature* **383**, 133.

Hamilton, T.H. & Armstrong, N.E. (1965) Environmental determination of insular variation in bird species abundance in the Gulf of Guinea. *Nature* **207**, 148–151.

Hammond, P.M. (1995) Described and estimated species numbers: an objective assess-ment of current knowledge. In: *Microbial Diversity and Ecosystem Function* (eds D. Allsopp, D.L. Hawksworth & R.R. Colwell), pp. 29–71. CAB International, Wallingford.

Hannah, L., Carr, J.L. & Lankerani, A. (1995) Human disturbance and natural habitat: a biome level analysis of a global data set. *Biodiversity and Conservation* **4**, 128–155.

Hansen, T.A. (1978) Larval dispersal and species longevity in Lower Tertiary gastropods. *Science* **199**, 885–887.

Hansen, T.A. (1980) Influence of larval dispersal and geographic distribution on species longevities in neogastropods. *Paleobiology* **6**, 193–207.

Hanski, I. (1982a) Dynamics of regional distribution: the core and satellite species hypothesis. *Oikos* **38**, 210–221.

Hanski, I. (1982b) Distributional ecology of anthropochorous plants in villages surrounded by forest. *Annales Botanici Fennici* **19**, 1–15.

Hanski, I. (1982c) Communities of bumblebees: testing the core-satellite species hypothesis. *Annales Zoologici Fennici* **19**, 65–73.

Hanski, I. (1991a) Single-species metapopulation dynamics: concepts, models and observations. *Biological Journal of the Linnean Society* **42**, 17–38.

Hanski, I. (1991b) Reply to Nee, Gregory and May. *Oikos* **62**, 88–89.

Hanski, I. (1994) A practical model of metapopulation dynamics. *Journal of Animal Ecology* **63**, 151–162.

Hanski, I. (1997a) Metapopulation dynamics: from concepts and observations to predictive models. In: *Metapopulation Biology: Ecology, Genetics, and Evolution* (eds I. Hanski & M.E. Gilpin), pp. 69–91. Academic Press, San Diego.

Hanski, I. (1997b) Predictive and practical metapopulation models: the incidence function approach. In: *Spatial Ecology: The Role of Space in Population Dynamics and*

Interspecific Interactions (eds D. Tilman & P. Kareiva), pp. 21–45. Princeton University Press, Princeton, New Jersey.

Hanski, I. & Gilpin, M. (1991) Metapopulation dynamics: brief history and conceptual domain. *Biological Journal of the Linnean Society* **42**, 3–16.

Hanski, I. & Gilpin, M.E. (1997) *Metapopulation Biology: Ecology, Genetics, and Evolution.* Academic Press, San Diego.

Hanski, I. & Gyllenberg, M. (1993) Two general metapopulation models and the core-satellite species hypothesis. *American Naturalist* **142**, 17–41.

Hanski, I. & Gyllenberg, M. (1997) Uniting two general patterns in the distribution of species. *Science* **275**, 397–400.

Hanski, I., Kouki, J. & Halkka, A. (1993) Three explanations of the positive relationship between distribution and abundance of species. In: *Species Diversity in Ecological Communities: Historical and Geographical Perspectives* (eds R.E. Ricklefs & D. Schluter), pp. 108–116. University of Chicago Press, Chicago.

Hanski, I., Pakkala, T., Kuussaari, M. & Lei, G. (1995) Metapopulation persistence of an endangered butterfly in a fragmented landscape. *Oikos* **72**, 21–28.

Harding, B.D. (1979) *Bedfordshire Bird Atlas.* Bedfordshire Natural History Society.

Harner, R.F. & Harper, K.T. (1976) The role of area, heterogeneity, and favorability in plant species diversity of pinyon–juniper ecosystems. *Ecology* **57**, 1254–1263.

Harris, S., Morris, P., Wray, S. & Yalden, D. (1995) *A Review of British Mammals: Population Estimates and Conservation Status of British Mammals other than Cetaceans.* JNCC, Peterborough.

Harrison, S. (1994) Metapopulations and conservation. In: *Large-Scale Ecology and Conservation Biology* (eds P.J. Edwards, R.M. May & N.R. Webb), pp. 111–128. Blackwell Scientific Publications, Oxford.

Harrison, S., Ross, S.J. & Lawton, J.H. (1992) Beta diversity on geographic gradients in Britain. *Journal of Animal Ecology* **61**, 151–158.

Harshman, J. (1994) Reweaving the tapestry: what can we learn from Sibley & Ahlquist (1990)? *Auk* **111**, 377–388.

Harte, J. & Kinzig, A.P. (1997) On the implications of species–area relationships for endemism, spatial turnover, and food web patterns. *Oikos* **80**, 417–427.

Harte, J., Kinzig, A. & Green, J. (1999) Self-similarity in the distribution and abundance of species. *Science* **284**, 334–336.

Hartley, S. (1998) A positive relationship between local abundance and regional occupancy is almost inevitable (but not all positive relationships are the same). *Journal of Animal Ecology* **67**, 992–994.

Harvey, P.H. (1996) Phylogenies for ecologists. *Journal of Animal Ecology* **65**, 255–263.

Harvey, P.H., Colwell, R.K., Silvertown, J. & May, R.M. (1983) Null models in ecology. *Annual Review of Ecology and Systematics* **14**, 189–211.

Harvey, P.H. & Godfray, H.C.J. (1987) How species divide resources. *American Naturalist* **129**, 318–320.

Harvey, P.H. & Lawton, J.H. (1986) Patterns in three dimensions. *Nature* **324**, 212.

Harvey, P.H. & Pagel, M.D. (1991) *The Comparative Method in Evolutionary Biology.* Oxford University Press, Oxford.

Harvey, P.H. & Rambaut, A. (1998) Phylogenetic extinction rates and comparative methodology. *Proceedings of the Royal Society, London, B* **265**, 1691–1696.

Hawkins, A.F.A. (1999) Altitudinal and latitudinal distribution of East Malagasy forest bird communities. *Journal of Biogeography* **26**, 447–458.

Hawkins, B.A. (1995) Latitudinal body-size gradients for the bees of the eastern United States. *Ecological Entomology* **20**, 195–198.

Hawkins, B.A. & Compton, S.G. (1992) African fig wasp communities: undersaturation and latitudinal gradients in species richness. *Journal of Animal Ecology* **61**, 361–372.

Hawkins, B.A. & DeVries, P.J. (1996) Altitudinal gradients in the body size of Costa Rican butterflies. *Acta Oecologica* **17**, 185–194.

Hawkins, B.A. & Lawton, J.H. (1995) Latitudinal gradients in butterfly body sizes: is there a general pattern? *Oecologia* **102**, 31–36.

Hawksworth, D.L. & Kalin-Arroyo, M.T. (1995) Magnitude and distribution of biodiversity. In: *Global Biodiversity Assessment* (ed. V.H. Heywood), pp. 107–191. Cambridge University Press, Cambridge.

He, F. & Legendre, P. (1996) On species–area relations. *American Naturalist* **148**, 719–737.

Heck, K.L., van Belle, G. & Simberloff, D. (1975) Explicit calculation of the rarefaction diversity measurement and the determination of sufficient sample size. *Ecology* **56**, 1459–1461.

Hecnar, S.J. (1999) Patterns of turtle species' geographic range size and a test of Rapoport's rule. *Ecography* **22**, 436–446.

Hecnar, S.J. & M'Closkey, R.T. (1998) Species richness patterns of amphibians in southwestern Ontario ponds. *Journal of Biogeography* **25**, 736–772.

Helle, P. & Mönkkönen, M. (1990) Forest succession and bird communities. In: *Biogeography and Ecology of Forest Bird Communities* (ed. A. Keast), pp. 299–318. SPB Academic Publishing, The Hague.

Helliwell, D.R. (1976) The effects of size and isolation on the conservation value of wooded sites in Britain. *Journal of Biogeography* **3**, 407–416.

Hemmingsen, A.M. (1934) A statistical analysis of the differences in body size of related species. *Videnskabelige Meddelelser Fra Dansk Naturhistorik Forening I Kobenhavn* **98**, 125–160.

Hengeveld, R. & Haeck, J. (1981) The distribution of abundance. II. Models and implications. *Proceedings of the Koninklijke Nederlandse Akademie Van Wetenschappen, C* **84**, 257–284.

Hengeveld, R. & Haeck, J. (1982) The distribution of abundance. I. Measurements. *Journal of Biogeography* **9**, 303–316.

Hengeveld, R., Kooijman, S.A.L.M. & Taillie, C. (1979) A spatial model explaining species–abundance curves. In: *Statistical Distributions in Ecological Work* (eds J.K. Ord, G.P. Patil & C. Taillie), pp. 333–347. International Co-operative Publishing House, Fairland, Maryland.

Herreid, C.F. & Kessel, B. (1967) Thermal conductance in birds and mammals. *Comparative Biochemistry and Physiology* **21**, 405–414.

Herrera, C.M. (1978) On the breeding distribution pattern of European migrant birds: MacArthur's theme reexamined. *Auk* **95**, 496–509.

Hill, D., Taylor, S., Thraxton, R., Amphlet, A. & Horn, W. (1990) Breeding bird communities of native pine forest, Scotland. *Bird Study* **37**, 133–141.

Hilty, S.L. & Brown, W.L. (1986) *A Guide to the Birds of Colombia.* Princeton University Press, Princeton.

Hinsley, S.A., Bellamy, P.E., Enoksson, B. *et al.* (1998) Geographical and land-use influences on bird species richness in small woods in agricultural landscapes. *Global Ecology and Biogeography Letters* **7**, 125–135.

Hinsley, S.A., Bellamy, P.E. & Newton, I. (1995) Bird species turnover and stochastic extinction in woodland fragments. *Ecography* **18**, 41–50.

Hinsley, S.A., Pakeman, R., Bellamy, P.E. & Newton, I. (1996) Influences of habitat fragmentation on bird species distributions and regional population size. *Proceedings of the Royal Society, London, B* **263**, 307–313.

Hodgson, J.G. (1993) Commonness and rarity in British butterflies. *Journal of Applied Ecology* **30**, 407–427.

Holdaway, R.N. (1999) Introduced predators and avifaunal extinction in New Zealand. In: *Extinctions in Near Time: Causes, Contexts, and Consequences* (eds R.D.E. MacPhee & H.-D. Sues), pp. 189–238. Kluwer Academic/Plenum, New York.

Holling, C.S. (1992) Cross-scale morphology, geometry, and dynamics of ecosystems. *Ecological Monographs* **62**, 447–502.

Holloway, S. (1996) *The Historical Atlas of Breeding Birds in Britain and Ireland: 1875–1900.* T. & A.D. Poyser, London.

Holmes, J., Marchant, J., Bucknell, N., Stroud, D. & Parkin, D.T. (1998) The British list: new categories and their relevance to conservation. *British Birds* **91**, 2–11.

Holmes, R.T., Sherry, T.W. & Sturges, F.W. (1986) Bird community dynamics in a temperate deciduous forest: long-term trends at Hubbard Brook. *Ecological Monographs* **56**, 201–220.

Holmes, R.T. & Sturges, F.W. (1975) Avian community dynamics and energetics in a northern hardwood ecosystem. *Journal of Animal Ecology* **44**, 175–200.

Holt, R.D. (1993) Ecology at the mesoscale: the influence of regional processes on local communities. In: *Species Diversity in Ecological Communities: Historical and Geographical Perspectives* (eds R.E. Ricklefs & D. Schluter), pp. 77–88. University of Chicago Press, Chicago.

Holt, R.D., Lawton, J.H., Gaston, K.J. & Blackburn, T.M. (1997) On the relationship between range size and local abundance: back to basics. *Oikos* **78**, 183–190.

Houde, P. (1987) Critical evaluation of DNA hybridization studies in avian systematics. *Auk* **104**, 17–32.

Howard, R. & Moore, A. (1991) *A Complete Checklist of the Birds of the World.* Academic Press, London.

Howe, R.W. (1984) Local dynamics of bird assemblages in small forest habitat islands in Australia and North America. *Ecology* **65**, 1585–1601.

Howell, S.N.G. & Webb, S. (1995) *A Guide to the Birds of Mexico and Northern Central America.* Oxford University Press, Oxford.

Hubbell, S.P. (1997) A unified theory of biogeography and relative species abundance and its application to tropical rain forests and coral reefs. *Coral Reefs* **16** (Suppl.), S9–S21.

Hubbell, S.P. (in press) *A Unified Theory of Biodiversity and Biogeography.* Princeton University Press, Princeton.

Hudec, K., Chytil, J., Stastny, K. & Bejcek, V. (1995) The birds of the Czech Republic. *Sylvia* **31**, 97–149.

Hughes, B. (1993) Stiff-tail threat. *BTO News* **185**, 14.

Hughes, L., Cawsey, E.M. & Westoby, M. (1996) Geographic and climatic range sizes of Australian eucalypts and a test of Rapoport's rule. *Global Ecology and Biogeography Letters* **5**, 128–142.

Hughes, R.G. (1986) Theories and models of species abundance. *American Naturalist* **128**, 879–899.

Hugueny, B., de Morais, L.T., de Mérigoux, S., Mérona, B. & Ponton, D. (1997) The relationship between local and regional species richness: comparing biotas with different evolutionary histories. *Oikos* **80**, 583–587.

Hurlbert, S.H. (1971) The nonconcept of species diversity: a critique and alternative parameters. *Ecology* **52**, 577–586.

Hurlbert, S.H. (1984) Pseudoreplication and the design of ecological field experiments. *Ecological Monographs* **54**, 187–211.

Hutchinson, G.E. (1957) Concluding remarks. *Cold Spring Harbor Symposia on Quantitative Biology* **22**, 415–427.

Hutchinson, G.E. & MacArthur, R.H. (1959) A theoretical ecological model of size distributions among species of animals. *American Naturalist* **93**, 117–125.

Inglis, I.R., Isaacson, A.J., Thearle, R.J.P. & Westwood, N.J. (1990) The effects of changing agricultural practice upon woodpigeon *Columba palumbus* numbers. *Ibis* **132**, 262–272.

Inkinen, P. (1994) Distribution and abundance in British noctuid moths revisited. *Annales Zoologici Fennici* **31**, 235–243.

Jablonski, D. (1986a) Background and mass extinctions: the alternation of macroevolutionary regimes. *Science* **231**, 129–133.

Jablonski, D. (1986b) Causes and consequences of mass extinctions: a comparative approach. In: *Dynamics of Extinction* (ed. D.K. Elliot), pp. 183–229. Wiley, New York.

Jablonski, D. (1987) Heritability at the species level: analysis of geographic ranges of cretaceous mollusks. *Science* **238**, 360–363.

Jablonski, D. (1996a) Mass extinctions: persistent problems and new directions. In: *The Cretaceous–Tertiary Event and Other Catastrophes in Earth History* (eds G. Ryder, D. Fastovsky & S. Gartner), pp. 1–9. Geological Society of America Special Paper 307, Boulder, Colorado.

Jablonski, D. (1996b) Body size and macroevolution. In: *Evolutionary Paleobiology* (eds D. Jablonski, D.H. Erwin & J.H. Lipps), pp. 256–289. University of Chicago Press, Chicago.

Jablonski, D. & Raup, D.M. (1995) Selectivity of end-Cretaceous marine bivalve extinctions. *Science* **268**, 389–391.

Jablonski, D. & Valentine, J.W. (1990) From regional to total geographic ranges: testing the relationship in Recent bivalves. *Paleobiology* **16**, 126–142.

Jackson, J.B.C. (1974) Biogeographic consequences of eurytopy and stenotopy among marine bivalves and their evolutionary consequences. *American Naturalist* **108**, 541–560.

Jaeger, E.C. (1949) Further observations on the hibernation of the poor-will. *Condor* **51**, 105–109.

Jaenike, J. (1990) Host specialization in phytophagous insects. *Annual Review of Ecology and Systematics* **21**, 243–273.

James, F.C. (1970) Geographic size variation in birds and its relationship to climate. *Ecology* **51**, 365–390.

James, F.C., McCulloch, C.E. & Wiedenfeld, D.A. (1996) New approaches to the analysis of population trends in land birds. *Ecology* **77**, 13–27.

James, F.C. & Rathbun, S. (1981) Rarefaction, relative abundance, and diversity of avian communities. *Auk* **98**, 785–800.

James, P. (1996) *Birds of Sussex.* Sussex Ornithological Society.

Janzen, D.H. (1981) The peak of North American ichneumonid species richness lies between 38° and 42°N. *Ecology* **62**, 532–537.

Järvinen, O. & Sammalisto, L. (1976) Regional trends in the avifauna of Finnish peatland bogs. *Annales Zoologici Fennici* **13**, 31–43.

Järvinen, O. & Ulfstrand, S. (1980) Species turnover of a continental bird fauna: northern Europe 1850–1970. *Oecologia* **46**, 186–195.

Johnson, C.N. (1998a) Species extinction and the relationship between distribution and abundance. *Nature* **394**, 272–274.

Johnson, C.N. (1998b) Rarity in the tropics: latitudinal gradients in distribution and abundance in Australian mammals. *Journal of Animal Ecology* **67**, 689–698.

Johnson, N.K. (1975) Controls of number of bird species on montane islands in the Great Basin. *Evolution* **29**, 545–567.

Johnson, T.H. & Stattersfield, A.J. (1990) A global review of island endemic birds. *Ibis* **132**, 167–180.

Johst, K. & Brandl, R. (1997) Body size and extinction risk in a stochastic environment. *Oikos* **78**, 612–617.

Jones, K.E. & Purvis, A. (1997) An optimum body size for mammals? Comparative evidence from bats. *Functional Ecology* **11**, 751–756.

Juanes, F. (1986) Population density and body size in birds. *American Naturalist* **128**, 921–929.

Kadmon, R. (1995) Nested subsets and geographic isolation: a case study. *Ecology* **76**, 458–465.

Kareiva, P. (1994) Higher order interactions as a foil to reductionist ecology. *Ecology* **75**, 1527–1528.

Kareiva, P. & Andersen, M. (1988) Spatial aspects of species interactions: the wedding of models and experiments. In: *Community Ecology* (ed. A. Hastings), pp. 35–50. Springer-Verlag, Berlin.

Karr, J.R. (1968) Habitat and avian diversity on strip-mined land in east-central Illinois. *Condor* **70**, 348–357.

Karr, J.R. (1982) Avian extinction on Barro Colorado Island, Panama: a reassessment. *American Naturalist* **119**, 220–239.

Karr, J.R. & Roth, R.R. (1971) Vegetation structure and avian diversity in several New World areas. *American Naturalist* **105**, 423–435.

Kattan, G.H. (1992) Rarity and vulnerability: the birds of the Cordillera Central of Colombia. *Conservation Biology* **6**, 64–70.

Kaufman, D.M. (1995) Diversity of New World mammals—universality of the latitudinal gradients of species and bauplans. *Journal of Mammalogy* **76**, 322–334.

Kaufman, D.M. (1998) *The Structure of Mammalian Faunas in the New World: From Continents to Communities*. PhD Thesis, University of New Mexico.

Kaufman, D.M. & Willig, M.R. (1998) Latitudinal patterns of mammalian species richness in the New World: the effects of sampling method and faunal group. *Journal of Biogeography* **25**, 795–805.

Kelsey, M.G., Green, G.H., Garnett, M.C. & Hayman, P.V. (1989) Marsh warblers in Britain. *British Birds* **82**, 239–256.

Kemp, W.P. (1992) Rangeland grasshopper (Orthoptera: Acrididae) community structure: a working hypothesis. *Environmental Entomology* **21**, 461–470.

Kendeigh, S.C. (1974) *Ecology with Special Reference to Animals and Man*. Prentice Hall, Englewood Cliffs, New Jersey.

Kennedy, C.R. & Guégan, J.-F. (1994) Regional versus local helminth parasite richness in British freshwater fish: saturated or unsaturated parasite communities? *Parasitology* **109**, 175–185.

Kerr, J.T. & Packer, L. (1997) Habitat heterogeneity as a determinant of mammal species richness in high-energy regions. *Nature* **385**, 252–254.

Kiester, A.R. (1971) Species density of North American amphibians and reptiles. *Systematic Zoology* **20**, 127–137.

Kindlmann, P., Dixon, A.F.G. & Dostálkova, I. (1999) Does body size optimization result in skewed body size distribution on a logarithmic scale? *American Naturalist* **153**, 445–447.

King, B. (1982) Turnstones feeding on gull excrement. *British Birds* **75**, 88.

Kirk, W.D.J. (1991) The size relationship between insects and their hosts. *Ecological Entomology* **16**, 351–359.

Kitchener, D.J., Chapman, A., Dell, J., Muir, B.G. & Palmer, M. (1980a) Lizard assemblage and reserve size and structure in the western Australian wheatbelt—some implications for conservation. *Biological Conservation* **17**, 25–62.

Kitchener, D.J., Chapman, A., Muir, B.G. & Palmer, M. (1980b) The conservation value for mammals of reserves in the Western Australian wheatbelt. *Biological Conservation* **18**, 179–207.

Kitchener, D.J., Dell, J. & Muir, B.G. (1982) Birds in western Australian wheatbelt reserves—implications for conservation. *Biological Conservation* **22**, 127–163.

Kleiber, M. (1962) *The Fire of Life*. Wiley, New York.

Klicka, J. & Zink, R.M. (1997) The importance of recent ice ages in speciation: a failed paradigm. *Science* **277**, 1666–1669.

Klicka, J. & Zink, R.M. (1999) Pleistocene effects on North American songbird evolution. *Proceedings of the Royal Society, London, B* **266**, 695–700.

Kluyver, H.N. & Tinbergen, L. (1953) Territory and regulation of density in titmice. *Archives Néerlandaises de Zoologie* **10**, 265–289.

Knight, T.W. & Morris, D.W. (1996) How many habitats do landscapes contain? *Ecology* **77**, 1756–1764.

Knox, A. (1990) The sympatric breeding of common and Scottish crossbills *Loxia curvirostra* and *L. scotica* and the evolution of crossbills. *Ibis* **132**, 454–466.

Knox, A. (1994) Lumping and splitting of species. *British Birds* **87**, 149–159.

Koch, C.F. (1980) Bivalve species duration, areal extent and population size in a Cretaceous sea. *Paleobiology* **6**, 184–192.

Kochmer, J.P. & Wagner, R.H. (1988) Why are there so many kinds of passerine birds? Because they are small. A reply to Raikow. *Systematic Zoology* **37**, 68–69.

Koenig, W.D. & Knops, J.M.H. (1998) Testing for spatial autocorrelation in ecological studies. *Ecography* **21**, 423–429.

Körner, C. (1998) A re-assessment of high elevation treeline positions and their explanation. *Oecologia* **115**, 445–459.

Kostrzewa, V.R. (1988) Die dichte des turmfalken (*Falco tinnunculus*) in Europa übersicht und kritische betrachtung. *Die Vogelwarte* **34**, 216–224.

Koteja, P. (1991) On the relation between basal and field metabolic rates in birds and mammals. *Functional Ecology* **5**, 56–64.

Kouki, J., Niemelä, P. & Viitasaari, M. (1994) Reversed latitudinal gradient in species richness of sawflies (Hymenoptera, Symphyta). *Annales Zoologici Fennici* **31**, 83–88.

Kozár, F. (1995) Geographical segregation of scale-insects (Homoptera: Coccoidea) on fruit trees and the role of host plant ranges. *Acta Zoologica Academiae Scientiarium Hungaricae* **41**, 315–325.

Kozłowski, J. (1996) Energetic definition of fitness? Yes, but not that one. *American Naturalist* **147**, 1087–1091.

Kozłowski, J. & Weiner, J. (1997) Interspecific allometries are by-products of body size optimization. *American Naturalist* **149**, 352–380.

Kunin, W.E. (1998) Extrapolating species abundance across spatial scales. *Science* **281**, 1513–1515.

Lack, D. (1947) The significance of clutch size. *Ibis* **89**, 302–352.

Lack, D. (1969) The numbers of bird species on islands. *Bird Study* **16**, 193–209.

Lack, D. (1976) *Island Biology, Illustrated by the Landbirds of Jamaica*. Blackwell, Oxford.

Lack, P. (1986) *The Atlas of Wintering Birds in Britain and Ireland*. T. & A.D. Poyser, Calton.

Lambshead, P.J.D. (1993) Recent developments in marine benthic biodiversity research. *Océanis* **19**, 5–24.

Latham, R.E. & Ricklefs, R.E. (1993) Global patterns of tree species richness in moist forests: energy-diversity theory does not account for variation in species richness. *Oikos* **67**, 325–333.

Laurance, W.F. (1991) Ecological correlates of extinction proneness in Australian tropical rain forest mammals. *Conservation Biology* **5**, 79–89.

Laurila, T. & Järvinen, O. (1989) Poor predictabilty of the threatened status of waterfowl by life-history traits. *Ornis Fennica* **66**, 165–167.

Lawton, J.H. (1982) Vacant niches and unsaturated communities: a comparison of bracken herbivores at sites on two continents. *Journal of Animal Ecology* **51**, 573–595.

Lawton, J.H. (1986) Surface availability and insect community structure: the effects of architecture and fractal dimension of plants. In: *Insects and Plant Surfaces* (eds B.E. Juniper & T.R.E. Southwood), pp. 317–331. Edward Arnold, London.

Lawton, J.H. (1989) What is the relationship between population density and body size in animals? *Oikos* **55**, 429–434.

Lawton, J.H. (1990) Species richness and population dynamics of animal assemblages. Patterns in body-size: abundance space. *Philosophical Transactions of the Royal Society, London, B* **330**, 283–291.

Lawton, J.H. (1993) Range, population abundance and conservation. *Trends in Ecology and Evolution* **8**, 409–413.

Lawton, J.H. (1995) Population dynamic principles. In: *Extinction Rates* (eds J.H. Lawton & R.M. May), pp. 147–163. Oxford University Press, Oxford.

Lawton, J.H. (1996a) Patterns in ecology. *Oikos* **75**, 145–147.

Lawton, J.H. (1996b) Population abundances, geographic ranges and conservation: 1994 Witherby Lecture. *Bird Study* **43**, 3–19.

Lawton, J.H. (1999) Are there general laws in ecology? *Oikos* **84**, 177–192.

Lawton, J.H., Lewinsohn, T.M. & Compton, S.G. (1993) Patterns of diversity for the insect herbivores on bracken. In: *Species Diversity in Ecological Communities: Historical and Geographical Perspectives* (eds R.E. Ricklefs & D. Schluter), pp. 178–184. University of Chicago Press, Chicago.

Lawton, J.H. & Strong, D.R. (1981) Community patterns and competition in folivorous insects. *American Naturalist* **118**, 317–338.

Leck, C.F. (1979) Avian extinctions in an isolated tropical wet-forest preserve, Ecuador. *Auk* **96**, 343–352.

Leck, C.F., Murray, B.G. & Swinebroad, J. (1988) Long-term changes in the breeding bird populations of a New Jersey forest. *Biological Conservation* **46**, 145–157.

Lees, D.C. (1996) The Périnet effect? Diversity gradients in an adaptive radiation of butterflies in Madagascar (Satyrinae: Mycalesina) compared with other rainforest taxa. In: *Biogéographie de Madagascar* (ed. W.R. Lourenço), pp. 479–490. Editions de l'ORSTOM, Paris.

Lees, D., Kremen, C. & Andriamampianina, L. (1999) A null model for species richness gradients: bounded range overlap of butterflies and other rainforest endemics in Madagascar. *Biological Journal of the Linnean Society* **67**, 529–584.

Leith, H. & Werger, M.J.A. (1989) *Ecosystems of the World 14B: Tropical Rain Forest Ecosystems.* Elsevier, Amsterdam.

Leitner, W.A. & Rosenzweig, M. (1997) Nested species–area curves and stochastic sampling: a new theory. *Oikos* **79**, 503–512.

Lennon, J.J., Turner, J.R.G. & Connell, D. (1997) A metapopulation model of species boundaries. *Oikos* **78**, 486–502.

Letcher, A.J. & Harvey, P.H. (1994) Variation in geographical range size among mammals of the Palearctic. *American Naturalist* **144**, 30–42.

Levins, R. (1969) Some demographic and genetic consequences of environmental heterogeneity for biological control. *Bulletin of the Entomological Society of America* **15**, 237–240.

Levins, R. (1970) Extinction. In: *Some Mathematical Problems in Ecology* (ed. M. Gesternhaber), pp. 77–107. American Mathematical Society, Providence, Rhode Island.

Lewin, R. (1989) Biologists disagree over bold signature of nature. *Science* **244**, 527–528.

Lindstedt, S.L. & Boyce, M.S. (1985) Seasonality, fasting endurance, and body size in mammals. *American Naturalist* **125**, 873–878.

Lloyd, P. & Palmer, A.R. (1998) Abiotic factors as predictors of distribution in southern African bulbuls. *Auk* **115**, 404–411.

Lockwood, J.L., Moulton, M.P. & Anderson, S.K. (1993) Morphological assortment and the assembly of communities of introduced passeriforms on oceanic islands: Tahiti versus Oahu. *American Naturalist* **141**, 398–408.

Loder, N. (1997) *Insect species—body size distributions.* PhD Thesis, University of Sheffield.

Loder, N., Blackburn, T.M. & Gaston, K.J. (1997) The slippery slope—towards an understanding of the body size frequency distribution. *Oikos* **78**, 195–201.

Lomolino, M.V. (1990) The target area hypothesis: the influence of island area on immigration rates of non-volant mammals. *Oikos* **57**, 297–300.

Lomolino, M.V. (1994) Species richness of mammals inhabiting archipelagoes: area, isolation, and immigration filters. *Journal of Mammalogy* **75**, 39–49.

Lomolino, M.V. (1996) Investigating causality of nestedness of insular communities: selective immigrations or extinctions? *Journal of Biogeography* **23**, 699–703.

Lomolino, M.V., Brown, J.H. & Davis, R. (1989) Island biogeography of montane forest mammals in the American southwest. *Ecology* **70**, 180–194.

Lundberg, S. & Persson, L. (1993) Optimal body size and resource density. *Journal of Theoretical Biology* **164**, 163–180.

Lutz, F.E. (1921) Geographic average, a suggested method for the study of distribution. *American Museum Novitates* **5**, 1–7.

Lyons, S.K. & Willig, M.R. (1997) Latitudinal patterns of range size: methodological concerns and empirical evaluations for New World bats and marsupials. *Oikos* **79**, 568–580.

McAllister, D.E., Platania, S.P., Schueler, F.W., Baldwin, M.E. & Lee, D.S. (1986) Ichthyofaunal patterns on a geographical grid. In: *Zoogeography of Freshwater Fishes of North America* (eds C.H. Hocutt & E.D. Wiley), pp. 17–51. Wiley, New York.

McArdle, B.H. (1996) Levels of evidence in studies of competition, predation, and disease. *New Zealand Journal of Ecology* **20**, 7–15.

McArdle, B.H. & Gaston, K.J. (1995) The temporal variability of densities: back to basics. *Oikos* **74**, 165–171.

McArdle, B.H., Gaston, K.J. & Lawton, J.H. (1990) Variation in the size of animal populations: patterns, problems and artefacts. *Journal of Animal Ecology* **59**, 439–454.

MacArthur, R.H. (1957) On the relative abundance of bird species. *Proceedings of the National Academy of Sciences, USA* **43**, 293–295.

MacArthur, R.H. (1960) On the relative abundance of species. *American Naturalist* **94**, 25–36.

MacArthur, R.H. (1969) Patterns of communities in the tropics. *Biological Journal of the Linnean Society* **1**, 19–30.

MacArthur, R.H. (1972) *Geographical Ecology: Patterns in the Distribution of Species.* Harper & Row, New York.

MacArthur, R.H. & Connell, J.H. (1966) *The Biology of Populations.* Wiley, New York.

MacArthur, R.H. & MacArthur, J.W. (1961) On bird species diversity. *Ecology* **42**, 594–598.

MacArthur, R.H., MacArthur, J.W. & Preer, J. (1962) On bird species diversity. II. Prediction of bird census from habitat measurements. *American Naturalist* **96**, 167–174.

MacArthur, R.H., Recher, H.F. & Cody, M.L. (1966) On the relation between habitat selection and species diversity. *American Naturalist* **100**, 319–327.

MacArthur, R.H. & Wilson, E.O. (1963) An equilibrium theory of insular zoogeography. *Evolution* **17**, 373–387.

MacArthur, R.H. & Wilson, E.O. (1967) *The Theory of Island Biogeography.* Princeton University Press, Princeton.

MacCall, A.D. (1990) *Dynamic Geography of Marine Fish Populations.* University of Washington Press, Seattle.

McClure, M.S. & Price, P.W. (1976) Ecotope characteristics of coexisting *Erythroneura* leafhoppers (Homoptera: Cicadellidae) on sycamore. *Ecology* **57**, 928–940.

McCollin, D. (1993) Avian distribution patterns in a fragmented wooded landscape (North Humberside, U.K.); the role of between-patch and within-patch structure. *Global Ecology and Biogeography Letters* **3**, 48–62.

McCoy, E.D. & Connor, E.F. (1980) Latitudinal gradients in the species diversity of North American mammals. *Evolution* **34**, 193–203.

McDowall, R.M. (1969) Extinction and endemism in New Zealand land birds. *Tuatara* **17**, 1–12.

Mace, G.M. & Kershaw, M. (1997) Extinction risk and rarity on an ecological timescale. In: *The Biology of Rarity* (eds W.E. Kunin & K.J. Gaston), pp. 130–149. Chapman & Hall, London.

McIntosh, R.P. (1962) Raunkiaer's 'law of frequency'. *Ecology* **43**, 533–535.

McIntosh, R.P. (1967) An index of diversity and the relation of certain concepts to diversity. *Ecology* **48**, 392–404.

McLain, D.K. (1993) Cope's rule, sexual selection, and the loss of ecological plasticity. *Oikos* **68**, 490–500.

McLain, D.K., Moulton, M.P. & Redfearn, T.P. (1995) Sexual selection and the risk of extinction of introduced birds on oceanic islands. *Oikos* **74**, 27–34.

McNab, B.K. (1971) On the ecological significance of Bergmann's rule. *Ecology* **52**, 845–854.

Mac Nally, R. (1997) Monitoring forest bird communities for impact assessment: the influence of sampling intensity and spatial scale. *Biological Conservation* **82**, 355–367.

McShea, D.W. (1994) Mechanisms of large-scale evolutionary trends. *Evolution* **48**, 1747–1763.

Magurran, A.E. (1988) *Ecological Diversity and its Measurement.* Croom Helm, London.

Maiorana, V.C. (1990) Evolutionary strategies and body size in a guild of mammals. In: *Body Size in Mammalian Paleobiology* (eds J. Damuth & B.J. MacFadden), pp. 69–102. Cambridge University Press, Cambridge.

Malmer, N. (1994) Ecological research at the beginning of the next century. *Oikos* **71**, 171–176.

Manly, B.F.J. (1996) Are there clumps in body-size distributions? *Ecology* **77**, 81–86.

Manne, L.L., Pimm, S.L., Diamond, J.D. & Reed, T.M. (1998) The form of the curves: a direct evaluation of MacArthur & Wilson's classic theory. *Journal of Animal Ecology* **67**, 784–794.

Marchant, J.H. & Gregory, R.D. (1994) Recent population changes among seed-eating passerines in the United Kingdom. In: *Bird Numbers 1992. Distribution, Monitoring and Ecological Aspects. Proceedings of the 12th International Conference of the IBCC and EOAC, Noordwijkerhout, The Netherlands* (eds W.J.M. Hagemeijer & T.J. Vestrael), pp. 87–95. Statistics Netherlands, Voorburg/Heerlen and SOVON, Beek-Ubbergen, The Netherlands.

Marchant, J.H., Hudson, R., Carter, S.P. & Whittington, P. (1990) *Population Trends in British Breeding Birds.* British Trust for Ornithology, Tring, Hertfordshire.

Markgraf, V., McGlone, M. & Hope, G. (1995) Neogene paleoenvironmental and paleo-climatic change in southern temperate ecosystems—a southern perspective. *Trends in Ecology and Evolution* **10**, 143–147.

Marquet, P.A. & Cofré, H. (1999) Large temporal and spatial scales in the structure of mammalian assemblages in South America. *Oikos* **85**, 299–309.

Marquet, P.A., Navarette, S.A. & Castilla, J.C. (1990) Scaling population density to body size in rocky intertidal communities. *Science* **250**, 1125–1127.

Martin, G. (1996) Birds in double trouble. *Nature* **380**, 666–667.

Martin, J. & Gurrea, P. (1990) The peninsular effect in Iberian butterflies (Lepidoptera: Papilionoidea and Hesperioidea). *Journal of Biogeography* **17**, 85–96.

Martin, J.-L. & Lepart, J. (1989) Impoverishment in the bird community of a Finnish archipelago: the role of island size, isolation and vegetation structure. *Journal of Biogeography* **16**, 159–172.

Martin, P.S. (1984) Prehistoric overkill: the global model. In: *Quaternary Extinctions: a Prehistoric Revolution* (eds P.S. Martin & R. Klein), pp. 354–403. University of Arizona Press, Tucson.

Martin, R.A. (1992) Generic richness and body mass in North American mammals: support for the inverse relationship of body size and speciation rate. *Historical Biology* **6**, 73–90.

Martins, E.P. & Garland, T.H. (1991) Phylogenetic analyses of the correlated evolution of continuous characters: a simulation study. *Evolution* **45**, 534–557.

Matthysen, E. (1997) Geographic variation in the occurrence of song types in nuthatch *Sitta europaea* populations. *Ibis* **139**, 102–106.

Maurer, B.A. (1985) Avian community dynamics in desert grasslands: observational scale and hierarchical structure. *Ecological Monographs* **27**, 351–384.

Maurer, B.A. (1990) The relationship between distribution and abundance in a patchy environment. *Oikos* **58**, 181–189.

Maurer, B.A. (1994) *Geographical Population Analysis: Tools for the Analysis of Biodiversity.* Blackwell Scientific Publications, Oxford.

Maurer, B.A. (1998a) The evolution of body size in birds. I. Evidence for non-random diversification. *Evolutionary Ecology* **12**, 925–934.

Maurer, B.A. (1998b) The evolution of body size in birds. II. The role of reproductive power. *Evolutionary Ecology* **12**, 935–944.

Maurer, B.A. (1999) *Untangling Ecological Complexity.* University of Chicago Press, Chicago.

Maurer, B.A. & Brown, J.H. (1988) Distribution of energy use and biomass among species of North American terrestrial birds. *Ecology,* **69**, 1923–1932.

Maurer, B.A., Brown, J.H. & Rusler, R.D. (1992) The micro and macro in body size evolution. *Evolution* **46**, 939–953.

Maurer, B.A., Ford, H.A. & Rapoport, E.H. (1991) Extinction rate, body size, and avifaunal diversity. *Acta XX Congressus Internationalis Ornithologici* 826–834.

Mawdsley, N. & Stork, N.E. (1995) Species extinctions in insects: ecological and biogeographical considerations. In: *Insects in a Changing Environment* (eds R. Harrington & N.E. Stork), pp. 321–369. Academic Press, New York.

May, R.M. (1975) Patterns of species abundance and diversity. In: *Ecology and Evolution of Communities* (eds M.L. Cody & J.M. Diamond), pp. 81–120. Harvard University Press, Cambridge, Massachusetts.

May, R.M. (1978) The dynamics and diversity of insect faunas. In: *Diversity of Insect Faunas* (eds L.A. Mound & N. Waloff), pp. 188–204. Blackwell Scientific Publications, Oxford.

May, R.M. (1986) The search for patterns in the balance of nature: advances and retreats. *Ecology* **67**, 1115–1126.

May, R.M. (1988) How many species are there on earth? *Science* **241**, 1441–1448.

May, R.M. (1990) How many species? *Philosophical Transactions of the Royal Society, London, B* **330**, 293–304.

May, R.M. (1994a) The effects of spatial scale on ecological questions and answers. In: *Large-Scale Ecology and Conservation Biology* (eds P.J. Edwards, R.M. May & N.R. Webb), pp. 1–17. Blackwell Scientific Publications, Oxford.

May, R.M. (1994b) Biological diversity—differences between land and sea. *Philosophical Transactions of the Royal Society, London, B* **343**, 105–111.

May, R.M., Lawton, J.H. & Stork, N.E. (1995) Assessing extinction rates. In: *Extinction Rates* (eds J.H. Lawton & R.M. May), pp. 1–24. Oxford University Press, Oxford.

Mayr, E. (1956) Geographical character gradients and climatic adaptation. *Evolution* **10**, 105–108.

Mehlman, D.W. (1994) Rarity in North American passerine birds. *Conservation Biology* **8**, 1141–1145.

Mercer, A.J. (1966) Turnstones feeding on human corpse. *British Birds* **59**, 307.

Michener, C.D. (1979) Biogeography of the bees. *Annals of the Missouri Botanical Garden* **66**, 277–347.

Mikkelson, G.M. (1993) How do food webs fall apart? A study of changes in trophic structure during relaxation on habitat fragments. *Oikos* **67**, 539–547.

Miller, W.E. (1991) Body size in North American Lepidoptera as related to geography. *Journal of the Lepidopterists' Society* **45**, 158–168.

Møller, A.P. (1987) Breeding birds in habitat patches: random distribution of species and individuals. *Journal of Biogeography* **14**, 225–236.

Mönkkönen, M. (1994) Diversity patterns in Palaearctic and Nearctic forest bird assemblages. *Journal of Biogeography* **21**, 183–195.

Mönkkönen, M. & Viro, P. (1997) Taxonomic diversity of the terrestrial bird and mammal fauna in temperate and boreal biomes of the northern hemisphere. *Journal of Biogeography* **24**, 603–612.

Montier, D. (1977) *Atlas of Breeding Birds of the London Area.* London Natural History Society.

Mooers, A.Ø. & Cotgreave, P. (1994) Sibley and Ahlquist's tapestry dusted off. *Trends in Ecology and Evolution* **9**, 458–459.

Moore, N.W. & Hooper, M.D. (1975) On the number of bird species in British woods. *Biological Conservation* **8**, 239–250.

Morse, D.R., Lawton, J.H., Dodson, M.M. & Williamson, M.H. (1985) Fractal dimensions of vegetation and the distribution of arthropod body lengths. *Nature* **314**, 731–733.

Morse, D.R., Stork, N.E. & Lawton, J.H. (1988) Species number, species abundance and body length relationships of arboreal beetles in Bornean lowland rain forest trees. *Ecological Entomology* **13**, 25–37.

Moss, D. (1978) Diversity of woodland song-bird populations. *Journal of Animal Ecology* **47**, 521–527.

Moss, S. (1998) Predictions of the effects of global climate change on Britain's birds. *British Birds* **91**, 307–325.

Moulton, M.P. & Pimm, S.L. (1983) The introduced Hawaiian avifauna: biogeographic evidence for competition. *American Naturalist* **121**, 669–690.

Moulton, M.P. & Pimm, S.L. (1986) The extent of competition in shaping an introduced avifauna. In: *Community Ecology* (eds J. Diamond & T.J. Case), pp. 80–97. Harper & Row, New York.

Mourelle, C. & Ezcurra, E. (1997) Rapoport's rule: a comparative analysis between South and North American columnar cacti. *American Naturalist* **150**, 131–142.

Mousseau, T.A. (1997) Ectotherms follow the converse to Bergmann's rule. *Evolution* **51**, 630–632.

Murray, B.R., Fonseca, C.R. & Westoby, M. (1998b) The macroecology of Australian frogs. *Journal of Animal Ecology* **67**, 567–579.

Murray, R.D., Holling, M., Dott, H.E.M. & Vandome, P. (1998a) *The Breeding Birds of South-East Scotland: A Tetrad Atlas 1988–94.* Scottish Ornithologists' Club, Edinburgh.

Naganuma, K.H. & Roughgarden, J.D. (1990) Optimal body size in Lesser Antillean *Anolis* lizards—a mechanistic approach. *Ecological Monographs* **60**, 239–256.

Nagy, K.A. (1987) Field metabolic rate and food requirement scaling in mammals and birds. *Ecological Monographs* **57**, 111–128.

Navarrete, S.A. & Menge, B.A. (1997) The body size–population density relationship in tropical rocky intertidal communities. *Journal of Animal Ecology* **66**, 557–566.

Nee, S., Gregory, R.D. & May, R.M. (1991a) Core and satellite species: theory and artefacts. *Oikos* **62**, 83–87.

Nee, S., Harvey, P.H. & Cotgreave, P. (1992a) Population persistence and the natural relationships between body size and abundance. In: *Conservation of Biodiversity for Sustainable Development* (eds O.T. Sandlund, K. Hindar & A.H.D. Brown), pp. 124–136. Scandinavian University Press, Oslo.

Nee, S., Harvey, P.H. & May, R.M. (1991b) Lifting the veil on abundance patterns. *Proceedings of the Royal Society, London, B* **243**, 161–163.

Nee, S. & Lawton, J.H. (1996) Body size and biodiversity. *Nature* **380**, 672–673.

Nee, S., Mooers, A.Ø. & Harvey, P.H. (1992b) Tempo and mode of evolution revealed from molecular phylogenies. *Proceedings of the National Academy of Sciences, USA* **89**, 8322–8326.

Nee, S., Read, A.F., Greenwood, J.J.D. & Harvey, P.H. (1991c) The relationship between abundance and body size in British birds. *Nature* **351**, 312–313.

Newmark, W.D. (1987) A land-bridge island perspective on mammalian extinctions in western North American parks. *Nature* **325**, 430–432.

Newton, I. (1993) Goldfinch. In: *The New Atlas of Breeding Birds in Britain and Ireland: 1988–1991* (eds D.W. Gibbons, J.B. Reid & R.A. Chapman), pp. 412–413. T. & A.D. Poyser, London.

Newton, I. (1995) The contribution of some recent research on birds to ecological understanding. *Journal of Animal Ecology* **64**, 675–696.

Newton, I. (1997) Links between abundance and distribution of birds. *Ecography* **20**, 137–145.

Newton, I. (1998) *Population Limitation in Birds*. Academic Press, San Diego.

Newton, I. & Dale, L. (1996) Relationship between migration and latitude among west European birds. *Journal of Animal Ecology* **65**, 137–146.

Newton, I. & Haas, M.B. (1984) The return of the sparrowhawk. *British Birds* **77**, 47–70.

Newton, I., Dale, L. & Rothery, P. (1997) Apparent lack of impact of sparrowhawks on the breeding densities of some woodland songbirds. *Bird Study* **44**, 129–135.

Newton, I., Rothery, P. & Dale, L.C. (1998) Density-dependence in the bird populations of an oak wood over 22 years. *Ibis* **140**, 131–136.

Nichols, W.F., Killingbeck, K.T. & August, P.V. (1998) The influence of geomorphological heterogeneity on biodiversity. II. A landscape perspective. *Conservation Biology* **12**, 371–379.

Nilsson, A.N., Elmberg, J. & Sjöberg, K. (1994) Abundance and species richness patterns of predaceous diving beetles (Coleoptera, Dytiscidae) in Swedish lakes. *Journal of Biogeography* **21**, 197–206.

Nilsson, S.G. (1977) Density compensation and competition among birds breeding on small islands in a south Swedish lake. *Oikos* **28**, 170–176.

Nilsson, S.G. (1986) Are bird communities in small biotype patches random samples from communities in large patches? *Biological Conservation* **38**, 179–204.

Nilsson, S.G., Bengtsson, J. & Ås, S. (1988) Habitat diversity or area *per se*? Species richness of woody plants, carabid beetles and land snails on islands. *Journal of Animal Ecology* **57**, 685–704.

Nores, M. (1995) Insular biogeography of birds on mountain-tops in north western Argentina. *Journal of Biogeography* **22**, 61–70.

Norris, C.A. (1947) Report on the distribution and status of the corn-crake. Part two. A consideration of the causes of decline. *British Birds* **40**, 226–244.

Norris, R.D. (1991) Biased extinction and evolutionary trends. *Paleobiology* **17**, 388–399.

Novotny, V. (1992) Community structure of Auchenorryncha (Homoptera) in montane rain forest in Vietnam. *Journal of Tropical Ecology* **8**, 169–179.

Oberdorff, T., Hugueny, B., Compin, A. & Belkessam, D. (1998) Non-interactive fish communities in the coastal streams of north-western France. *Journal of Animal Ecology* **67**, 472–484.

Obeso, J.R. (1992) Geographic distribution and community structure of bumblebees in the northern Iberian peninsula. *Oecologia* **89**, 244–252.

O'Connor, R.J. (1981) Habitat correlates of bird distribution in British census plots. *Studies in Avian Biology* **6**, 533–537.

O'Connor, R.J. (1982) Habitat occupancy and regulation of clutch size in the European kestrel *Falco tinnunculus*. *Bird Study* **29**, 17–26.

O'Connor, R.J. (1987) Organisation of avian assemblages—the influence of intraspecific habitat dynamics. In: *The Organisation of Communities: Past and Present* (eds J.H.R. Gee & P.S. Giller), pp. 163–183. Blackwell Scientific Publications, Oxford.

O'Connor, R.J. & Shrubb, M. (1986) *Farming and Birds*. Cambridge University Press, Cambridge.

Odum, E.P. (1950) Bird populations of the Highlands (North Carolina) Plateau in relation to plant succession and avian invasion. *Ecology* **31**, 587–605.

Opdam, P. & Schotman, A. (1987) Small woods in a rural landscape as habitat islands for woodland birds. *Acta Oecologica* **8**, 269–274.

Opdam, P., van Dorp, D. & ter Braak, C.J.F. (1984) The effect of isolation on the number of woodland birds in small woods in the Netherlands. *Journal of Biogeography* **11**, 473–478.

Oro, D. & Pradel, R. (1999) Recruitment of Audouin's gull to the Ebro Delta colony at metapopulation level in the western Mediterranean. *Marine Ecology Progress Series* **180**, 267–273.

Ortega, J. & Arita, H.T. (1998) Neotropical–Nearctic limits in middle America as determined by distributions of bats. *Journal of Mammalogy* **79**, 772–783.

Osborne, P.E. & Tigar, B.J. (1992) Priorities for bird conservation in Lesotho, southern Africa. *Biological Conservation* **61**, 159–169.

Osman, R.W. & Whitlatch, R.B. (1978) Patterns of species diversity: fact or artifact? *Paleobiology* **4**, 41–54.

Owen, D.F. & Owen, J. (1974) Species diversity in temperate and tropical Ichneumonidae. *Nature* **249**, 583–584.

Owen, J.G. (1988) On productivity as a predictor of rodent and carnivore diversity. *Ecology* **69**, 1161–1165.

Owen, J.G. (1990) Patterns of mammalian species richness in relation to temperature. *Journal of Mammalogy* **71**, 1–13.

Owen, J. & Gilbert, F.S. (1989) On the abundance of hoverflies (Syrphidae). *Oikos* **55**, 183–193.

Pagel, M.D., Harvey, P.H. & Godfray, H.C.J. (1991b) Species abundance, biomass and resource use distributions. *American Naturalist* **138**, 836–850.

Pagel, M.D., May, R.M. & Collie, A.R. (1991a) Ecological aspects of the geographical distribution and diversity of mammalian species. *American Naturalist* **137**, 791–815.

Palmer, M.W. & White, P.S. (1994) Scale dependence and the species–area relationship. *American Naturalist* **144**, 717–740.

Papp, L. & Izsák, J. (1997) Bimodality in occurrence classes: a direct consequence of lognormal or logarithmic series distribution of abundances—a numerical experimentation. *Oikos* **79**, 191–194.

Paradis, E., Baillie, S.R., Sutherland, W.J. & Gregory, R.D. (1998) Patterns of natal and breeding dispersal in birds. *Journal of Animal Ecology* **67**, 518–536.

Paradis, E., Baillie, S.R., Sutherland, W.J. & Gregory, R.D. (in press) Spatial synchrony in populations of birds: effects of habitat, population trend, and spatial scale. *Ecology*.

Parkin, D.T. & Knox, A.G. (1994) Occurrence patterns of rare passerines in Britain and Ireland. *British Birds* **87**, 585–592.

Parr, D. (1972) *Birds in Surrey 1900–70*. B.T. Batsford Ltd, London.

Pärtel, M., Zobel, M., Zobel, K. & van der Maarel, E. (1996) The species pool and its relation to species richness: evidence from Estonian plant communities. *Oikos* **75**, 111–117.

Partridge, L. & Coyne, J.A. (1997) Bergmann's rule in ectotherms: is it adaptive? *Evolution* **51**, 632–635.

Paterson, J.D. (1990) Comment—Bergmann's rule is invalid: a reply to V. Geist. *Canadian Journal of Zoology* **68**, 1610–1612.

Patterson, B.D. (1984) Mammalian extinction and biogeography in the southern Rocky Mountains. In: *Extinctions* (ed. M.H. Nitecki), pp. 247–293. University of Chicago Press, Chicago.

Patterson, B.D. (1990) On the temporal development of nested subset patterns of species composition. *Oikos* **59**, 330–342.

Patterson, B.D. (1994) Accumulating knowledge on the dimensions of biodiversity: systematic perspectives on Neotropical mammals. *Biodiversity Letters* **2**, 79–86.

Patterson, B.D. (1999) Contingency and determinism in mammalian biogeography: the role of history. *Journal of Mammalogy* **80**, 345–360.

Patterson, B.D. & Atmar, W. (1986) Nested subsets and the structure of insular mammalian faunas and archipelagoes. *Biological Journal of the Linnean Society* **28**, 65–82.

Patterson, B.D., Stotz, D.F., Solari, S. & Fitzpatrick, J.W. (1998) Contrasting patterns of elevational zonation for birds and mammals in the Andes of southeastern Peru. *Journal of Biogeography* **25**, 593–607.

Pearson, D.L. & Cassola, F. (1992) World-wide species richness patterns of tiger beetles (Coleoptera: Cicindelidae): indicator taxon for biodiversity and conservation studies. *Conservation Biology* **6**, 376–391.

Peck, K.M. (1989) Tree species preferences shown by foraging birds in forest plantations in northern England. *Biological Conservation* **48**, 41–57.

Perrin, N. (1998) On body size, energy and fitness. *Functional Ecology* **12**, 500–502.

Perrins, C.M. (1979) *British Tits.* Collins, London.

Perry, J.N. (1988) Some models for spatial variability of animal species. *Oikos* **51**, 124–130.

Peters, R.H. (1983) *The Ecological Implications of Body Size.* Cambridge University Press, Cambridge.

Peters, R.H. & Raelson, J.V. (1984) Relations between individual size and mammalian population density. *American Naturalist* **124**, 498–517.

Peters, R.H. & Wassenberg, K. (1983) The effect of body size on animal abundance. *Oecologia* **60**, 89–96.

Peterson, S.R. (1975) Ecological distribution of breeding birds. In: *The Proceedings of the Symposium on Management of Forest and Range Habitats for Nongame Birds* (ed. D.R. Smith), pp. 22–38. USDA Forest Service, Washington DC.

Petit, J.R., Jouzel, J., Raynaud, D. *et al.* (1999) Climate and atmospheric history of the past 420 000 years from the Vostok ice core, Antarctica. *Nature* **399**, 429–436.

Pettersson, R.B. (1997) Lichens, invertebrates and birds in spruce canopies: impacts of forestry. *Acta Universitatis Agriculturae Sueciae: Silvestria 16.*

Pianka, E.R. (1966) Latitudinal gradients in species diversity: a review of the concepts. *American Naturalist* **100**, 33–46.

Pianka, E.R. (1989) Latitudinal gradients in species diversity. *Trends in Ecology and Evolution* **4**, 223.

Pielou, E.C. (1975) *Ecological Diversity.* John Wiley & Sons, New York.

Pielou, E.C. (1977a) The latitudinal spans of seaweed species and their patterns of overlap. *Journal of Biogeography* **4**, 299–311.

Pielou, E.C. (1977b) *Mathematical Ecology.* John Wiley & Sons, New York.

Pimm, S.L. (1986) Putting the species back into community ecology. *Trends in Ecology and Evolution* **1**, 51–52.

Pimm, S.L., Jones, H.L. & Diamond, J. (1988) On the risk of extinction. *American Naturalist* **132**, 757–785.

Platnick, N.I. (1991) Patterns of biodiversity: tropical vs temperate. *Journal of Natural History* **25**, 1083–1088.

Platnick, N.I. (1992) Patterns of biodiversity. In: *Systematics, Ecology and the Biodiversity Crisis* (ed. N. Eldredge), pp. 15–24. Columbia University Press, New York.

Pollard, E., Hooper, M.D. & Moore, N.W. (1974) *Hedges*. Collins, London.

Polo, V. & Carrascal, L.M. (1999) Shaping the body mass distribution of Passeriformes: habitat use and body mass are evolutionarily and ecologically related. *Journal of Animal Ecology* **68**, 324–337.

Pomeroy, D. & Ssekabiira, D. (1990) An analysis of the distributions of terrestrial birds in Africa. *African Journal of Ecology* **28**, 1–13.

Potts, G.R. (1998) Global dispersion of nesting hen harriers *Circus cyaneus*: implications for grouse moors in the UK. *Ibis* **140**, 76–88.

Poulin, R. (1998) *Evolutionary Ecology of Parasites*. Chapman & Hall, London.

Poulin, R. (1999) The intra- and interspecific relationships between abundance and distribution in helminth parasites of birds. *Journal of Animal Ecology* **68**, 719–725.

Poulsen, B.O. & Krabbe, N. (1997) Avian rarity in ten cloud-forest communities in the Andes of Ecuador: implications for conservation. *Biodiversity and Conservation* **6**, 1365–1375.

Preston, F.W. (1948) The commonness, and rarity, of species. *Ecology* **29**, 254–283.

Preston, F.W. (1960) Time and space and the variation of species. *Ecology* **41**, 612–627.

Preston, F.W. (1962) The canonical distribution of commonness and rarity. *Ecology* **43**, 185–215, 410–432.

Price, P.W. (1980) *Evolutionary Biology of Parasites*. Princeton University Press, Princeton, New Jersey.

Price, T.D., Helbig, A.J. & Richman, A.D. (1997) Evolution of breeding distributions in the Old World leaf warblers (Genus *Phylloscopus*). *Evolution* **51**, 552–561.

Promislow, D.E.L. & Harvey, P.H. (1989) Living fast and dying young: a comparative analysis of life history variation among mammals. *Journal of Zoology, London* **220**, 417–437.

Purvis, A., Gittleman, J.L. & Luh, H.-K. (1994) Truth or consequences: effects of phylogenetic accuracy on two comparative methods. *Journal of Theoretical Biology* **167**, 293–300.

Purvis, A. & Harvey, P.H. (1997) The right size for a mammal. *Nature* **386**, 332–333.

Purvis, A. & Rambaut, A. (1994) *Comparative Analysis by Independent Contrasts (CAIC)*, Version 2. Oxford University, Oxford.

Purvis, A. & Rambaut, A. (1995) Comparative analysis by independent contrasts (CAIC): an Apple Macintosh application for analysing comparative data. *Computer Applications in the Biosciences* **11**, 247–251.

Quinn, R.M., Gaston, K.J. & Arnold, H.R. (1996) Relative measures of geographic range size: empirical comparisons. *Oecologia* **107**, 179–188.

Quinn, R.M., Gaston, K.J., Blackburn, T.M. & Eversham, B.E. (1997a) Abundance–range size relationships of macrolepidoptera in Britain: the effects of taxonomy and life history variables. *Ecological Entomology* **22**, 453–461.

Quinn, R.M., Gaston, K.J. & Roy, D.B. (1997b) Coincidence between consumer and host occurrence: macrolepidoptera in Britain. *Ecological Entomology* **22**, 197–208.

Quinn, R.M., Gaston, K.J. & Roy, D.B. (1998) Coincidence in the distributions of butterflies and their foodplants. *Ecography* **21**, 279–288.

Quinn, S.L., Wilson, J.B. & Mark, A.F. (1987) The island biogeography of Lake Manapouri, New Zealand. *Journal of Biogeography* **14**, 569–581.

Rafe, R.W., Usher, M.B. & Jefferson, R.G. (1985) Birds on reserves: the influence of area and habitat on species richness. *Journal of Applied Ecology* **22**, 327–335.

Rahbek, C. (1995) The elevational gradient of species richness: a uniform pattern? *Ecography* **18**, 200–205.

Rahbek, C. (1997) The relationship among area, elevation, and regional species richness in Neotropical birds. *American Naturalist* **149**, 875–902.

Ralls, K. & Harvey, P.H. (1985) Geographic variation in size and sexual dimorphism of North American weasels. *Biological Journal of the Linnean Society* **25**, 119–167.

Rapoport, E.H. (1982) *Areography: Geographical Strategies of Species.* Pergamon, Oxford.

Rapoport, E.H. (1994) Remarks on marine and continental biogeography: an areographical viewpoint. *Philosophical Transactions of the Royal Society, London, B* **343**, 71–78.

Raunkiaer, C. (1934) *The Life Forms of Plants and Statistical Plant Geography.* Clarendon, Oxford.

Raup, D.M. (1991) *Extinction: Bad Genes or Bad Luck?* Oxford University Press, Oxford.

Raup, D.M. (1994) The role of extinction in evolution. *Proceedings of the National Academy of Sciences, USA* **91**, 6758–6763.

Ray, C. (1960) The application of Bergmann's and Allen's rules to the poikilotherms. *Journal of Morphology* **106**, 85–108.

Read, A.F. & Harvey, P.H. (1989) Life history differences among the eutherian radiations. *Journal of Zoology, London* **219**, 329–353.

Reavey, D. (1992) Egg size, first instar behaviour and the ecology of Lepidoptera. *Journal of Zoology, London* **227**, 277–297.

Reavey, D. (1993) Why body size matters to caterpillars. In: *Caterpillars* (eds N.E. Stamp & T.M. Casey), pp. 248–279. Chapman & Hall, London.

Rebelo, A.G. (1992) Red Data Book species in the Cape Floristic Region: threats, priorities and target species. *Transactions of the Royal Society of South Africa* **48**, 55–86.

Recher, H.F. (1969) Bird species diversity and habitat diversity in Australia and North America. *American Naturalist* **103**, 75–80.

Reed, T.M. (1980) Turnover frequency in island birds. *Journal of Biogeography* **7**, 329–335.

Reed, T. (1981) The number of breeding landbird species on British islands. *Journal of Animal Ecology* **50**, 613–624.

Reed, T.M. (1983) The role of species–area relationships in reserve choice: a British example. *Biological Conservation* **25**, 263–271.

Reed, T.M. (1984) The number of landbird species on the Isles of Scilly. *Biological Journal of the Linnean Society* **21**, 431–437.

Reed, T.M. (1987) Island birds and isolation: lack revisited. *Biological Journal of the Linnean Society* **30**, 25–29.

Reiss, M.J. (1989) *The Allometry of Growth and Reproduction.* Cambridge University Press, Cambridge.

Rensch, B. (1938) Some problems of geographical variation and species-formation. *Proceedings of the Linnean Society of London* **150**, 275–285.

Repasky, R.R. (1991) Temperature and the northern distributions of wintering birds. *Ecology* **72**, 2274–2285.

Restrepo, C., Renjifo, L.M. & Marples, P. (1997) Frugivorous birds in fragmented Neotropical montane forests: landscape pattern and body mass distribution. In: *Tropical Forest Remnants: Ecology, Management, and Conservation of Fragmented Com-*

munities (eds W.F. Laurance & R.O. Bierregaard), pp. 171–189. University of Chicago Press, Chicago.

Rex, M.A., Stuart, C.T., Hessler, R.R., Allen, J.A., Sanders, H.L. & Wilson, G.D.F. (1993) Global-scale latitudinal patterns of species diversity in the deep-sea benthos. *Nature* **365**, 636–639.

Richards, J.F. (1990) Land transformation. In: *The Earth as Transformed by Human Action* (eds B.L. Turner, W.C. Clark, R.W. Kates, J.F. Richards, J.T. Mathews & W.B. Meyer), pp. 163–178. Cambridge University Press, New York.

Ricklefs, R.E. (1979) *Ecology.* Chiron, New York.

Ricklefs, R.E. (1987) Community diversity: relative roles of local and regional processes. *Science* **235**, 167–171.

Ricklefs, R.E. & Cox, G.W. (1972) Taxon cycles in the West Indian avifauna. *American Naturalist* **106**, 195–219.

Ricklefs, R.E. & Cox, G.W. (1978) Stage of the taxon cycle, habitat distribution and population density in the avifauna of the West Indies. *American Naturalist* **112**, 875–895.

Ricklefs, R.E., Konarzewski, M. & Daan, S. (1996) The relationship between basal metabolic rate and daily energy expenditure. *American Naturalist* **147**, 1047–1071.

Ricklefs, R.E. & Latham, R.E. (1992) Intercontinental correlation of geographic ranges suggests stasis in ecological traits of relict genera of temperate perennial herbs. *American Naturalist* **139**, 1305–1321.

Ricklefs, R.E. & Schluter, D. (eds) (1993) *Species Diversity in Ecological Communities.* University of Chicago Press, Chicago.

Ricklefs, R.E. & Starck, J.M. (1996) Applications of phylogenetically independent contrasts: a mixed progress report. *Oikos* **77**, 167–172.

Riddle, B.R. (1998) The historical assembly of continental biotas: Late Quaternary range-shifting, areas of endemism, and biogeographic structure in the North American mammal fauna. *Ecography* **21**, 437–446.

Rigby, C. & Lawton, J.H. (1981) Species–area relationships of arthropods on host plants: herbivores on bracken. *Journal of Biogeography* **8**, 125–133.

van Riper, C. III (1991) The impact of introduced vectors and avian malaria on insular passeriform bird populations in Hawaii. *Bulletin of the Society for Vector Ecology* **16**, 59–83.

Robbins, C.S., Bystrak, D. & Geissler, P.H. (1986) *The Breeding Bird Survey: Its First Fifteen Years, 1965–79.* US Department of the Interior Fish and Wildlife Service, Resource Publication 157, Washington DC.

Robbins, C.S., Sauer, J.R., Greenberg, R.S. & Droege, S. (1989) Population declines in North American birds that migrate to the Neotropics. *Proceedings of the National Academy of Sciences, USA* **86**, 7658–7662.

Robinson, G.R. & Quinn, J.F. (1988) Extinction, turnover and species diversity in an experimentally fragmented California annual grassland. *Oecologia* **76**, 71–82.

Roff, D.R. (1981) On being the right size. *American Naturalist* **118**, 405–422.

Rogovin, K.A. & Shenbrot, G.I. (1995) Geographical ecology of Mongolian desert rodent communities. *Journal of Biogeography* **22**, 111–128.

Rohde, K. (1978) Latitudinal gradients in species diversity and their causes. II. Marine parasitological evidence for a time hypothesis. *Biologisches Zentralblatt* **97**, 405–418.

Rohde, K. (1986) Differences in species diversity of Monogenea between the Pacific and Atlantic oceans. *Hydrobiologia* **137**, 21–28.

Rohde, K. (1992) Latitudinal gradients in species diversity: the search for the primary cause. *Oikos* **65**, 514–527.

Rohde, K. (1996) Rapoport's rule is a local phenomenon and cannot explain latitudinal gradients in species diversity. *Biodiversity Letters* **3**, 10–13.

Rohde, K. (1997) The larger area of the tropics does not explain latitudinal gradients in species diversity. *Oikos* **79**, 169–172.

Rohde, K. & Heap, M. (1996) Latitudinal ranges of teleost fish in the Atlantic and Indo-Pacific oceans. *American Naturalist* **147**, 659–665.

Rohde, K., Heap, M. & Heap, D. (1993) Rapoport's rule does not apply to marine teleosts and cannot explain latitudinal gradients in species richness. *American Naturalist* **142**, 1–16.

Rohde, K., Worthen, W.B., Heap, M., Hugueny, B. & Guégan, J.-F. (1998) Nestedness in assemblages of metazoan ecto- and endoparasites of marine fish. *International Journal for Parasitology* **28**, 543–549.

Root, T. (1988a) *Atlas of Wintering North American Birds*. University of Chicago Press, Chicago.

Root, T. (1988b) Environmental factors associated with avian distributional boundaries. *Journal of Biogeography* **15**, 489–505.

Root, T. (1988c) Energy constraints on avian distributions and abundances. *Ecology* **69**, 330–339.

Root, T. (1989) Energy constraints on avian distributions: a reply to Castro. *Ecology* **70**, 1183–1185.

Rosenzweig, M.L. (1975) On continental steady states of species diversity. In: *Ecology and Evolution of Communities* (eds M.L. Cody & J.M. Diamond), pp. 124–140. Harvard University Press, Cambridge, Massachusetts.

Rosenzweig, M.L. (1978) Geographical speciation: on range size and the probability of isolate formation. In: *Proceedings of the Washington State University Conference on Biomathematics and Biostatistics* (ed. D. Wollkind), pp. 124–140. Washington State University, Washington.

Rosenzweig, M.L. (1991) Habitat selection and population interactions: the search for mechanism. *American Naturalist* **137**, S5–S28.

Rosenzweig, M.L. (1992) Species diversity gradients: we know more and less than we thought. *Journal of Mammalogy* **73**, 715–730.

Rosenzweig, M.L. (1995) *Species Diversity in Space and Time*. Cambridge University Press, Cambridge.

Rosenzweig, M.L. & Clark, C.W. (1994) Island extinction rates from regular censuses. *Conservation Biology* **8**, 491–494.

Rosenzweig, M.L. & Sandlin, E.A. (1997) Species diversity and latitudes: listening to area's signal. *Oikos* **80**, 172–176.

Roubik, D.W. (1992) Loose niches in tropical communities: why are there so few bees and so many trees?. In: *Effects of Resource Distribution on Animal–Plant Interactions* (eds M.D. Hunter, T. Ohgushi & P.W. Price), pp. 327–354. Academic Press, London.

Rousch, W. (1995) When rigor meets reality. *Science* **269**, 313–315.

Routledge, R.D. (1977) On Whittaker's components of diversity. *Ecology* **58**, 1120–1127.

Routledge, R.D. & Swartz, T.B. (1991) Taylor's power law re-examined. *Oikos* **60**, 107–112.

Roy, K. (1994) Effects of the Mesozoic marine revolution on the taxonomic, morphologic, and biogeographic evolution of a group: aporrhaid gastropods during the Mesozoic. *Paleobiology* **20**, 274–296.

Roy, K., Jablonski, D. & Valentine, J.W. (1994) Eastern Pacific molluscan provinces and latitudinal diversity gradient: no evidence for Rapoport's Rule. *Proceedings of the National Academy of Sciences, USA* **91**, 8871–8874.

Roy, K., Jablonski, D. & Valentine, J.W. (1995) Thermally anomalous assemblages revisited; patterns in the extraprovincial latitudinal range shifts of Pleistocene marine mollusks. *Geology* **23**, 1071–1074.

Ruggiero, A. (1994) Latitudinal correlates of the sizes of mammalian geographical ranges in South America. *Journal of Biogeography* **21**, 545–559.

Ruggiero, A. & Lawton, J.H. (1998) Are there latitudinal and altitudinal Rapoport effects in the geographic ranges of Andean passerine birds? *Biological Journal of the Linnean Society* **63**, 283–304.

Ruggiero, A., Lawton, J.H. & Blackburn, T.M. (1998) The geographic ranges of mammalian species in South America: spatial patterns in environmental resistance and anisotropy. *Journal of Biogeography* **25**, 1093–1103.

Russell, G.J., Diamond, J.M., Pimm, S.L. & Reed, T.M. (1995) A century of turnover: community dynamics at three timescales. *Journal of Animal Ecology* **64**, 628–641.

Russell, M.P. & Lindberg, D.R. (1988a) Real and random patterns associated with molluscan spatial and temporal distributions. *Paleobiology* **14**, 322–330.

Russell, M.P. & Lindberg, D.R. (1988b) Estimates of species duration. *Science* **240**, 969.

Ryti, R.T. & Gilpin, M.E. (1987) The comparative analysis of species occurrence patterns on archipelagos. *Oecologia* **73**, 282–287.

Sabelis, M. (1992) Predatory arthropods. In: *Natural Enemies: The Population Biology of Predators, Parasites and Diseases* (ed. M.J. Crawley), pp. 225–264. Blackwell Scientific, Oxford.

Sabo, S.R. (1980) Niche and habitat relations in subalpine bird communities of the White Mountains of New Hampshire. *Ecological Monographs* **50**, 241–259.

Sæther, B.-E. (1989) Survival rates in relation to body weight in European birds. *Ornis Scandinavica* **20**, 13–21.

Safriel, U.N., Volis, S. & Kark, S. (1994) Core and peripheral populations and global climate change. *Israel Journal of Plant Sciences* **42**, 331–345.

Salt, G.W. (1952) The relation of metabolism to climate and distribution in three finches of the genus *Carpodacus*. *Ecological Monographs* **22**, 121–152.

Sanders, H.L. (1968) Marine benthic diversity: a comparative study. *American Naturalist* **102**, 243–282.

Sanderson, W.G. (1996) *Rare Marine Benthic Flora and Fauna in Great Britain: The Development of Criteria for Assessment.* JNCC Report no. 240, Peterborough.

Sarich, V.M., Schmid, C.W. & Marks, J. (1989) DNA hybridization as a guide to phylogenies: a critical evaluation. *Cladistics* **5**, 3–12.

Schall, J.J. & Pianka, E.R. (1978) Geographical trends in numbers of species. *Science* **201**, 679–686.

Scharf, F.S., Juanes, F. & Sutherland, M. (1998) Inferring ecological relationships from the edges of scatter diagrams: comparison of regression techniques. *Ecology* **79**, 448–460.

Schmidt-Nielsen, K. (1984) *Scaling: Why Is Animal Size So Important?* Cambridge University Press, Cambridge.

Schoener, T.W. (1987) The geographical distribution of rarity. *Oecologia* **74**, 161–173.

Schoener, T.W. & Janzen, D.H. (1968) Notes on environmental determinants of tropical versus temperate insect size patterns. *American Naturalist* **102**, 207–224.

Scholander, P.F. (1955) Evolution of climatic adaptation in homeotherms. *Evolution* **9**, 15–26.

Scholander, P.F., Walters, V., Hock, R. & Irving, L. (1950) Body insulation of some arctic and tropical mammals and birds. *Biological Bulletin* **99**, 225–236.

Schonewald-Cox, C., Azari, R. & Blume, S. (1991) Scale, variable density, and conservation planning for mammalian carnivores. *Conservation Biology* **5**, 491–495.

Schopf, T.J.M., Fisher, J.B. & Smith, C.A.F. (1977) Is the marine latitudinal gradient merely another example of the species area curve? In: *Marine Organisms: Genetics, Ecology, and Evolution* (eds B. Battaglia & J.A. Beardmore), pp. 365–386. Plenum Press, New York.

Seib, R.L. (1980) Baja California: a peninsula for rodents but not for reptiles. *American Naturalist* **115**, 613–620.

Sfenthourakis, S. (1996) The species–area relationship of terrestrial isopods (Isopoda; Oniscidea) from the Aegean archipelago (Greece); a comparative study. *Global Ecology and Biogeography Letters* **5**, 149–157.

Sharrock, J.T.R. (1976) *The Atlas of Breeding Birds in Britain and Ireland*. T. & A.D. Poyser, Berkhamstead.

Sharrock, J.T.R. (1999) Editorial: Britain and Ireland. *British Birds* **92**, 62–63.

Sharrock, J.T.R. & Sharrock, E.M. (1976) *Rare Birds in Britain and Ireland*. T. & A.D. Poyser, Berkhamsted.

Shelford, V.E. (1911) Physiological animal geography. *Journal of Morphology* **22**, 551–618.

Shepherd, U.L. (1998) A comparison of species diversity and morphological diversity across the North American latitudinal gradient. *Journal of Biogeography* **25**, 19–29.

Shmida, A. & Wilson, M.V. (1985) Biological determinants of species diversity. *Journal of Biogeography* **12**, 1–20.

Shorrocks, B. (1993) Trends in the *Journal of Animal Ecology*: 1932–92. *Journal of Animal Ecology* **62**, 599–605.

Shorrocks, B., Marsters, J., Ward, I. & Evennett, P.J. (1991) The fractal dimension of lichens and the distribution of arthropod body lengths. *Functional Ecology* **5**, 457–460.

Short, J.J. (1979) Patterns of alpha-diversity and abundance in breeding bird communities across North America. *Condor* **81**, 21–27.

Shrader-Frechette, K.S. & McCoy, E.D. (1993) *Method in Ecology*. Cambridge University Press, Cambridge.

Shugart, H.H. & Patten, B.C. (1972) Niche quantification and the concept of niche pattern. In: *Systems Analysis and Simulation Ecology* (ed. B.C. Patten), pp. 283–327. Academic Press, New York.

Sibley, C.G. & Ahlquist, J.E. (1990) *Phylogeny and Classification of Birds: A Study in Molecular Evolution*. Yale University Press, New Haven.

Sibley, C.G. & Monroe, B.L., Jr (1990) *Distribution and Taxonomy of Birds of the World*. Yale University Press, New Haven.

Sibley, C.G. & Monroe, B.L., Jr (1993) *Supplement to the Distribution and Taxonomy of Birds of the World*. Yale University Press, New Haven.

Sibly, R.M. & Calow, P. (1986) *Physiological Ecology of Animals: An Evolutionary Approach*. Blackwell Scientific Publications, Oxford.

Siemann, E., Tilman, D. & Haarstad, J. (1996) Insect species diversity, abundance and body size relationships. *Nature* **380**, 704–706.

Siemann, E., Tilman, D. & Haarstad, J. (1999) Abundance, diversity and body size: patterns from a grassland arthropod community. *Journal of Animal Ecology* **68**, 824–835.

Sieving, K.E. & Karr, J.R. (1997) Avian extinction and persistence mechanisms in lowland Panama. In: *Tropical Forest Remnants: Ecology, Management, and Conservation of Fragmented Communities* (eds W.F. Laurance & R.O. Bierregaard), pp. 156–170. University of Chicago Press, Chicago.

Silva, M. & Downing, J.A. (1994) Allometric scaling of minimal mammal densities. *Conservation Biology* **8**, 732–743.

Silva, M. & Downing, J.A. (1995) The allometric scaling of density and body mass: a non-linear relationship for terrestrial mammals. *American Naturalist* **145**, 704–727.

Simberloff, D. (1976) Experimental zoogeography of islands: effects of island size. *Ecology* **57**, 629–648.

Simberloff, D. (1979) Rarefaction as a distribution free method of expressing and estimating diversity. In: *Ecological Diversity in Theory and Practice* (eds J.F. Grassle, G.P. Patil, W.K. Smith & C. Taillie), pp. 159–176. International Co-operative Publishing House, Fairland, Maryland.

Simberloff, D. (1983) When is an island community in equilibrium? *Science* **220**, 1275–1276.

Simberloff, D. & Martin, J.-L. (1991) Nestedness of insular avifaunas: simple summary statistics masking complex species patterns. *Ornis Fennica* **68**, 178–192.

Simpson, G.G. (1964) Species density of North American recent mammals. *Systematic Zoology* **13**, 57–73.

Simpson, R. (1991) The problem of escapes and potential vagrants. *Birding World* **3**, 421–422.

Siriwardena, G.M., Baillie, S.R., Buckland, S.T., Fewster, R.M., Marchant, J.H. & Wilson, J.D. (1998) Trends in the abundance of farmland birds: a quantitative comparison of smoothed Common Birds Census indices. *Journal of Applied Ecology* **35**, 24–43.

Sitters, H.P. (1988) *Tetrad Atlas of the Breeding Birds of Devon.* Devon Bird Watching and Preservation Society.

Slud, P. (1976) Geographic and climatic relationships of avifaunas with special reference to comparative distribution in the Neotropics. *Smithsonian Contributions to Zoology* **212**, 1–149.

Smallwood, K.S. (1995) Scaling Swainson's hawk population density for assessing habitat use across an agricultural landscape. *Journal of Raptor Research* **29**, 172–178.

Smallwood, K.S., Jones, G. & Schonewald, C. (1996) Spatial scaling of allometry for terrestrial, mammalian carnivores. *Oecologia* **107**, 588–594.

Smallwood, K.S. & Schonewald, C. (1996) Scaling population density and spatial pattern for terrestrial mammalian carnivores. *Oecologia* **105**, 329–335.

Smith, F.D.M., May, R.M. & Harvey, P.H. (1994) Geographical ranges of Australian mammals. *Journal of Animal Ecology* **63**, 441–450.

Smith, K.W., Dee, C.W., Fearnside, J.D., Fletcher, E.W. & Smith, R.N. (1993) *The Breeding Birds of Hertfordshire.* Hertfordshire Natural History Society.

Sokal, R.R. & Rohlf, F.J. (1995) *Biometry*, 3rd edn. W.H. Freeman, New York.

Solonen, T. (1994) Finnish bird fauna—species dynamics and adaptive constraints. *Ornis Fennica* **71**, 81–94.

Soulé, M.E. (ed.) (1987) *Viable Populations for Conservation.* Cambridge University Press, Cambridge.

Soulé, M.E., Bolger, D.T., Alberts, A.C. *et al.* (1988) Reconstructed dynamics of rapid extinctions of chaparral-requiring birds in urban habitat islands. *Conservation Biology* **2**, 75–92.

Southwood, T.R.E., Brown, V.K., Reader, P.M. & Green, E.E. (1986) The use of different stages of a secondary succession. *Bird Study* **33**, 159–163.

Spicer, J.I. & Gaston, K.J. (1999) *Physiological Diversity and its Ecological Implications.* Blackwell Science, Oxford.

Srivastava, D. (1999) Using local–regional richness plots to test for species saturation: pitfalls and potentials. *Journal of Animal Ecology* **68**, 1–16.

Standley, P., Bucknell, N.J., Swash, A. & Collins, I.D. (1996) *The Birds of Berkshire.* Berkshire Atlas Group, Reading.

Stanley, S.M. (1973) An explanation for Cope's Rule. *Evolution* **27**, 1–26.

Stanley, S.M. (1979) *Macroevolution.* W.H. Freeman, New York.

Stanley, S.M. (1986) Population size, extinction, and speciation: the fission effect in Neogene Bivalvia. *Paleobiology* **12**, 89–110.

Stearns, S.C. (1992) *The Evolution of Life Histories.* Oxford University Press, Oxford.

Stenseth, N.C. (1979) Where have all the species gone? On the nature of extinction and the Red Queen hypothesis. *Oikos* **33**, 196–227.

Stevens, G.C. (1989) The latitudinal gradient in geographical range: how so many species co-exist in the tropics. *American Naturalist* **133**, 240–256.

Stevens, G.C. (1992) The elevational gradient in altitudinal range: an extension of Rapoport's latitudinal rule to altitude. *American Naturalist* **140**, 893–911.

Stevens, G.C. (1996) Extending Rapoport's rule to Pacific marine fishes. *Journal of Biogeography* **23**, 149–154.

Stevens, G.C. & Enquist, B.J. (1997) Macroecological limits to the abundance and distribution of *Pinus*. In: *Ecology and Biogeography of Pinus* (ed. D.M. Richardson), pp. 183–190. Cambridge University Press, Cambridge.

Stone, B.H., Sears, J., Cranswick, P.A. *et al.* (1997) Population estimates of birds in Britain and in the United Kingdom. *British Birds* **90**, 1–22.

Stork, N.E. (1997) Measuring global biodiversity and its decline. In: *Biodiversity II* (eds M.L. Reaka-Kudla, D.E. Wilson & E.O. Wilson), pp. 41–68. Joseph Henry Press, Washington DC.

Stotz, D.F., Fitzpatrick, J.W., Parker, T.A. & Moskovits, D.K. (1996) *Neotropical Birds. Ecology and Conservation.* University of Chicago Press, Chicago.

Strayer, D.L. (1994) Body size and abundance of benthic animals in Mirror Lake, New Hampshire. *Freshwater Biology* **32**, 83–90.

Strong, D.R. (1980) Null hypotheses in ecology. *Synthese* **43**, 271–285.

Sugihara, G. (1980) Minimal community structure: an explanation of species abundance patterns. *American Naturalist* **116**, 770–787.

Sugihara, G. (1989) How *do* species divide resources? *American Naturalist* **133**, 458–463.

Summers-Smith, J.D. (1989) A history of the status of the Tree Sparrow *Passer montanus* in the British Isles. *Bird Study* **36**, 23–31.

Summers-Smith, J.D. (1999) Current status of the House Sparrow in Britain. *British Wildlife* **10**, 381–386.

Sutherland, W.J. & Baillie, S.R. (1993) Patterns in the distribution, abundance and variation of bird populations. *Ibis* **135**, 209–210.

Svensson, B.W. (1992) Changes in occupancy, niche breadth and abundance of three *Gyrinus* species as their respective range limits are approached. *Oikos* **63**, 147–156.

Symonds, M.R.E. (1999) Insectivore life histories: further evidence against an optimum body size for mammals. *Functional Ecology* **13**, 508–513.

Taper, M.L. & Marquet, P.A. (1996) How do species really divide resources? *American Naturalist* **147**, 1072–1086.

Taylor, C.M. & Gotelli, N.J. (1994) The macroecology of *Cyprinella*: correlates of phylogeny, body size and geographical range. *American Naturalist* **144**, 549–569.

Taylor, D.W., Davenport, D.L. & Flegg, J.J.M. (1981) *The Birds of Kent. A Review of their Status and Distribution*. The Kent Ornithological Society.

Taylor, L.R., Woiwod, I.P. & Perry, J.N. (1978) The density dependence of spatial behaviour and the rarity of randomness. *Journal of Animal Ecology* **47**, 383–406.

Taylor, L.R., Woiwod, I.P. & Perry, J.N. (1979) The negative binomial as an ecological model and the density-dependence of *k*. *Journal of Animal Ecology* **48**, 289–304.

Taylor, R.J. & Regal, P.J. (1978) The peninsular effect on species diversity and the biogeography of Baja California. *American Naturalist* **112**, 583–593.

Tellería, J.L. & Santos, T. (1993) Distributional patterns of insectivorous passerines in the Iberian forests: does abundance decrease near the border? *Journal of Biogeography* **20**, 235–240.

Tellería, J.L. & Santos, T. (1999) Distribution of birds in fragments of Mediterranean forests: the role of ecological densities. *Ecography* **22**, 13–19.

Tellería, J.L., Santos, T., Sanchez, A. & Galarza, A. (1992) Habitat structure predicts bird diversity distribution in Iberian forests better than climate. *Bird Study* **39**, 63–68.

Terborgh, J. (1973) On the notion of favorableness in plant ecology. *American Naturalist* **107**, 481–501.

Terborgh, J. (1977) Bird species diversity on an Andean elevational gradient. *Ecology* **58**, 1007–1019.

Terborgh, J. (1985) The role of ecotones in the distribution of Andean birds. *Ecology* **66**, 1237–1246.

Terborgh, J. (1989) *Where Have All the Birds Gone?* Princeton University Press, Princeton.

Terborgh, J.W. & Faaborg, J. (1980) Saturation of bird communities in the West Indies. *American Naturalist* **116**, 178–195.

Terborgh, J., Robinson, S.K., Parker, T.A., Munn, C.A. & Pierpont, N. (1990) Structure and organization of an Amazonian forest bird community. *Ecological Monographs* **60**, 213–238.

Terborgh, J. & Winter, B. (1980) Some causes of extinction. In: *Conservation Biology: An Evolutionary-Ecological Perspective* (eds M.E. Soule & B.A. Wilcox), pp. 119–133. Sinauer Associates, Sunderland, Massachusetts.

Terres, J.K. (1980) *The Audubon Society Encyclopedia of North American Birds*. Alfred A. Knopf, New York.

Thiollay, J.-M. (1990) Comparative diversity of temperate and tropical forest bird communities: the influence of habitat heterogeneity. *Acta Oecologica* **11**, 887–911.

Thiollay, J.-M. (1994) Structure, density and rarity in an Amazonian rainforest bird community. *Journal of Tropical Ecology* **10**, 449–481.

Thiollay, J.-M. (1997) Distribution and abundance patterns of bird community and raptor populations in the Andaman archipelago. *Ecography* **20**, 67–82.

Thiollay, J.-M. (1998) Distribution patterns and insular biogeography of south Asian raptor communities. *Journal of Biogeography* **25**, 57–72.

Thomas, D.K. (1992) *An Atlas of Breeding Birds in West Glamorgan*. Gower Ornithological Society.

Thompson, J.N. (1998) Rapid evolution as an ecological process. *Trends in Ecology and Evolution* **13**, 329–332.

Thomson, J.D., Weiblen, G., Thomson, B.A., Alfaro, S. & Legendre, P. (1996) Untangling multiple factors in spatial distributions: lilies, gophers, and rocks. *Ecology* **77**, 1698–1715.

Tilman, D. (1982) *Resource Competition and Community Structure*. Princeton University Press, Princeton, New Jersey.

Tilman, D. (1989) Ecological experimentation: strengths and conceptual problems. In: *Long-Term Studies in Ecology* (ed. G.E. Likens), pp. 136–157. Springer, New York.

Tilman, D. & Pacala, S. (1993) The maintenance of species richness in plant communities. In: *Species Diversity in Ecological Communities: Historical and Geographical Perspectives* (eds R.E. Ricklefs & D. Schluter), pp. 13–25. University of Chicago Press, Chicago.

Tokeshi, M. (1990) Niche apportionment or random assortment: species abundance patterns revisited. *Journal of Animal Ecology* **59**, 1129–1146.

Tokeshi, M. (1992) Dynamics and distribution in animal communities: theory and analysis. *Researches in Population Ecology* **34**, 249–273.

Tokeshi, M. (1996) Power fraction: a new explanation of relative abundance patterns in species-rich assemblages. *Oikos* **75**, 543–550.

Tokeshi, M. (1999) *Species Coexistence*. Blackwell Science, Oxford.

Tonn, W.M. & Magnuson, J.J. (1982) Patterns in the species composition and richness of fish assemblages in northern Wisconsin lakes. *Ecology* **63**, 1149–1166.

Tracy, C.R. & George, T.L. (1992) On the determinants of extinction. *American Naturalist* **139**, 102–122.

Tramer, E.J. (1974) On latitudinal gradients in avian diversity. *Condor* **76**, 123–130.

Trnka, A. (1997) *Current List of Birds of Slovakia*. Trnava University, Trnava.

Tucker, G.M. & Heath, M.F. (1994) *Birds in Europe: Their Conservation Status*. BirdLife International, Cambridge.

Turner, J.R.G., Gatehouse, C.M. & Corey, C.A. (1987) Does solar energy control organic diversity? Butterflies, moths and the British climate. *Oikos* **48**, 195–205.

Turner, J.R.G., Lennon, J.J. & Greenwood, J.J.D. (1996) Does climate cause the global biodiversity gradient?. In: *Aspects of the Genesis and Maintenance of Biological Diversity* (eds M.E. Hochberg, J. Clobert & R. Barbault), pp. 199–220. Oxford University Press, Oxford.

Turner, J.R.G., Lennon, J.J. & Lawrenson, J.A. (1988) British bird species distributions and energy theory. *Nature* **335**, 539–541.

Tyler, S., Lewis, J., Venables, A. & Walton, J. (1987) *The Gwent Atlas of Breeding Birds*. Gwent Ornithological Society.

Valentine, J.W. & Jablonski, D. (1993) Fossil communities: compositional variation at many time scales. In: *Species Diversity in Ecological Communities: Historical and Geographical Perspectives* (eds R.E. Ricklefs & D. Schluter), pp. 341–349. University of Chicago Press, Chicago.

Van den Bosch, F., Hengeveld, R. & Metz, J.A.J. (1992) Analysing the velocity of animal range expansion. *Journal of Biogeography* **19**, 135–150.

Van Valen, L. (1973a) A new evolutionary law. *Evolutionary Theory* **1**, 1–30.

Van Valen, L. (1973b) Body size and numbers of plants and animals. *Evolution* **27**, 27–35.

Van Valen, L. (1975) Group selection, sex, and fossils. *Evolution* **29**, 87–94.

Van Valen, L. & Sloan, R.E. (1977) Ecology and extinction of the dinosaurs. *Evolutionary Theory* **2**, 37–64.

Van Voorhies, W.A. (1996) Bergmann size clines: a simple explanation for their occurrence in ectotherms. *Evolution* **50**, 1259–1264.

Van Voorhies, W.A. (1997) On the adaptive nature of Bergmann size clines: a reply to Mousseau, Partridge and Coyne. *Evolution* **51**, 635–640.

Vaughn, C.C. (1997) Regional patterns of mussel species distributions in North American rivers. *Ecography* **20**, 207–215.

Veltman, C.J., Nee, S. & Crawley, M.J. (1996) Correlates of introduction success in exotic New Zealand birds. *American Naturalist* **147**, 542–557.

Venier, L.A. & Fahrig, L. (1998) Intra-specific abundance–distribution relationships. *Oikos* **82**, 483–490.

Venier, L.A., McKenney, D.W., Wang, Y. & McKee, J. (1999) Models of large-scale breeding-bird distribution as a function of macro-climate in Ontario, Canada. *Journal of Biogeography* **26**, 315–328.

Vezina, A.F. (1985) Empirical relationships between predator and prey size among terrestrial vertebrate predators. *Oecologia* **67**, 555–565.

Village, A. (1984) Problems in estimating kestrel breeding density. *Bird Study* **31**, 121–125.

Village, A. (1990) *The Kestrel.* Poyser, London.

Vinicombe, K., Marchant, J. & Knox, A. (1993) Review of status and categorisation of feral birds on the British list. *British Birds* **86**, 605–614.

Virkkala, R. (1993) Ranges of northern forest passerines: a fractal analysis. *Oikos* **67**, 218–226.

Vitousek, P.M., D'Antonio, C.M., Loope, L.L., Rejmánek, M. & Westbrooks, R. (1997) Introduced species: a significant component of human-caused global change. *New Zealand Journal of Ecology* **21**, 1–16.

Voous, K.H. (1960) *Atlas of European Birds.* Nelson, Edinburgh.

Vuilleumier, F. (1970) Insular biogeography in continental regions. I. The northern Andes of South America. *American Naturalist* **104**, 373–388.

Warburton, N.H. (1997) Structure and conservation of forest avifauna in isolated rainforest remnants in tropical Australia. In: *Tropical Forest Remnants: Ecology, Management, and Conservation of Fragmented Communities* (eds W.F. Laurance & R.O. Bierregaard), pp. 190–206. University of Chicago Press, Chicago.

Warren, P.H. & Gaston, K.J. (1992) Predator–prey ratios: a special case of a general pattern? *Philosophical Transactions of the Royal Society, London, B* **338**, 113–130.

Warren, P.H. & Gaston, K.J. (1997) Interspecific abundance–occupancy relationships: a test of mechanisms using microcosms. *Journal of Animal Ecology* **66**, 730–742.

Warren, P.H. & Lawton, J.H. (1987) Invertebrate predator–prey body size relationships: an explanation for upper triangular food webs and patterns in food web structure? *Oecologia* **74**, 231–235.

Watts, D.J. & Strogatz, S.H. (1998) Collective dynamics of 'small-world' networks. *Nature* **393**, 440–442.

Weatherhead, P.J. (1986) How unusual are unusual events? *American Naturalist* **128**, 150–154.

Webb, T.J., Kershaw, M. & Gaston, K.J. (2000) Rarity and phylogeny in birds. In: *Biotic Homogenization: The Loss of Diversity through Invasion and Extinction* (eds J.L. Lockwood & M.L. McKinney), in press. Academic/Plenum Publishers, New York.

West, G.B., Brown, J.H. & Enquist, B.J. (1999) A general model for the structure and allometry of plant vascular systems. *Nature* **400**, 664–667.

Westoby, M. (1993) Biodiversity in Australia compared with other continents. In: *Species Diversity in Ecological Communities: Historical and Geographical Perspectives* (eds R.E. Ricklefs & D. Schluter), pp. 170–177. University of Chicago Press, Chicago.

Westoby, M., Leishman, M.R. & Lord, J.M. (1995a) On misinterpreting the 'phylogenetic correction'. *Journal of Ecology* **83**, 531–534.

Westoby, M., Leishman, M.R. & Lord, J. (1995b) Further remarks on phylogenetic correction. *Journal of Ecology* **83**, 727–729.

Westoby, M., Leishman, M.R. & Lord, J. (1995c) Issues of interpretation after relating comparative datasets to phylogeny. *Journal of Ecology* **83**, 892–893.

Whiteman, P. & Millington, R. (1991) The British list and rare birds in the eighties. *Birding World* **3**, 429–434.

Whittaker, R.J. (1998) *Island Biogeography.* Oxford University Press, Oxford.

Whittaker, R.M. (1960) Vegetation of the Siskiyou Mountains, Oregon and California. *Ecological Monographs* **30**, 279–338.

Whittaker, R.M. (1972) Evolution and measurement of species diversity. *Taxon* **21**, 213–251.

Wiegert, R.G. (1988) Holism and reductionism in ecology: hypotheses, scale and systems. *Oikos* **53**, 267–269.

Wiens, J.A. (1973) Pattern and process in grassland bird communities. *Ecological Monographs* **43**, 237–270.

Wiens, J.A. (1981) Single-sample surveys of communities: are the revealed patterns real? *American Naturalist* **117**, 90–98.

Wiens, J.A. (1989) *The Ecology of Bird Communities*, Vol. 1. *Foundations and Patterns.* Cambridge University Press, Cambridge.

Wiens, J.A. (1991) Ecological similarity of shrub-desert avifaunas of Australia and North America. *Ecology* **72**, 479–495.

Wilber, K. (1979) *No Boundary.* Shambhala Publications, Boston.

Williams, C.B. (1943) Area and the number of species. *Nature* **152**, 264–267.

Williams, C.B. (1950) The application of the logarithmic series to the frequency of occurrence of plant species in quadrats. *Journal of Ecology* **38**, 107–138.

Williams, C.B. (1964) *Patterns in the Balance of Nature.* Academic Press, London.

Williams, G.R. (1973) Birds. In: *The Natural History of New Zealand* (ed. G.R. Williams), pp. 304–333. Reed, Wellington.

Williams, P.H. (1988) Habitat use by bumblebees (*Bombus* spp.). *Ecological Entomology* **13**, 223–237.

Williams, P.H. (1992) *WORLDMAP. Priority Areas for Biodiversity. Using Version 3.* Privately distributed computer software and manual, London.

Williams, P.H. (1993) Measuring more of biodiversity for choosing conservation areas, using taxonomic relatedness. In: *International Symposium on Biodiversity and Conservation* (ed. T.-Y. Moon), pp. 194–227. Korean Entomological Institute, Seoul.

Williams, P.H. (1996a) Measuring biodiversity value. *World Conservation* **1**, 12–14.

Williams, P.H. (1996b) Mapping variations in the strength and breadth of biogeographic transition zones using species turnover. *Proceedings of the Royal Society, London, B* **263**, 579–588.

Williams, P.H. & Gaston, K.J. (1998) Biodiversity indicators: graphical techniques, smoothing and searching for what makes relationships work. *Ecography* **21**, 551–560.

Williams, P., Gibbons, D., Margules, C., Rebelo, A., Humphries, C. & Pressey, R. (1996) A comparison of richness hotspots, rarity hotspots, and complementary areas for conserving diversity of British birds. *Conservation Biology* **10**, 155–174.

Williamson, K. (1969) Habitat preferences of the wren on English farmland. *Bird Study* **16**, 53–59.

Williamson, K. (1972) The relevance of the mapping census technique to the conservation of migratory bird populations. *Population Ecology of Migratory Birds: a Symposium. US Department of the Interior Wildlife Research Report* **2**, 27–40.

Williamson, K. (1974) Breeding birds in the deciduous woodlands of mid-Argyll, Scotland. *Bird Study* **21**, 29–44.

Williamson, M. (1981) *Island Populations.* Oxford University Press, Oxford.

Williamson, M. (1987) Are communities ever stable? In: *Colonization, Succession and Stability* (eds A.J. Gray, M.J. Crawley & P.J. Edwards), pp. 353–371. Blackwell Scientific Publications, Oxford.

Williamson, M. (1988) Relationship of species number to area, distance and other variables. In: *Analytical Biogeography: An Integrated Approach to the Study of Animal and Plant Distributions* (eds A.A. Myers & P.S. Giller), pp. 91–115. Chapman & Hall, London.

Williamson, M. (1996) *Biological Invasions.* Chapman & Hall, London.

Williamson, M. (1997) Marine biodiversity in its global context. In: *Marine Biodiversity: Patterns and Processes* (eds R.F.G. Ormond, J.D. Gage & M.V. Angel), pp. 1–17. Cambridge University Press, Cambridge.

Williamson, M. & Gaston, K.J. (1999) A simple transformation for sets of range sizes. *Ecography* **22**, 674–680.

Williamson, M.H. & Lawton, J.H. (1991) Fractal geometry of ecological habitats. In: *Habitat Structure: the Physical Arrangement of Objects in Space* (eds S.S. Bell, E.D. McCoy & H.R. Mushinsky), pp. 69–81. Chapman & Hall, London.

Willig, M.R. & Gannon, M.R. (1997) Gradients of species density and turnover in marsupials: a hemispheric perspective. *Journal of Mammalogy* **78**, 756–765.

Willig, M.R. & Lyons, S.K. (1998) An analytical model of latitudinal gradients of species richness with an empirical test for marsupials and bats in the New World. *Oikos* **81**, 93–98.

Willig, M.R. & Sandlin, E.A. (1992) Gradients of species density and species turnover in New World bats: a comparison of quadrat and band methodologies. In: *Latin American Mammalogy* (eds M.A. Mares & D.J. Schmidly), pp. 81–96. University of Oklahoma Press, Norman, Oklahoma.

Willig, M.R. & Selcer, K.W. (1989) Bat species density gradients in the New World: a statistical assessment. *Journal of Biogeography* **16**, 189–195.

Willis, J.C. (1922) *Age and Area.* Cambridge University Press, Cambridge.

Willson, M.F. & Comet, T.A. (1996) Bird communities of northern forests: ecological correlates of diversity and abundance in the understorey. *Condor* **98**, 350–362.

Wilson, D.S. (1988) Holism and reductionism in evolutionary ecology. *Oikos* **53**, 269–273.

Wilson, E.O. (1961) The nature of the taxon cycle in the Melanesian ant fauna. *American Naturalist* **95**, 169–193.

Wilson, E.O. (1992) *The Diversity of Life.* Penguin Books, London.

Wilson, J.B. (1991) Methods for fitting dominance/diversity curves. *Journal of Vegetation Science* **2**, 35–46.

Wilson, J.B. (1993) Would we recognise a broken-stick community if we found one? *Oikos* **67**, 181–183.

Wilson, J.B., Gitay, H., Steel, J.B. & King, W.M. (1998) Relative abundance distributions in plant communities: effects of species richness and of spatial scale. *Journal of Vegetation Science* **9**, 213–220.

Wilson, M.V. & Shmida, A. (1984) Measuring beta diversity with presence–absence data. *Journal of Ecology* **72**, 1055–1064.

Winstanley, D., Spencer, R. & Williamson, K. (1974) Where have all the whitethroats gone? *Bird Study* **21**, 1–14.

Woolhouse, M.E.J. (1983) The theory and practice of the species–area effect, applied to the breeding birds of British woods. *Biological Conservation* **27**, 315–332.

Wright, D.H. (1983) Species-energy theory: an extension of species area theory. *Oikos* **41**, 496–506.

Wright, D.H. (1991) Correlations between incidence and abundance are expected by chance. *Journal of Biogeography* **18**, 463–466.

Wright, D.H., Currie, D.J. & Maurer, B.A. (1993) Energy supply and patterns of species richness on local and regional scales. In: *Species Diversity in Ecological Communities* (eds R.E. Ricklefs & D. Schluter), pp. 66–74. University of Chicago Press, Chicago.

Wright, D.H., Patterson, B.D., Mikkelson, G.M., Cutler, A. & Atmar, W. (1998) A comparative analysis of nested subset patterns of species composition. *Oecologia* **113**, 1–20.

Wright, D.H. & Reeves, J.H. (1992) On the meaning and measurement of nestedness of species assemblages. *Oecologia* **92**, 416–428.

Wright, S.J. (1981) Intra-archipelago vertebrate distributions: the slope of the species–area relation. *American Naturalist* **118**, 726–748.

Wylie, J.L. & Currie, D.J. (1993a) Species-energy theory and patterns of species richness: I. Patterns of bird, angiosperm, and mammal species richness on islands. *Biological Conservation* **63**, 137–144.

Wylie, J.L. & Currie, D.J. (1993b) Species-energy theory and patterns of species richness: II. Predicting mammal species richness on isolated nature reserves. *Biological Conservation* **63**, 145–148.

Yamagishi, S. & Eguchi, K. (1996) Comparative foraging ecology of Madagascar vangids (Vangidae). *Ibis* **138**, 283–290.

Zeveloff, S.I. & Boyce, M.S. (1988) Body size patterns in North American mammal faunas. In: *Evolution of Life Histories of Mammals* (ed. M.S. Boyce), pp. 123–146. Yale University Press, New Haven.

Zimmerman, D., Turner, D., Pearson, D., Willis, I. & Pratt, D. (1996). *The Birds of Kenya and Northern Tanzania.* Christopher Helm, London.

Zink, R.M. (1996) Bird species diversity. *Nature* **381**, 566.

Zobel, M. (1997) The relative role of species pools in determining plant species richness; an alternative explanation of species coexistence? *Trends in Ecology and Evolution* **12**, 266–269.

List of Common and Scientific Bird Names

List of common and scientific names of bird species referred to in the book. Scientific names follow Sibley and Monroe (1990, 1993), but for the purposes of this list species are grouped into families, and families are ordered following the more traditional taxonomy given in Howard and Moore (1991).

Family	Common name	Scientific name
Gaviidae	Red-throated diver	*Gavia stellata*
	Black-throated diver	*Gavia arctica*
	Great northern diver	*Gavia immer*
Podicipedidae	Little grebe	*Tachybaptus ruficollis*
	Great-crested grebe	*Podiceps cristatus*
	Red-necked grebe	*Podiceps grisegena*
	Slavonian grebe	*Podiceps auritus*
	Black-necked grebe	*Podiceps nigricollis*
Procellariidae	Fulmar	*Fulmarus glacialis*
	Manx shearwater	*Puffinus puffinus*
Hydrobatidae	Storm petrel	*Hydrobates pelagicus*
	Leach's petrel	*Oceanodroma leucorhoa*
Pelecanoididea	Common diving-petrel	*Pelecanoides urinatrix*
Pelecanidae	Dalmatian pelican	*Pelecanus crispus*
Sulidae	Gannet	*Morus bassanus*
Phalacrocoracidae	Cormorant	*Phalacrocorax carbo*
	Shag	*Phalacrocorax aristotelis*
Ardeidae	Little bittern	*Ixobrychus minutus*
	Bittern	*Botaurus stellaris*
	Grey heron	*Ardea cinerea*
Anatidae	Mute swan	*Cygnus olor*
	Bewick's swan	*Cygnus columbianus*
	Whooper swan	*Cygnus cygnus*
	Bean goose	*Anser fabalis*
	Pink-footed goose	*Anser brachyrhynchus*
	White-fronted goose	*Anser albifrons*
	Greylag	*Anser anser*
	Canada goose	*Branta canadensis*
	Barnacle goose	*Branta leucopsis*
	Brent goose	*Branta bernicla*

(continued on p. 350)

Family	Common name	Scientific name
	Egyptian goose	*Alopochen aegyptiacus*
	Shelduck	*Tadorna tadorna*
	Mandarin	*Aix galericulata*
	Wood duck	*Aix sponsa*
	Wigeon	*Anas penelope*
	Gadwall	*Anas strepera*
	Teal	*Anas crecca*
	Mallard	*Anas platyrhynchos*
	Pintail	*Anas acuta*
	Garganey	*Anas querquedula*
	Shoveler	*Anas clypeata*
	Red-crested pochard	*Netta rufina*
	Pochard	*Aythya ferina*
	Tufted duck	*Aythya fuligula*
	Greater scaup	*Aythya marila*
	Eider	*Somateria mollissima*
	Long-tailed duck	*Clangula hyemalis*
	Common scoter	*Melanitta nigra*
	Velvet scoter	*Melanitta fusca*
	Goldeneye	*Bucephala clangula*
	Smew	*Mergellus albellus*
	Red-breasted merganser	*Mergus serrator*
	Goosander	*Mergus merganser*
	Ruddy duck	*Oxyura jamaicensis*
	White-headed duck	*Oxyura leucocephala*
Pandionidae	Osprey	*Pandion haliaetus*
Accipitridae	Honey buzzard	*Pernis apivorus*
	Red kite	*Milvus milvus*
	White-tailed sea-eagle	*Haliaeetus albicilla*
	Marsh harrier	*Circus aeruginosus*
	Hen harrier	*Circus cyaneus*
	Montagu's harrier	*Circus pygargus*
	Goshawk	*Accipiter gentilis*
	Sparrowhawk	*Accipiter nisus*
	Buzzard	*Buteo buteo*
	Rough-legged buzzard	*Buteo lagopus*
	Golden eagle	*Aquila chrysaetos*
Falconidae	Kestrel	*Falco tinnunculus*
	American kestrel	*Falco sparverius*
	Merlin	*Falco columbarius*
	Hobby	*Falco subbuteo*
	Peregrine	*Falco peregrinus*
Phasianidae	Red grouse	*Lagopus lagopus*
	Ptarmigan	*Lagopus mutus*
	Black grouse	*Tetrao tetrix*
	Capercaillie	*Tetrao urogallus*
	Red-legged partridge	*Alectoris rufa*
	Grey partridge	*Perdix perdix*
	Quail	*Coturnix coturnix*

Family	Common name	Scientific name
	Ring-necked pheasant	*Phasianus colchicus*
	Golden pheasant	*Chrysolophus pictus*
	Lady Amherst's pheasant	*Chrysolophus amherstiae*
Gruidae	Crane	*Grus grus*
Rallidae	Water rail	*Rallus aquaticus*
	Spotted crake	*Porzana porzana*
	Baillon's crake	*Porzana pusilla*
	Corncrake	*Crex crex*
	Moorhen	*Gallinula chloropus*
	Coot	*Fulica atra*
Otididae	Great bustard	*Otis tarda*
Haematopodidae	Oystercatcher	*Haematopus ostralegus*
Recurvirostridae	Black-winged stilt	*Himantopus himantopus*
	Avocet	*Recurvirostra avosetta*
Burhunidae	Stone-curlew	*Burhinus oedicnemus*
Charadriidae	Little ringed plover	*Charadrius dubius*
	Ringed plover	*Charadrius hiaticula*
	Dotterel	*Eudromias morinellus*
	Golden plover	*Pluvialis apricaria*
	Grey plover	*Pluvialis squatarola*
	Lapwing	*Vanellus vanellus*
Scolopacidae	Knot	*Calidris canutus*
	Sanderling	*Calidris alba*
	Temminck's stint	*Calidris temminckii*
	Purple sandpiper	*Calidris maritima*
	Dunlin	*Calidris alpina*
	Ruff	*Philomachus pugnax*
	Jack snipe	*Lymnocryptes minimus*
	Snipe	*Gallinago gallinago*
	Woodcock	*Scolopax rusticola*
	Black-tailed godwit	*Limosa limosa*
	Bar-tailed godwit	*Limosa lapponica*
	Whimbrel	*Numenius phaeopus*
	Curlew	*Numenius arquata*
	Spotted redshank	*Tringa erythropus*
	Redshank	*Tringa totanus*
	Greenshank	*Tringa nebularia*
	Green sandpiper	*Tringa ochropus*
	Wood sandpiper	*Tringa glareola*
	Common sandpiper	*Tringa hypoleucos*
	Spotted sandpiper	*Tringa macularia*
	Turnstone	*Arenaria interpres*
	Red-necked phalarope	*Phalaropus lobatus*
Stercorariidae	Arctic skua	*Stercorarius parasiticus*
	Great skua	*Catharacta skua*
Laridae	Little gull	*Larus minutus*
	Mediterranean gull	*Larus melanocephalus*

(continued on p. 352)

Family	Common name	Scientific name
	Black-headed gull	*Larus ridibundus*
	Common gull	*Larus canus*
	Lesser black-backed gull	*Larus fuscus*
	Herring gull	*Larus argentatus*
	Great black-backed gull	*Larus marinus*
	Kittiwake	*Rissa tridactyla*
	Sandwich tern	*Sterna sandvicensis*
	Roseate tern	*Sterna dougallii*
	Common tern	*Sterna hirundo*
	Arctic tern	*Sterna paradisaea*
	Little tern	*Sterna albifrons*
	Black tern	*Chlidonias niger*
Alcidae	Great auk	*Pinguinus impennis*
	Guillemot	*Uria aalge*
	Razorbill	*Alca torda*
	Black guillemot	*Cepphus grylle*
	Puffin	*Fratercula arctica*
Columbidae	Stock dove	*Columba oenas*
	Feral pigeon/rock dove	*Columba livia*
	Woodpigeon	*Columba palumbus*
	Collared dove	*Streptopelia decaocto*
	Turtle dove	*Streptopelia turtur*
Psittacidae	Rose-ringed parakeet	*Psittacula krameri*
Cuculidae	Cuckoo	*Cuculus canorus*
Tytonidae	Barn owl	*Tyto alba*
Strigidae	Snowy owl	*Nyctea scandiaca*
	Little owl	*Athene noctua*
	Tawny owl	*Strix aluco*
	Long-eared owl	*Asio otus*
	Short-eared owl	*Asio flammeus*
Caprimulgidae	Common poorwill	*Phalaenoptilus nuttallii*
	Nightjar	*Caprimulgus europaeus*
Apodidae	Swift	*Apus apus*
Trochilidae	Anna's hummingbird	*Calypte anna*
	Ruby-throated hummingbird	*Archilochus colubris*
Alcedinidae	Kingfisher	*Alcedo atthis*
Upupidae	Hoopoe	*Upupa epops*
Picidae	Wryneck	*Jynx torquilla*
	Green woodpecker	*Picus viridis*
	Great spotted woodpecker	*Dendrocopos major*
	Lesser spotted woodpecker	*Dendrocopos minor*
Alaudidae	Woodlark	*Lullula arborea*
	Skylark	*Alauda arvensis*
	Shore lark	*Eremophila alpestris*
Hirundinidae	Sand martin	*Riparia riparia*
	Swallow	*Hirundo rustica*
	House martin	*Delichon urbica*
Motacillidae	Tree pipit	*Anthus trivialis*
	Meadow pipit	*Anthus pratensis*

Family	Common name	Scientific name
	Rock pipit	*Anthus petrosus*
	Water pipit	*Anthus spinoletta*
	Yellow wagtail	*Motacilla flava*
	Grey wagtail	*Motacilla cinerea*
	Pied wagtail	*Motacilla alba*
Laniidae	Red-backed shrike	*Lanius collurio*
	Great grey shrike	*Lanius excubitor*
Bombycillidae	Waxwing	*Bombycilla garrulus*
Cinclidae	Dipper	*Cinclus cinclus*
Troglodytidae	Wren	*Troglodytes troglodytes*
Prunellidae	Dunnock	*Prunella modularis*
Turdidae	Robin	*Erithacus rubecula*
	Bluethroat	*Luscinia svecica*
	Nightingale	*Luscinia megarhynchos*
	Black redstart	*Phoenicurus ochruros*
	Redstart	*Phoenicurus phoenicurus*
	Whinchat	*Saxicola rubetra*
	Stonechat	*Saxicola torquata*
	Wheatear	*Oenanthe oenanthe*
	Ring ouzel	*Turdus torquatus*
	Blackbird	*Turdus merula*
	Fieldfare	*Turdus pilaris*
	Song thrush	*Turdus philomelos*
	Redwing	*Turdus iliacus*
	Mistle thrush	*Turdus viscivorus*
Panuridae	Bearded tit	*Panurus biarmicus*
Sylviidae	Cetti's warbler	*Cettia cetti*
	Grasshopper warbler	*Locustella naevia*
	Savi's warbler	*Locustella luscinioides*
	Sedge warbler	*Acrocephalus schoenobaenus*
	Marsh warbler	*Acrocephalus palustris*
	Reed warbler	*Acrocephalus scirpaceus*
	Icterine warbler	*Hippolais icterina*
	Dartford warbler	*Sylvia undata*
	Lesser whitethroat	*Sylvia curruca*
	Whitethroat	*Sylvia communis*
	Garden warbler	*Sylvia borin*
	Blackcap	*Sylvia atricapilla*
	Wood warbler	*Phylloscopus sibilatrix*
	Chiffchaff	*Phylloscopus collybita*
	Willow warbler	*Phylloscopus trochilus*
	Pallas's warbler	*Phylloscopus proregulus*
	Goldcrest	*Regulus regulus*
	Firecrest	*Regulus ignicapillus*
Muscicapidae	Spotted flycatcher	*Muscicapa striata*
	Pied flycatcher	*Ficedula hypoleuca*
Aegithalidae	Long-tailed tit	*Aegithalos caudatus*

(continued on p. 354)

Family	Common name	Scientific name
Paridae	Marsh tit	*Parus palustris*
	Willow tit	*Parus montanus*
	Crested tit	*Parus cristatus*
	Coal tit	*Parus ater*
	Blue tit	*Parus caeruleus*
	Great tit	*Parus major*
Sittidae	Nuthatch	*Sitta europaea*
Certhiidae	Treecreeper	*Certhia familiaris*
Meliphagidae	Stitchbird	*Notiomystis cincta*
Emberizidae	Lapland bunting	*Calcarius lapponicus*
	Snow bunting	*Plectrophenax nivalis*
	Yellowhammer	*Emberiza citrinella*
	Cirl bunting	*Emberiza cirlus*
	Reed bunting	*Emberiza schoeniclus*
	Corn bunting	*Miliaria calandra*
Fringillidae	Chaffinch	*Fringilla coelebs*
	Brambling	*Fringilla montifringilla*
	Serin	*Serinus serinus*
	Greenfinch	*Carduelis chloris*
	Goldfinch	*Carduelis carduelis*
	Siskin	*Carduelis spinus*
	Linnet	*Carduelis cannabina*
	Twite	*Carduelis flavirostris*
	Lesser redpoll	*Carduelis flammea*
	Common crossbill	*Loxia curvirostra*
	Scottish crossbill	*Loxia scotica*
	Parrot crossbill	*Loxia pytyopsittacus*
	Common rosefinch	*Carpodacus erythrinus*
	Bullfinch	*Pyrrhula pyrrhula*
	Hawfinch	*Coccothraustes coccothraustes*
Ploceidae	House sparrow	*Passer domesticus*
	Tree sparrow	*Passer montanus*
	Red-billed quelea	*Quelea quelea*
Sturnidae	Starling	*Sturnus vulgaris*
Oriolidae	Golden oriole	*Oriolus oriolus*
Corvidae	Jay	*Garrulus glandarius*
	Magpie	*Pica pica*
	Chough	*Pyrrhocorax pyrrhocorax*
	Jackdaw	*Corvus monedula*
	Rook	*Corvus frugilegus*
	Carrion/hooded crow	*Corvus corone*
	Raven	*Corvus corax*

Eastern Wood Breeding Bird Data

The number of territories of all bird species recorded breeding in Eastern Wood in the period 1949–79. NC = not censused. Data on species abundance from 1949 are used throughout the book, even though the number of territories for that year will be an underestimate. From data in Beven (1976) and Williamson (1987).

Species	1949	1950	1951	1952	1953	1954	1955	1956	1958	1959	1960	1961	1962
Mandarin	0	0	0	0	0	0	0	0	0	0	0	0	0
Sparrowhawk	1	1	1	0	1	1	1	1	1	1	0	0	0
Pheasant	0	1	1	1	1	1	0	0	0	0	1	0	1
Woodcock	0	0	0	1	1	0	0	0	1	0	0	0	1
Stock dove	0	0	0	0	0	0	0	0	0	0	0	0	0
Woodpigeon	2	2	5	2	5	4	3	1	3	4	4	3	2
Turtle dove	0	0	0	1	1	0	0	2	2	0	0	0	0
Cuckoo	0	2	0	2	1	2	1	1	0	1	0	2	1
Tawny owl	0	1	1	1	1	1	1	1	1	1	1	1	1
Green woodpecker	1	2	2	3	2	2	0	0	1	1	1	1	0
Great spotted woodpecker	1	2	2	2	3	3	2	2	2	1	2	2	2
Lesser spotted woodpecker	0	0	0	0	0	1	1	0	0	0	0	0	0
Carrion crow	1	1	1	1	1	1	1	1	1	1	1	1	1
Magpie	1	1	1	1	1	1	1	1	1	1	2	1	1
Jay	2	5	6	6	5	7	4	3	4	4	4	4	3
Great tit	6	8	8	8.5	11	11	10	12	4.5	11	12	12	11
Blue tit	NC	6.5	13.5	9.5	13	16	10	14	13	10.5	16	18	16.5
Coal tit	1	4	3	3	2	2	1	2	2	3	2	3	3
Marsh tit	2	2	2	4	5	4	2	4	2	3	4	2	5
Willow tit	1	1	0	1	1	1	1	2	1	0	1	1	0
Long-tailed tit	1	1	2	2	1	2	2	2	2	0	1	1	2
Nuthatch	0	1	3	1	2	2	3	4	1	4	4	3	3
Treecreeper	1	2	0	2	2	1	2	1	1	2	3	1	3
Wren	17	20	20.5	18	10.5	14	11.5	7	16	13	14.5	17	12
Mistle thrush	0	0	0	1	1	2	0	0	0	0	0	2	2
Song thrush	1	1	7	5	3	5	4	5	4	2	3	6	5
Blackbird	5	9	9.5	11.5	8	12.5	8	7.5	11.5	9	7	9	8
Nightingale	1	0	0	0	1	1	2	0	0	0	0	0	0
Robin	22.5	28	28.5	30.5	33	30.5	23	24	32	29	23.5	27	32
Blackcap	3	3	4	3	2	3	2	3	3	4	2	4	3
Garden warbler	3	3	3	3	4	2	4	4	7	3	3	2	4
Whitethroat	1	2	1	2	2	2	0	1	1	1	1	0	1
Lesser whitethroat	0	1	1	0	0	0	0	0	0	0	0	0	0
Willow warbler	16.5	21	16	15	15	11	8.5	17.5	12	5	2.5	2	1.5
Chiffchaff	1	2.5	4.5	2.5	2	5.5	4.5	4.5	1.5	2	3	1	1.5
Goldcrest	0	1	0	0	0	0	0	0	0	0	1	0	0
Spotted flycatcher	0	0	1	0	0	0	1	0	0	0	0	0	0
Dunnock	NC	2	2	2	4	1	2	4	1	2	0	1	4.5
Starling	0	0	0	0	0	0	0	0	0	1	2	3	2
Hawfinch	0	0	0	1	0	0	0	0	0	0	0	0	0
Greenfinch	0	0	0	0	0	0	0	0	0	0	0	0	0
Redpoll	0	0	0	0	0	0	0	0	0	0	0	0	0
Bullfinch	2	2	2	4	4	5	3	4	2	3	2	2	1
Chaffinch	9.5	11	12.5	9.5	9	8	8	6	4.5	2	2.5	4.5	7
House sparrow	0	0	0	0	0	0	0	0	0	0	0	0	0

1963	1964	1965	1966	1967	1968	1969	1970	1971	1972	1973	1974	1975	1976	1977	1978	1979
0	0	0	0	0	0	1	2	0	1	0	0	1	0	1	0	1
0	0	0	0	0	0	0	0	0	0	1	0	1	0	0	0	0
0	0	0	1	0	1	3	4	4	5	0	1	0	1	2	0	1
0	1	1	0	1	0	0	1	0	0	1	0	1	1	1	0	0
0	0	0	0	0	0	0	0	0	1	0	0	1	1	0	1	1
3	5	5	7	8	6	13	15	15	10	12	8	10	11	14	6	12
2	2	2	1	1	1	1	4	5	1	1	1	0	1	0	0	0
1	1	2	2	1	2	1	2	2	2	2	2	1	1	2	1	2
0	1	1	1	1	1	1	1	1	1	1	1	1	1	1	1	1
1	0	0	1	1	1	1	2	1	2	1	1	1	0	2	1	1
2	2	2	2	3	2	3	3	3	3	3	2	3	3	3	3	3
1	1	1	0	1	1	1	0	1	1	2	1	1	1	2	1	1
1	1	1	2	2	1	1	1	2	3	2	2	2	1	2	1	2
1	1	1	0	1	1	1	3	1	2	3	2	2	1	2	2	3
4	7	4	5	5	3	4	7	7	8	5	4	5	5	5	4	5
12	17	17	12	16	19	14	13	13	12	18	19	17	15	14	13	16
19	19	22	17	16	20	19	20	15	16	19	26	19	18	15	14	19
2	5	4	3	8	6	7	5	4	7	9	5	6	11	7	7	5
3	1	1	3	3	3	3	3	2	2	4	2	1	3	2	1	2
1	0	1	1	1	1	2	1	1	2	2	1	1	1	1	1	0
0	0	1	2	2	1	1	2	3	2	3	2	2	2	2	1	1
2	5	5	5	7	5	3	5	5	4	6	5	4	4	5	3	5
1	1	2	2	2	4	4	3	5	3	1	2	2	2	1	3	2
1	5.5	11	17	25	26	24	25	17	26	27	27	30	22	23	33	18
2	2	3	2	1	1	2	2	2	2	2	1	1	2	1	2	2
4	7	7	8	10	7	7	7	9	6	6	6	4	4	8	5	4
8	10	12	11	12	11	13	9	14	12	6	12	12	10	11	11	13
0	1	0	0	0	0	0	0	0	1	1	0	0	0	0	0	0
21.5	32	37	37	42	43	44	36	44	37	35	38	33	33	34	40	41
5	3	4	7	6	3	6	1	3	4	3	2	2	4	5	4	4
2	2	4	1	1	1	1	1	0	0	0	1	1	2	2	0	1
1	1	0	1	0	1	0	0	0	0	0	0	0	0	0	0	0
0	0	0	0	0	0	0	0	0	0	0	0	0	0	0	0	0
2.5	1.5	4	4	2	6	2	2	0	0	0	0	2	1	1	1	2
2	5	7	6	6	3	3	4	1	1	4	4	1	4	3	5	4
0	0	0	0	0	3	0	1	2	1	3	2	2	1	0	1	0
1	0	1	0	1	0	0	0	1	0	0	0	1	0	0	0	0
4	5	5	4	8	6	6	3	2	1	0	3	1	2	2	4	4
5	5.5	10	6	8	12	7	8	8	6	7	9	9	6	7	7	7
0	0	0	0	0	1	0	1	0	0	0	0	0	0	0	0	0
0	0	0	1	0	0	0	0	0	1	0	0	0	0	0	0	1
0	0	0	0	0	0	0	0	0	2	0	0	0	0	1	0	0
2	2	2	2	2	2	3	4	1	4	2	2	2	2	2	0	2
5.5	7	8	7	6	4	4	3	6	8	3	1	3	3	3	3	2
0	0	0	0	0	0	0	0	0	0	0	0	0	1	1	0	0

British Bird Assemblage Data

This appendix presents data on the abundance, distribution and body mass of those bird species we consider to be part of the current British assemblage, together with information on their status. These data are used for all analyses of these variables throughout the book. Details of their derivation are as follows.

The assemblage: We assumed that a species was part of the British avifauna if both the following criteria were met:

(i) either a breeding or a wintering population estimate was available in Stone *et al.* (1997);

(ii) an estimate of geographical extent in Britain was available for the same season in either the most recent BTO atlas of British bird breeding distributions (Gibbons *et al.* 1993) or the BTO atlas of British bird winter distributions (Lack 1986).

Three species (black-winged stilt, little gull and icterine warbler) met the first criterion but not the second, while a further two species (wood duck and feral pigeon/rock dove) met the second criterion but not the first. These were excluded from the assemblage. Whimbrel met both criteria for its breeding population but not its wintering population, and so was included in the breeding assemblage only.

We freely admit that our definition of the British bird assemblage is arbitrary. However, that will be true of any such definition. It is rare to find two analyses of the British bird fauna that utilize exactly the same set of species, unless the studies are by the same authors, or the data for one are taken directly from another. No one list will satisfy all readers, and so we make no apologies for the one chosen.

Breeding population size (number of individuals; BREPOP): The source of these data is Stone *et al.* (1997). We used the population size estimates referring to Great Britain only, or the arithmetic mean of the highest and lowest values where a range was indicated. Stone *et al.* report the population sizes using a variety of units (individuals, adult individuals, pairs, nests, territories, wild pairs, males, females). We converted all these to number of individuals on the assumption that the number of individuals equals the number of adults, or twice the estimated number of pairs, territories, nests, males, females, or wild pairs. We ignored greater than or less than signs. Stone *et al.* do not give a breeding population size for one species (mandarin). We assumed the breeding population size for this resident species equals its wintering population size.

Wintering population size (number of individuals; WINPOP): The principal source of these data is Stone *et al.* (1997), with selection and conversion criteria as for breeding population size. Winter population sizes for seabirds were not calculated (see below). For a number of species that both breed and winter in Britain, Stone *et al.* do not report wintering population size (e.g. meadow pipit, robin, chaffinch). These species are indicated by a 'W' in the wintering population size column, to indicate that they are a part of the British wintering bird species assemblage.

Breeding range size (number of 10 × 10-km squares occupied; BREGR): These figures are taken from Gibbons *et al.* (1993), and refer to the total number of 10 × 10-km squares of the National Grid in Britain (excluding Ireland) in which the species was recorded, whether or not there was evidence of breeding in the square, over the course of the fieldwork for this atlas.

Wintering range size (number of 10 × 10-km squares occupied; WINGR): These figures are taken from Lack (1986), and refer to the total number of 10 × 10-km squares of the National Grid in Britain (excluding Ireland) in which the species was recorded wintering over the course of the fieldwork for this atlas. Note that since both breeding and wintering range sizes were calculated from data compiled over several years, whereas the equivalent population sizes are estimates of annual figures, the former are sometimes larger than the latter for rare species.

Body mass (grams): Data were taken from Cramp and Simmons (1977, 1980, 1983), Cramp (1985, 1988, 1992), and Cramp and Perrins (1993, 1994a, 1994b), with the exception of those for golden pheasant, Lady Amherst's pheasant and Mediterranean gull, which were taken from Dunning (1992) (Cramp *et al.* provide no data on female mass for the first two, which presents problems for comparison with the ring-necked pheasant, and no mass data at all for the last). We used mid-winter masses from unstressed British female birds where available, but in many cases had to settle for data that did not fulfil one or more (or indeed any) of these criteria. In such cases, particular attention was paid to the comparability of estimates among closely related species. We avoided using extreme mass values wherever possible.

Origin: This column distinguishes among three groups of species; wild: those species with British populations that are principally of wild origin; captive: species whose populations are principally of captive origin; [Captive]: species whose *breeding* populations are principally of captive origin. The last two categories of species have established British populations after either deliberate introduction (pheasant) or escaping from collections (ruddy duck). Arguably, they may be considered not to be natural components of the British avifauna, albeit that certain of the introductions have been of native species previously driven extinct by humans (e.g. white-tailed eagle, capercaillie). We identify them here for this reason, so that they can be excluded from certain analyses.

Pelagics: Although most species in the British avifauna are firmly terrestrial in habits, a number make use of the surrounding seas. These can be separated into those principally found in coastal waters, and those of more pelagic habits. Species associated with coasts are generally tied somewhat to the land area of Britain. Many of these species breed inland and use the coastal waters only for wintering. Other species winter partly inland and partly at the coast. At the very least, the extents of their distributions will depend on the land area of Britain, which relates to the amount of coastline. Pelagic species, in contrast, present a special problem for analyses of the British bird community. In effect, they are only making use of Britain as a nesting platform. Away from the nest, they range widely across an area that is difficult to estimate or define, and is certainly very different to that used by all other British birds. Thus, it is not clear that populations of pelagic species should be limited in the same way as are populations of other British species. For this reason, we identify here species that we consider to be pelagic, so that they can be excluded from certain analyses.

As for our definition of the composition of the British bird assemblage, we are also aware that data other than those we present here could have been used. For example, some may have preferred estimates of breeding distribution to include only those 10 × 10-km squares in which the species had been proven to breed. The estimates of body mass provided could have been calculated in a range of different ways, and the lack of confidence intervals around the values belies a significant degree of intraspecific variability. Nevertheless, we believe that the need for a consistent body of data for as many of the analyses relating to the British bird assemblage as possible is clear, and that those we use are no worse than any other set for which ultimately some definitions must be somewhat arbitrary.

Species	BREPOP	WINPOP	BREGR	WINGR	Mass	Origin	Pelagics
Red-throated diver	2435	4850	379	536	1144.0	Wild	
Black-throated diver	344	700	199	299	1688.0	Wild	
Great northern diver		3000		486	4250.0	Wild	
Little grebe	15 000	3290	1275	1405	187.0	Wild	
Great-crested grebe	8000	9800	892	919	830.0	Wild	
Red-necked grebe	4	150	8	198	476.0	Wild	
Slavonian grebe	148	400	24	283	364.0	Wild	
Black-necked grebe	71	120	35	107	280.0	Wild	
Fulmar	1 078 000		550		706.0	Wild	✓
Manx shearwater	470 000		22		424.0	Wild	✓
Storm petrel	170 000		48		23.7	Wild	✓
Leach's petrel	110 000		10		39.9	Wild	✓

Species	BREPOP	WINPOP	BREGR	WINGR	Mass	Origin	Pelagics
Gannet	402 000		18		3067.0	Wild	✓
Cormorant	14 000	13 200	174	1486	2127.0	Wild	
Shag	75 000	W	386	771	1760.0	Wild	
Bittern	40	100	13	184	1008.5	Wild	
Grey heron	20 000	W	2335	2387	1361.0	Wild	
Mute swan	25 750	25 750	1579	1577	9700.0	Captive	
Bewick's swan		7200		423	5700.0	Wild	
Whooper swan	4	5600	38	849	8750.0	Wild	
Bean goose		450		121	2843.0	Wild	
Pink-footed goose		192 000		516	2520.0	Wild	
White-fronted goose		19 800		381	2180.0	Wild	
Greylag	14 300	124 150	718	972	3170.0	[Captive]	
Canada goose	46 700	61 000	1196	1004	3550.0	Captive	
Barnacle goose	730	39 890	43	420	1702.0	[Captive]	
Brent goose		105 730		303	1377.0	Wild	
Egyptian goose	700	910	87	73	2040.0	Captive	
Shelduck	21 200	73 500	959	946	813.0	Wild	
Mandarin	7000	7000	218	128	512.0	Captive	
Wigeon	800	277 800	360	1423	557.5	Wild	
Gadwall	1540	8200	357	585	737.0	Wild	
Teal	4100	135 800	1147	1851	294.0	Wild	
Mallard	230 000	500 000	2596	2507	944.0	Wild	
Pintail	50	27 800	85	568	655.0	Wild	
Garganey	140		138		320.0	Wild	
Shoveler	2500	10 000	454	734	432.0	Wild	
Red-crested pochard	100	W	12	75	967.0	Captive	
Pochard	657	43 700	511	1399	807.0	Wild	
Tufted duck	15 000	60 600	1484	1614	680.0	Wild	
Greater scaup	3	11 000	18	380	1183.0	Wild	
Eider	63 000	77 500	488	626	2142.0	Wild	
Long-tailed duck		23 500		418	705.0	Wild	
Common scoter	165	34 500	51	369	1059.0	Wild	
Velvet scoter		3000		162	1730.0	Wild	
Goldeneye	192	17 000	173	1614	787.0	Wild	
Smew		250		231	814.0	Wild	
Red-breasted merganser	4400	10 000	674	825	984.0	Wild	
Goosander	5200	8900	674	1004	1390.0	Wild	
Ruddy duck	1140	3500	292	238	510.0	Captive	
Honey buzzard	27		27		620.0	Wild	
Red kite	320	W	85	70	1175.0	Wild	
White-tailed sea-eagle	20	W	9	7	5572.0	Captive	
Marsh harrier	317	W	114	51	669.0	Wild	
Hen harrier	1260	W	498	975	527.0	Wild	
Montagu's harrier	14		32		370.0	Wild	
Goshawk	850	W	236	110	1206.0	Captive	
Sparrowhawk	64 000	W	2178	2076	264.0	Wild	
Buzzard	29 000	W	1544	1508	1018.0	Wild	

(continued on p. 362)

Species	BREPOP	WINPOP	BREGR	WINGR	Mass	Origin	Pelagics
Rough-legged buzzard		43		93	1086.0	Wild	
Golden eagle	844	W	408	349	5194.0	Wild	
Osprey	198		168		1627.0	Wild	
Kestrel	100 000	W	2481	2415	252.0	Wild	
Merlin	2600	W	693	1008	212.0	Wild	
Hobby	1400		625		240.0	Wild	
Peregrine	2370	W	1048	1006	1112.5	Wild	
Red grouse	500 000	W	945	749	600.0	Wild	
Ptarmigan	20 000	W	173	82	449.0	Wild	
Black grouse	50 540	W	432	316	945.0	Wild	
Capercaillie	2200	W	66	73	1985.0	Captive	
Red-legged partridge	340 000	W	1214	882	439.0	Captive	
Grey partridge	290 000	W	1629	1513	390.0	Wild	
Quail	300		804		88.6	Wild	
Ring-necked pheasant	3 100 000	W	2269	2096	850.0	Captive	
Golden pheasant	1500	W	47	47	607.5	Captive	
Lady Amherst's pheasant	150	W	9	7	714.0	Captive	
Water rail	1350	W	420	808	107.0	Wild	
Spotted crake	21		26		90.0	Wild	
Corncrake	960		161		158.0	Wild	
Moorhen	480 000	W	2032	1921	289.0	Wild	
Coot	46 000	114 100	1603	1605	720.0	Wild	
Crane	6	W	2	10	5895.0	Wild	
Oystercatcher	76 000	359 000	1702	1176	587.0	Wild	
Avocet	942	1270	28	51	275.0	Wild	
Stone-curlew	346		54		461.0	Wild	
Little ringed plover	1895		421		39.2	Wild	
Ringed plover	17 000	28 600	1025	772	65.4	Wild	
Dotterel	1790		99		108.0	Wild	
Golden plover	45 200	250 000	784	1509	256.0	Wild	
Grey plover		43 200		464	278.0	Wild	
Lapwing	430 000	1 750 000	2340	2248	189.0	Wild	
Knot		291 000		366	130.0	Wild	
Sanderling		23 200		287	60.5	Wild	
Temminck's stint	4		3		19.0	Wild	
Purple sandpiper	4	21 300	3	401	65.8	Wild	
Dunlin	19 050	532 000	569	815	56.2	Wild	
Ruff	26	700	42	222	99.0	Wild	
Jack snipe		55 000		636	46.7	Wild	
Snipe	110 000	100 000	1806	2084	107.0	Wild	
Woodcock	30 000	W	1204	1762	313.0	Wild	
Black-tailed godwit	82	7410	59	384	247.0	Wild	
Bar-tailed godwit		52 500		151	330.0	Wild	
Whimbrel	1060		83		440.0	Wild	
Curlew	71 000	115 000	1893	1430	1001.0	Wild	
Spotted redshank		120		123	161.0	Wild	
Redshank	64 200	114 000	1473	1210	146.0	Wild	
Greenshank	2700	380	243	247	209.0	Wild	

Species	BREPOP	WINPOP	BREGR	WINGR	Mass	Origin	Pelagics
Green sandpiper		750		545	72.5	Wild	
Wood sandpiper	6		8		65.6	Wild	
Common sandpiper	31 600	100	1424	193	44.8	Wild	
Turnstone		64 400		671	112.0	Wild	
Red-necked phalarope	72		9		31.3	Wild	
Arctic skua	6400		113		491.0	Wild	
Great skua	17 000		97		1525.0	Wild	
Mediterranean gull	35	W	7	142	256.0	Wild	
Black-headed gull	334 000	1 900 000	671	2274	268.0	Wild	
Common gull	136 000	900 000	577	2217	360.0	Wild	
Lesser black-backed gull	166 000	500 000	434	1447	755.0	Wild	
Herring gull	320 000	450 000	729	2293	813.0	Wild	
Great black-backed gull	38 000	40 000	486	2006	1486.0	Wild	
Kittiwake	980 000		252		393.0	Wild	✓
Sandwich tern	28 000		43		229.0	Wild	
Roseate tern	128		19		123.5	Wild	
Common tern	24 600		426		126.2	Wild	
Arctic tern	88 000		303		107.0	Wild	
Little tern	4800		110		57.0	Wild	
Guillemot	1 050 000		212		670.0	Wild	✓
Razorbill	148 000		233		453.0	Wild	✓
Black guillemot	36 500	W	383	344	380.0	Wild	
Puffin	898 000		151		368.0	Wild	✓
Stock dove	480 000	W	1821	1663	299.0	Wild	
Woodpigeon	4 700 000	W	2510	2369	480.0	Wild	
Collared dove	400 000	W	2210	1880	200.0	Wild	
Turtle dove	150 000		940		152.0	Wild	
Rose-ringed parakeet	3500	W	63	68	92.0	Captive	
Cuckoo	39 000		2418		98.0	Wild	
Barn owl	8800	W	1110	1140	298.0	Wild	
Little owl	18 000	W	1228	1021	206.0	Captive	
Tawny owl	40 000	W	2054	1683	530.0	Wild	
Long-eared owl	4700	W	445	263	308.0	Wild	
Short-eared owl	4500	W	679	973	346.0	Wild	
Nightjar	6800		274		66.3	Wild	
Swift	160 000		2215		39.0	Wild	
Kingfisher	8800	W	1224	1037	43.7	Wild	
Wryneck	7		6		32.5	Wild	
Green woodpecker	30 000	W	1555	1342	198.0	Wild	
Great spotted woodpecker	55 000	W	1959	1731	89.6	Wild	
Lesser spotted woodpecker	9000	W	790	748	19.8	Wild	
Woodlark	1200	W	73	56	32.1	Wild	
Skylark	4 000 000	W	2729	2094	35.8	Wild	
Shore lark		300		53	36.9	Wild	
Sand martin	327 500		1559		12.8	Wild	
Swallow	1 140 000		2626		19.2	Wild	
House martin	750 000		2393		17.6	Wild	

(continued on p. 364)

Species	BREPOP	WINPOP	BREGR	WINGR	Mass	Origin	Pelagics
Tree pipit	240 000		1524		21.3	Wild	
Meadow pipit	3 800 000	W	2539	2314	17.7	Wild	
Rock pipit	68 000	W	654	866	21.8	Wild	
Water pipit		100		93	21.8	Wild	
Yellow wagtail	100 000		1047		18.3	Wild	
Grey wagtail	68 000	W	1979	1492	18.4	Wild	
Pied wagtail	600 000	W	2669	2065	23.6	Wild	
Waxwing		100		149	65.8	Wild	
Dipper	28 000	W	1309	1152	59.0	Wild	
Wren	14 200 000	W	2747	2582	9.2	Wild	
Dunnock	4 000 000	W	2511	2418	21.7	Wild	
Robin	8 400 000	W	2629	2563	19.9	Wild	
Nightingale	11 000		457		19.3	Wild	
Black redstart	101		103		16.5	Wild	
Redstart	180 000		1327		13.4	Wild	
Whinchat	42 000		1404		18.7	Wild	
Stonechat	30 500	W	1034	908	15.1	Wild	
Wheatear	110 000		1738		25.7	Wild	
Ring ouzel	16 500		544		114.4	Wild	
Blackbird	8 800 000	W	2664	2634	108.2	Wild	
Fieldfare	50	750 000	104	2452	116.6	Wild	
Song thrush	1 980 000	W	2620	2357	84.8	Wild	
Redwing	120	750 000	136	2391	67.9	Wild	
Mistle thrush	460 000	W	2397	2277	139.1	Wild	
Cetti's warbler	299	W	86	64	11.7	Wild	
Grasshopper warbler	2100		1189		12.9	Wild	
Savi's warbler	16		27		15.2	Wild	
Sedge warbler	500 000		1887		11.5	Wild	
Marsh warbler	45		15		12.2	Wild	
Reed warbler	120 000		790		10.8	Wild	
Dartford warbler	3490	W	45	35	9.4	Wild	
Lesser whitethroat	160 000		1271		10.7	Wild	
Whitethroat	1 320 000		2186		14.7	Wild	
Garden warbler	400 000		1867		18.4	Wild	
Blackcap	1 160 000	W	2048	857	18.5	Wild	
Wood warbler	34 400		1270		11.0	Wild	
Chiffchaff	1 280 000	W	2100	576	7.8	Wild	
Willow warbler	4 600 000		2602		9.2	Wild	
Goldcrest	1 120 000	W	2327	2230	5.6	Wild	
Firecrest	330	W	99	244	5.3	Wild	
Spotted flycatcher	240 000		2378		14.8	Wild	
Pied flycatcher	75 000		732		12.3	Wild	
Bearded tit	747	W	60	121	15.0	Wild	
Long-tailed tit	420 000	W	2106	2050	7.7	Wild	
Marsh tit	120 000	W	1133	1208	10.5	Wild	
Willow tit	50 000	W	1100	1152	10.0	Wild	
Crested tit	1800	W	51	46	10.8	Wild	
Coal tit	1 220 000	W	2315	2261	9.5	Wild	

Species	BREPOP	WINPOP	BREGR	WINGR	Mass	Origin	Pelagics
Blue tit	6 600 000	W	2480	2050	10.7	Wild	
Great tit	3 200 000	W	2443	2404	19.2	Wild	
Nuthatch	260 000	W	1270	1150	23.4	Wild	
Treecreeper	400 000	W	2120	2055	9.0	Wild	
Golden oriole	36		45		69.2	Wild	
Red-backed shrike	10		15		26.9	Wild	
Great grey shrike		60		238	63.4	Wild	
Jay	320 000	W	1713	1756	158.5	Wild	
Magpie	1 180 000	W	1958	1909	206.5	Wild	
Chough	680	W	88	87	265.5	Wild	
Jackdaw	780 000	W	2344	2312	231.0	Wild	
Rook	1 710 000	W	2237	2266	418.0	Wild	
Carrion/hooded crow	1 940 000	W	2762	2369	517.0	Wild	
Raven	14 000	W	1131	1096	1147.1	Wild	
Starling	2 200 000	W	2620	2498	79.9	Wild	
House sparrow	7 200 000	W	2525	2407	30.2	Wild	
Tree sparrow	220 000	W	1346	1403	20.8	Wild	
Chaffinch	10 800 000	W	2602	2544	20.7	Wild	
Brambling	2	922 500	13	1611	25.7	Wild	
Serin	2		8		11.8	Wild	
Greenfinch	1 060 000	W	2323	2145	28.3	Wild	
Goldfinch	440 000	W	2209	1984	15.5	Wild	
Siskin	600 000	W	1158	1666	12.7	Wild	
Linnet	1 040 000	W	2268	1693	17.3	Wild	
Twite	130 000	W	651	493	16.0	Wild	
Lesser redpoll	320 000	W	1754	1531	11.2	Wild	
Common crossbill	21 000	W	763	439	40.5	Wild	
Scottish crossbill	1550	W	59	68	41.8	Wild	
Parrot crossbill	1		2		50.3	Wild	
Common rosefinch	8		5		23.6	Wild	
Bullfinch	380 000	W	2173	2114	25.8	Wild	
Hawfinch	9500	W	315	252	51.9	Wild	
Lapland bunting		350		86	22.5	Wild	
Snow bunting	170	11 250	42	636	31.4	Wild	
Yellowhammer	2 400 000	W	2224	2011	27.4	Wild	
Cirl bunting	760	W	29	26	25.9	Wild	
Reed bunting	440 000	W	2188	1777	19.2	Wild	
Corn bunting	39 000	W	921	737	43.9	Wild	

Berkshire Breeding Bird Data

The distribution of species in the breeding avifauna of Berkshire documented by Standley *et al.* (1996) in a sample of 25 of the 391 tetrads (2 × 2-km squares) by which the county is covered. The sample tetrads are those comprising the 10 × 10-km grid square SU87. Tetrad letters follow Gibbons *et al.* (1993). The final two columns give the total number of tetrads occupied by species in the sample and in Berkshire as a whole.

		Range size in:																									Sample	Berkshire
Species	A	B	C	D	E	F	G	H	I	J	K	L	M	N	P	Q	R	S	T	U	V	W	X	Y	Z	Sample	Berkshire
Little grebe	X									X		X		X				X			X				X	8	121
Great-crested grebe																	X	X		X	X		X	X	X	2	98
Grey heron	X	X	X	X	X	X	X		X		X	X	X	X	X	X	X	X	X	X				X	X	17	166
Mute swan			X										X	X		X	X								X	4	152
Greylag goose																										0	29
Canada goose	X						X				X	X		X		X		X		X	X		X	X	X	12	192
Barnacle goose																										0	1
Egyptian goose																										0	7
Shelduck																										0	13
Mandarin														X		X	X	X			X					4	51
Gadwall																	X									1	23
Teal																	X									1	25
Mallard	X	X		X			X	X	X	X	X	X	X	X		X		X	X	X	X			X	X	19	319
Garganey																										0	2
Shoveler																										0	7
Pochard																										0	20
Tufted duck				X			X		X				X	X							X			X		7	129
Ruddy duck																										0	3
Sparrowhawk	X		X	X	X		X			X	X	X	X	X		X	X	X		X	X	X		X	X	21	294
Buzzard																										0	27
Kestrel	X		X	X	X		X	X		X	X	X	X	X		X	X	X	X	X	X	X	X	X	X	24	342
Hobby																						X				2	68
Red-legged partridge	X				X					X	X	X	X	X		X	X	X	X		X	X	X			14	233
Grey partridge	X										X		X	X		X	X	X	X	X			X	X		11	251
Quail																										0	25
Pheasant	X		X	X	X		X		X	X	X	X	X	X		X	X	X	X	X	X		X		X	24	363
Golden pheasant																										0	5
Water rail							X	X																		0	22
Moorhen	X	X	X		X		X	X		X	X	X	X	X		X	X	X	X	X	X			X	X	21	311

(continued on p. 368)

Species	A	B	C	D	E	F	G	H	I	J	K	L	M	N	P	Q	R	S	T	U	V	W	X	Y	Z	Sample	Berkshire
Coot	X					X					X	X	X	X	X	X	X	X	X		X	X	X	X	X	13	204
Little ringed plover																										0	39
Ringed plover																										0	28
Lapwing	X	X	X		X	X			X	X	X	X	X		X	X	X	X	X		X	X	X	X		19	298
Snipe												X								X						2	57
Woodcock				X								X	X	X											X	4	137
Curlew																										0	8
Redshank																										0	49
Black-headed gull								X	X		X				X	X	X	X	X				X		X	9	87
Common tern																				X						0	36
Stock dove	X	X	X	X	X	X				X	X	X	X	X	X	X	X	X	X	X	X	X	X		X	19	307
Woodpigeon	X	X		X	X	X	X	X	X	X	X	X	X	X	X	X	X	X	X	X	X	X	X		X	25	390
Collared dove				X	X			X	X	X	X	X	X	X	X	X	X	X	X	X	X	X	X		X	21	369
Turtle dove	X			X	X					X	X	X	X	X	X	X	X	X	X	X	X	X	X		X	18	228
Ring-necked parakeet																										0	16
Cuckoo	X	X		X	X		X		X	X	X	X	X		X	X	X	X	X	X	X	X	X	X	X	21	327
Barn owl	X																									1	54
Little owl	X		X	X			X			X	X		X		X	X	X	X	X		X	X	X		X	16	226
Tawny owl	X	X		X	X				X	X	X	X					X	X	X	X	X	X	X	X	X	13	270
Nightjar				X											X											1	25
Swift	X	X		X	X		X	X	X	X	X	X	X		X	X	X	X	X	X	X	X	X		X	23	329
Kingfisher								X	X	X	X	X	X			X	X	X	X	X	X	X	X	X	X	9	145
Green woodpecker	X	X	X	X	X	X	X		X	X	X	X	X		X	X	X	X	X	X	X	X	X	X	X	23	300
Great spotted woodpecker	X	X	X	X	X	X	X	X	X	X	X	X	X		X	X	X	X	X	X	X	X	X	X	X	25	352
Lesser spotted woodpecker				X							X		X	X		X				X	X	X	X			8	116
Woodlark																										0	11

Species			
Skylark		25	342
Sand martin		1	74
Swallow		23	369
House martin		22	372
Tree pipit		1	87
Meadow pipit		4	135
Yellow wagtail		5	93
Pied wagtail		21	361
Grey wagtail		11	154
Wren		25	391
Dunnock		24	387
Robin		25	391
Nightingale		3	83
Black redstart		1	3
Redstart		2	34
Stonechat		0	15
Blackbird		25	390
Fieldfare		2	10
Song thrush		22	382
Mistle thrush		24	373
Cetti's warbler		0	8
Grasshopper warbler		1	26
Sedge warbler		0	128
Reed warbler		1	145
Lesser whitethroat		12	207
Whitethroat		19	324
Garden warbler		14	290
Blackcap		23	372
Wood warbler		1	42
Chiffchaff		25	365
Willow warbler		22	374
Goldcrest		17	317
Firecrest		0	20
Spotted flycatcher		16	276
Bearded tit		0	1

(continued on p. 370)

Species	A	B	C	D	E	F	G	H	I	J	K	L	M	N	P	Q	R	S	T	U	V	W	X	Y	Z	Sample	Berkshire
Long-tailed tit	X	X	X	X	X	X	X	X	X	X	X	X	X	X	X	X	X	X	X	X	X	X	X	X	X	25	367
Marsh tit		X			X					X		X		X		X	X	X	X		X	X			X	10	198
Willow tit							X													X			X			3	124
Coal tit	X		X	X	X	X	X	X	X	X	X	X	X	X	X	X	X	X	X	X	X	X	X	X		19	293
Blue tit	X	X	X	X	X	X	X	X	X	X	X	X	X	X	X	X	X	X	X	X	X	X	X	X	X	25	391
Great tit	X	X	X	X	X	X	X	X	X	X	X	X	X	X	X	X	X	X	X	X	X	X	X	X	X	25	389
Nuthatch	X			X	X	X	X	X	X	X	X	X		X	X	X	X		X	X	X	X	X	X	X	18	306
Treecreeper	X			X	X	X	X	X	X	X	X	X	X	X	X	X	X	X	X	X	X	X	X	X	X	21	311
Jay	X	X	X	X	X	X	X	X	X	X	X	X	X	X	X	X	X	X	X	X	X	X	X	X	X	25	347
Magpie	X	X	X	X	X	X	X	X	X	X	X	X	X	X	X	X	X	X	X	X	X	X	X	X	X	25	388
Jackdaw	X	X	X	X	X	X	X	X	X	X	X	X	X	X	X	X	X		X	X	X	X	X	X	X	19	359
Rook	X	X	X	X	X	X	X	X	X	X	X	X	X	X	X	X	X	X	X	X	X	X	X	X	X	23	319
Carrion crow	X	X	X	X	X	X	X	X	X	X	X	X	X	X	X	X	X	X	X	X	X	X	X	X	X	25	375
Starling	X	X	X	X	X	X	X	X	X	X	X	X	X	X	X	X	X	X	X	X	X	X	X	X	X	24	388
House sparrow	X	X	X	X	X	X	X	X	X	X	X	X	X	X	X	X	X	X	X	X	X	X	X	X	X	24	380
Tree sparrow											X	X				X										3	61
Chaffinch	X	X	X	X	X	X	X	X	X	X	X	X	X	X	X	X	X	X	X	X	X	X	X	X	X	25	391
Greenfinch	X	X	X	X	X	X	X	X	X	X	X	X	X	X	X	X	X	X	X	X	X	X	X	X	X	22	384
Goldfinch	X		X	X	X				X	X	X	X	X	X	X	X	X	X	X	X	X	X	X	X	X	21	341
Siskin																				X				X	X	2	38
Linnet	X	X			X		X		X	X	X	X	X	X	X	X	X	X	X	X	X	X	X	X	X	18	291
Redpoll	X	X			X												X				X					4	46
Crossbill																										0	14
Bullfinch	X	X		X	X	X	X	X	X	X	X	X	X	X	X		X	X	X	X	X	X	X	X	X	21	340
Hawfinch																				X						1	16
Yellowhammer	X	X		X	X	X	X	X	X	X	X	X	X	X	X	X	X	X	X	X	X	X	X	X	X	24	344
Reed bunting				X	X	X	X				X	X	X	X	X	X	X	X	X	X	X	X	X	X	X	15	195
Corn bunting										X	X			X		X	X	X	X	X	X	X				11	138
Species richness	62	46	34	47	60	36	50	30	46	55	55	68	58	62	52	64	68	52	60	65	63	56	57	52	59		

Index